Lecture Notes in Computer Science   16230

Founding Editors

Gerhard Goos
Juris Hartmanis

Editorial Board Members

Elisa Bertino, *Purdue University, West Lafayette, IN, USA*
Wen Gao, *Peking University, Beijing, China*
Bernhard Steffen, *TU Dortmund University, Dortmund, Germany*
Moti Yung, *Columbia University, New York, NY, USA*

The series Lecture Notes in Computer Science (LNCS), including its subseries Lecture Notes in Artificial Intelligence (LNAI) and Lecture Notes in Bioinformatics (LNBI), has established itself as a medium for the publication of new developments in computer science and information technology research, teaching, and education.

LNCS enjoys close cooperation with the computer science R & D community, the series counts many renowned academics among its volume editors and paper authors, and collaborates with prestigious societies. Its mission is to serve this international community by providing an invaluable service, mainly focused on the publication of conference and workshop proceedings and postproceedings. LNCS commenced publication in 1973.

Pierre Ganty · Alessio Mansutti
Editors

Reachability Problems

19th International Conference, RP 2025
Madrid, Spain, October 1–3, 2025
Proceedings

Editors
Pierre Ganty
IMDEA Software Institute
Madrid, Spain

Alessio Mansutti
IMDEA Software Institute
Madrid, Spain

ISSN 0302-9743   ISSN 1611-3349 (electronic)
Lecture Notes in Computer Science
ISBN 978-3-032-09523-7   ISBN 978-3-032-09524-4 (eBook)
https://doi.org/10.1007/978-3-032-09524-4

© The Editor(s) (if applicable) and The Author(s), under exclusive license to Springer Nature Switzerland AG 2026

This work is subject to copyright. All rights are solely and exclusively licensed by the Publisher, whether the whole or part of the material is concerned, specifically the rights of translation, reprinting, reuse of illustrations, recitation, broadcasting, reproduction on microfilms or in any other physical way, and transmission or information storage and retrieval, electronic adaptation, computer software, or by similar or dissimilar methodology now known or hereafter developed.
The use of general descriptive names, registered names, trademarks, service marks, etc. in this publication does not imply, even in the absence of a specific statement, that such names are exempt from the relevant protective laws and regulations and therefore free for general use.
The publisher, the authors and the editors are safe to assume that the advice and information in this book are believed to be true and accurate at the date of publication. Neither the publisher nor the authors or the editors give a warranty, expressed or implied, with respect to the material contained herein or for any errors or omissions that may have been made. The publisher remains neutral with regard to jurisdictional claims in published maps and institutional affiliations.

This Springer imprint is published by the registered company Springer Nature Switzerland AG
The registered company address is: Gewerbestrasse 11, 6330 Cham, Switzerland

If disposing of this product, please recycle the paper.

Preface

This volume contains the papers presented at the 19th International Conference on Reachability Problems (RP 2025), organized by the IMDEA Software Institute, Madrid, Spain.

The RP 2025 conference took place as a physical event during October 1–3, 2025 at the IMDEA Software Institute, Madrid, Spain. Previous events in the RP conference series were located at the TU Wien, Vienna, Austria (2024); the Laboratoire d'Informatique, Signaux et Systèmes de Sophia Antipolis of the Université Côte d'Azur, France (2023); the University of Kaiserslautern, Germany (2022); the University of Liverpool, UK (2021); Université Paris Cité, France (2020); Université Libre de Bruxelles, Belgium (2019); Aix-Marseille University, France (2018); Royal Holloway, University of London, UK (2017); Aalborg University, Denmark (2016); the University of Warsaw, Poland (2015); the University of Oxford, UK (2014); Uppsala University, Sweden (2013); the University of Bordeaux, France (2012); the University of Genoa, Italy (2011); Masaryk University, Czech Republic (2010); École Polytechnique, France (2009); the University of Liverpool, UK (2008); and Turku University, Finland (2007). The RP conference aims to gather together scholars from diverse disciplines and backgrounds interested in reachability problems that appear, among others, in algebraic structures, automata theory and formal languages, computational game theory, concurrency and distributed computation, decision procedures in computational models, hybrid dynamical systems, logic and model checking, and verification of finite- and infinite-state systems. In addition, the conference promotes the exploration and combination of new approaches for the modeling and analysis of computational processes by combining mathematical, algorithmic, and computational techniques. As such, the RP conference provides an active forum for discussion and networking for researchers interested in diverse fields of reachability analysis. RP 2025 received 35 submissions, consisting of 18 regular research papers, 3 invited papers, and 14 presentation-only submissions. All regular paper submissions to RP 2025 were reviewed using a single-blind reviewing process. Each submission received three reviews. All reviews were completed by Program Committee (PC) members, except for two reviews. Using the review reports, the PC had a thorough discussion on each regular paper, which, in some cases, changed the PC support on the respective paper. Submissions with predominantly supportive reviews were agreed upon to be accepted by the PC: in total, the PC accepted 12 regular papers and 13 presentation-only submissions, in addition to the 3 invited paper contributions.

The RP 2025 conference featured 5 invited talks by

- **Albert Atserias**, Universitat Politècnica de Catalunya, Spain
    Title: Local-vs-Global Consistency of Annotated Relations
- **Alastair F. Donaldson**, Imperial College London, UK
    Title: When You Have a Fuzzer, Everything Looks Like a Reachability Problem
- **Zachary Kincaid**, Princeton University, USA

Title: Reachability Problems and Program Analysis
- **Anthony W. Lin**, University of Kaiserslautern, Germany
  Title: The Role of Logic and Automata in Understanding Transformers
- **Mickael Randour**, Université de Mons, Belgium
  Title: Simplicity Lies in the Eye of the Beholder: A Strategic Perspective on Controllers in Reactive Synthesis

In addition, the RP 2025 conference featured one invited tutorial

Niki Vazou, IMDEA Software Institute, Spain
  Title: LiquidHaskell: Theorem Proving with Refinement Types

This volume contains the 12 regular papers accepted at RP 2025, the 3 invited papers, and abstracts of the 13 presentation-only submissions, of the remaining invited talks, and of the invited tutorial.

Springer Nature sponsored the best paper award, selected by the program committee. The **best paper award** was given to:

Joel D. Day and Matthew Konefal
  Title: Word Equations with Length Constraints via Weak Arithmetics and Matrix Reachability Problems.

We would like to thank everyone who helped to make RP 2025 successful. We thank the authors for submitting their papers to RP 2025. The PC members and additional reviewers did an excellent job in reviewing papers: they provided detailed reports and engaged in the PC discussions. We thank the RP steering committee, and especially Igor Potapov for his valuable advice. We are grateful to members of our local organizing committee, and in particular to María Alcaraz, Tania Rodríguez, Ana Cecilia Monteverde, and Helena del Río (responsible for local arrangements). We acknowledge the financial support provided by Amazon Web Services, the Délégation générale Wallonie-Bruxelles in Spain, and Springer Nature. We are grateful for the institutional support RP 2025 received from the IMDEA Software Institute. Finally, a big thank you goes to all our RP 2025 authors, invited speakers, and invited tutorial speaker for their high-quality contributions, and to the participants for making RP 2025 a success!

September 2025

Pierre Ganty
Alessio Mansutti

Organization

Program Committee Chairs

Pierre Ganty	IMDEA Software Institute, Spain
Alessio Mansutti	IMDEA Software Institute, Spain

Steering Committee

Parosh Aziz Abdulla	Uppsala University, Sweden
Olivier Bournez	École Polytechnique, France
Vesa Halava	University of Turku, Finland
Alain Finkel	ENS Paris-Saclay, France
Oscar Ibarra	UC Santa Barbara, USA
Juhani Karhumaki	University of Turku, Finland
Jérôme Leroux	CNRS, University of Bordeaux, France
Joël Ouaknine	Max Planck Institute for Software Systems, Germany
Igor Potapov	University of Liverpool, UK
James Worrell	University of Oxford, UK

Program Committee

Mohamed Faouzi Atig	Uppsala University, Sweden
Laura Bozzelli	University of Naples "Federico II", Italy
Michaël Cadilhac	DePaul University, USA
Dmitry Chistikov	University of Warwick, UK
Rayna Dimitrova	CISPA Helmholtz Center for Information Security, Germany
Kyveli Doveri	University of Warsaw, Poland
Cezara Drăgoi	AWS, France
Hadar Frenkel	BarIlan University, Israel
Moses Ganardi	MPI-SWS, Germany
Piotrek Hofman	University of Warsaw, Poland
Lukáš Holík	Aalborg University, Denmark & Brno University of Technology, Czechia
George Kenison	Liverpool John Moores University, UK

Sandra Kiefer — University of Oxford, UK
Rupak Majumdar — MPI for Software Systems, Germany
Kaushik Mallik — IMDEA Software Institute, Spain
Joshua Moerman — Open Universiteit, Netherlands
Guillermo Perez — University of Antwerp, Belgium
Igor Potapov — University of Liverpool, UK
Gabriele Puppis — University of Udine, Italy
Andrew Ryzhikov — University of Warsaw, Poland
Mahsa Shirmohammadi — CNRS & IRIF, France
Mikhail R. Starchak — Max Planck Institute for Software Systems, Germany
Andrea Turrini — Institute of Software, Chinese Academy of Sciences, China
Chana Weil-Kennedy — CEA List, France
Sarah Winkler — Free University of Bozen-Bolzano, Italy
Sarah Winter — IRIF, France
Martin Zimmerman — Aalborg University, Denmark

Additional Reviewers

Mahsa Naraghi
Neha Rino

Abstract of Invited Talks and the Invited Tutorial

Local-vs-Global Consistency of Annotated Relations

Albert Atserias

Universitat Politècnica de Catalunya, Spain

Abstract. In modern relational database theory, annotated relations constitute an expressive formal extension of the classical framework to model a large variety of problems. When the tuples of the relations come annotated with values from the tropical semiring, annotated relations can be used to model reachability and shortest paths problems on weighted graphs and cost optimization problems. When the tuples come annotated with positive real numbers, annotated relations can be used to model statistical data and its associated data analysis problems. A third, very different, example is that of relations whose tuples come annotated with positive linear operators of a Hilbert space, which can be used to model the joint measurability problem in quantum mechanical systems.

A fundamental challenge in each of these fields is determining the necessary and sufficient conditions under which local components given by local annotated relations admit a global realization as a single annotated relation that projects to the local constituents. This is an instance of the local-vs-global consistency problem for annotated relations. After reviewing the formal framework of K-relations, where K is a positive commutative semiring or monoid and each tuple in each relation comes annotated with a value from K, we adapt the standard notions of consistency for classical relations to the setting of K-relations. We highlight both the parallels with the classical framework of standard relations and the key differences that make this theory particularly rich and interesting.

When You Have a Fuzzer, Everything Looks Like a Reachability Problem

Alastair F. Donaldson

Imperial College London, UK

Abstract. We provide an overview of three projects that explore the idea of using coverage-guided fuzzing, a technique traditionally used for finding bugs in software, in unconventional domains: (1) efficiently solving SMT formulas that use floating-point constraints; (2) achieving fast SMT sampling for such formulas; and (3) simulating operational memory models. In each case, the idea is to reduce the problem at hand to a *reachability problem*: transforming a problem instance into a program equipped with a special error location, such that finding an input that reaches the error location equates to finding a solution to the problem instance. Coverage-guided fuzzing, which excels at mutating a corpus of inputs to achieve increasing statement coverage of a system under test, can then be used to search for an input that reaches the error location—i.e., for a solution to the problem instance. We hope this overview will inspire other researchers to consider recasting search problems into a reachability problem form where coverage-guided fuzzing may prove effective.

Reachability Problems and Program Analysis

Zachary Kincaid

Princeton University, US

Abstract. Since essentially all decision problems concerning the semantics of computer programs are undecidable, the goal in program analysis is to compute approximate semantics. While approximation heuristics are often effective in practice, they can also be brittle and unpredictable. In this talk, I will summarize a line of work that aims to create more robust program analyses by exploiting reachability techniques for sub-Turing models of computation.

The Role of Logic and Automata in Understanding Transformers

Anthony W. Lin

Max-Planck Institute for Software Systems, Kaiserslautern, Germany
& University of Kaiserslautern-Landau, Kaiserslautern, Germany

Abstract. The advent of transformers has in recent years led to powerful and revolutionary Large Language Models (LLMs). Despite this, our understanding of the capability of transformers is still meager. In this invited contribution, we recount the rapid progress in the last few years on the question of what transformers can do. In particular, we will see the integral role of logic and automata (also with some help from circuit complexity) in answering this question. We also mention several open problems at the intersection of logic, automata, verification and transformers.

Simplicity Lies in the Eye of the Beholder: A Strategic Perspective on Controllers in Reactive Synthesis

Mickael Randour

F.R.S.-FNRS & UMONS – Université de Mons, Belgium

Abstract. In the game-theoretic approach to controller synthesis, we model the interaction between a system to be controlled and its environment as a game between these entities, and we seek an appropriate (e.g., winning or optimal) strategy for the system. This strategy then serves as a formal blueprint for a real-world controller. A common belief is that *simple* (e.g., using limited memory) *strategies are better*: corresponding controllers are easier to conceive and understand, and cheaper to produce and maintain.

This invited contribution focuses on the complexity of strategies in a variety of synthesis contexts. We discuss recent results concerning memory and randomness, and take a brief look at what lies beyond our traditional notions of complexity for strategies.

LiquidHaskell: Theorem Proving with Refinement Types

Niki Vazou

IMDEA Software Institute, Spain

Abstract. Liquid Haskell is an extension to Haskell that adds refinement types to the language, which are then checked via an external theorem prover such as Z3. With refinement types, one can express many interesting properties of programs that are normally out of reach of Haskell's type system or only achievable via quite substantial encoding efforts and advanced type system constructs. On the other hand, the overhead for checking refinement types is often rather small, because the external solver is quite powerful.

Liquid Haskell used to be an external, standalone executable, but is now available as a GHC plugin, making it much more convenient to use.

In this tutorial, we'll discuss how refinement types work, give many examples of their use and learn how to work with Liquid Haskell productively.

Contents

Invited Papers

When You Have a Fuzzer, Everything Looks Like a Reachability Problem 3
   Alastair F. Donaldson, Cristian Cadar, Manuel Carrasco, Dan Iorga,
   Daniel Liew, and John Wickerson

The Role of Logic and Automata in Understanding Transformers 17
   Anthony W. Lin and Pablo Barcelo

Simplicity Lies in the Eye of the Beholder: A Strategic Perspective
on Controllers in Reactive Synthesis 31
   Mickael Randour

Regular Papers

Word Equations with Length Constraints via Weak Arithmetics and Matrix
Reachability Problems 51
   Joel D. Day and Matthew Konefal

Word Chain Generators for Prefix Normal Words 68
   Duncan Adamson, Moritz Dudey, Pamela Fleischmann, and Annika Huch

Reachability and Mortality for Two-Dimensional RHPCD Systems Are
co-NP-hard 83
   Olga Tveretina

UPPAAL COSHY: Automatic Synthesis of Compact Shields for Hybrid
Systems 97
   Asger Horn Brorholt, Andreas Holck Høeg-Petersen, Peter Gjøl Jensen,
   Kim Guldstrand Larsen, Marius Mikučionis, Christian Schilling,
   and Andrzej Wasowski

Weighing Obese Timed Languages 112
   Eugene Asarin, Aldric Degorre, Cătălin Dima,
   and Bernardo Jacobo Inclán

Box-Reachability in Vector Addition Systems 126
   Shaull Almagor, Itay Hasson, Michał Pilipczuk, and Michael Zaslavski

Knowing-How Reasoning with Budgets Recasted: Universal Reachability
Problem on VASS 140
    Stéphane Demri, Laurent Doyen, and Raul Fervari

Nets-Within-Nets Through the Lens of Data Nets 156
    Francesco Di Cosmo, Soumodev Mal, and Tephilla Prince

Compositional Verification of Almost-Sure Büchi Objectives in MDPs 171
    Marck van der Vegt, Kazuki Watanabe, Ichiro Hasuo,
    and Sebastian Junges

DTMC Model Checking by Path Abstraction Revisited 186
    Arnd Hartmanns and Robert Modderman

Counterexample-Guided Abstraction Refinement for Star-Based Neural
Network Verification 202
    László Antal, Franz Link, and Erika Ábrahám

Maximum Path Sets in Trees 217
    A. Subramani, K. Subramani, and Jacob Restanio

Presentation-Only Contributions

Stationary Regimes of Piecewise Linear Dynamical Systems with Priorities 233
    Xavier Allamigeon, Pascal Capetillo, and Stéphane Gaubert

Membership and Conjugacy in Inverse Semigroups 234
    Lukas Fleischer, Florian Stober, Alexander Thumm, and Armin Weiß

The Ultimate Signs of Second-Order Holonomic Sequences 236
    FugenHagihara, and Akitoshi Kawamura

Quantitative Language Automata 237
    Thomas A. Henzinger, Pavol Kebis, Nicolas Mazzocchi, and N. Ege Saraç

Robust Identification of Hybrid Automata from Noisy Data 238
    Niklas Kochdumper, Mohammed Aristide Foughali, Peter Habermehl,
    and Eugene Asarin

Regular Model Checking for Systems with Effectively Regular
Reachability Relation 239
    Javier Esparza, and Valentin Krasotin

Galois Energy Games to Solve All Kinds of Quantitative Reachability
Problems 240
    Caroline Lemke, and Benjamin Bisping

Learning Deterministic One-Counter Automata 241
    Prince Mathew, Vincent Penelle, and A. V. Sreejith

BT2Automata: Expressing Behavior Trees as Automata for Formal
Control Synthesis 242
    Ryan Matheu, Aniruddh G. Puranic, John S. Baras, and Calin Belta

Model Checking as Program Verification by Abstract Interpretation 243
    Paolo Baldan, Roberto Bruni, Francesco Ranzato, and Diletta Rigo

A Complexity Dichotomy for Semilinear Target Sets in Automata
with One Counter 244
    Yousef Shakiba, Henry Sinclair-Banks, and Georg Zetzsche

Quantifier Elimination for Regular Integer Linear-Exponential
Programming 245
    Mikhail R. Starchak

Flexible Catalysis 246
    Mate Weisz, and Sergii Strelchuk

Author Index 247

Invited Papers

When You Have a Fuzzer, Everything Looks Like a Reachability Problem

Alastair F. Donaldson(✉), Cristian Cadar, Manuel Carrasco, Dan Iorga, Daniel Liew, and John Wickerson

Imperial College London, London, UK
alastair.donaldson@imperial.ac.uk

Abstract. We provide an overview of three projects that explore the idea of using coverage-guided fuzzing, a technique traditionally used for finding bugs in software, in unconventional domains: (1) efficiently solving SMT formulas that use floating-point constraints; (2) achieving fast SMT sampling for such formulas; and (3) simulating operational memory models. In each case, the idea is to reduce the problem at hand into a *reachability problem*: transforming a problem instance into a program equipped with a special error location, such that finding an input that reaches the error location equates to finding a solution to the problem instance. Coverage-guided fuzzing, which excels at mutating a corpus of inputs to achieve increasing statement coverage of a system under test, can then be used to search for an input that reaches the error location—i.e., for a solution to the problem instance. We hope this overview will inspire other researchers to consider recasting search problems into a reachability problem form where coverage-guided fuzzing may prove effective.

Keywords: Coverage-guided fuzzing · constraint solving · floating point · memory models

1 Introduction

Coverage-guided mutation-based fuzzing is a randomised testing technique for automatically finding software bugs, and has been widely adopted through tools such as AFL [36], AFL++ [14] and libFuzzer [25]. Building on the basic idea of fuzzing—testing a software system on randomised inputs [28]—coverage-guided mutation-based fuzzing is a search-based test case generation technique [3], using ideas from evolutionary algorithms [18] to guide the randomised testing process.

The technique starts with a corpus of *seed* inputs: inputs that already exercise the software under test (SUT) to some extent. Further inputs are then

---

This invited paper is associated with Alastair Donaldson's invited talk at RP 2025. The other authors are the main authors of the three existing papers on which this overview paper is based [8,21,26] and are listed alphabetically. Imperial College London was the affiliation of all authors when they contributed to these papers.

© The Author(s), under exclusive license to Springer Nature Switzerland AG 2026
P. Ganty and A. Mansutti (Eds.): RP 2025, LNCS 16230, pp. 3–16, 2026.
https://doi.org/10.1007/978-3-032-09524-4_1

obtained by mutating and combining existing inputs drawn from the corpus. This is what makes the technique "mutation-based", and the hypothesis behind mutation-based fuzzing is that inputs obtained via mutation are more likely to further exercise the SUT compared with inputs generated from scratch in a naïve manner.

To decide whether a mutated input is interesting enough to be added to the corpus (so that it will be considered for further mutation), a check is made to see whether the input covers code in the SUT that is not covered by any existing input in the corpus. Intuitively, favouring inputs that reach new parts of the SUT is a good strategy for guiding the fuzzing process towards finding bugs. This use of coverage information is what makes the technique "coverage-guided". In the context of evolutionary algorithms, code coverage is used as a measure of fitness.

The *raison d'être* of coverage-guided mutation-based fuzzing (henceforth referred to as coverage-guided fuzzing for brevity) is to find bugs that cause a program to crash. The technique has been very successful in this regard: ClusterFuzz, Google's continuous fuzzing infrastructure for Chrome [17], and OSS-Fuzz [35], a deployment of ClusterFuzz targeting open-source projects, are reported to have found tens of thousands of bugs [17].

However, from a more abstract viewpoint, coverage-guiding fuzzing can be seen as a technique for solving program reachability problems: a coverage-guided fuzzer demonstrates that a program can crash by synthesising an input that *reaches* an error location. With this viewpoint it is interesting to consider applications of coverage-guided fuzzing to reachability programs that go beyond finding bugs in programs, by *dressing up* said reachability problems as bug-finding problems.

In this invited paper we survey three pieces of work from the authors that leverage coverage-guided fuzzing for other kinds of reachability problems:

1. JFS, where coverage-guided fuzzing is used to find solutions to SMT formulas that feature floating-point constraints [26] (Sect. 2).
2. JFSAMPLER, an extension to JFS concerned with *SMT sampling*—finding a diverse range of solutions to an SMT formula [8] (Sect. 3).
3. A project on the use of program analysis tools—including coverage-guided fuzzing—for the simulation of operational memory models, to determine whether behaviours of concurrent programs characterised by litmus tests are allowed according to a given memory model [21] (Sect. 4).

We with a discussion of the conditions under which coverage-guided fuzzing may be an effective reachability analysis in these application domains, and the pros and cons of other program reachability analyses (Sect. 5). Key related work is discussed throughout, and we refer the reader to the related work sections of the original papers on these projects for a broader discussion of relevant literature [8,21,26].

Relationship to Existing Papers. The article draws on material from the original papers about these works [8,21,26]. The authors of the relevant material are also authors on this article and the material is reused here with their consent.

2 Just Fuzz it: Solving Floating-Point SMT Formulas

Satisfiability modulo theories (SMT) solvers have found application in many domains including software testing and software verification (see e.g. [5,7,9,15, 16,23,24,31]). For example, symbolic execution techniques involve gathering the constraints on a program input that must hold for the program to reach a given location or trigger a particular error, and then using an SMT solver to solve for inputs that satisfy these constraints [7,15,16,31].

A limitation to the utility of SMT solvers in these domains can be a lack of scalability; e.g. solver timeouts cause symbolic executors to grind to a halt and lead to inclusive results from program verifiers. Scalability is a particular problem when reasoning about formulas that use the floating-point SMT theory (QF_FP) [33] or the combination of floating-point and bitvector theories (QF_BVFP), to the extent that it can be impractical to apply SMT-based analysis methods to numeric applications that operate on floating-point numbers [27].

```
1  (declare-fun a () Float64)
2  (declare-fun b () Float64)
3  (define-fun div_rne () Float64 (fp.div RNE a b))
4  (define-fun div_rtp () Float64 (fp.div RTP a b))
5  (assert (not (fp.isNaN a)))
6  (assert (not (fp.isNaN b)))
7  (assert (not (fp.isNaN div_rne)))
8  (assert (not (fp.isNaN div_rtp)))
9  (assert (not (fp.eq div_rne div_rtp)))
10 (check-sat)
```

Fig. 1. Example QF_FP formula

```
1  int FuzzOneInput(const uint8_t* data, size_t size) {
2    double a = makeFloatFrom(data, size, 0, 63);
3    double b = makeFloatFrom(data, size, 64, 127);
4    if (!isnan(a)) {} else return 0;
5    if (!isnan(b)) {} else return 0;
6    double a_b_rne = div_rne(a, b);
7    double a_b_rtp = div_rtp(a, b);
8    if (!isnan(a_b_rne)) {} else return 0;
9    if (!isnan(a_b_rtp)) {} else return 0;
10   if (a_b_rne != a_b_rtp) {} else return 0;
11   abort(); // TARGET REACHED
12 }
```

Fig. 2. C++ program generated by JFS for Fig. 1 using the *fail-fast* encoding

The limited scalability of floating-point SMT solvers inspired the first project that we survey here: the Just Fuzz It Solver (JFS).

Overview of JFS. The idea behind JFS is to transform the problem of finding a satisfying assignment to an SMT formula into a program reachability problem. Specifically, JFS transforms an SMT formula into a program such that (1) a program input corresponds to an assignment to the free variables of the formula, and (2) the program contains a special *target* statement that is reachable if and only if the input corresponds to a satisfying assignment to the formula.

A coverage-guided fuzzer aims to find inputs that maximise coverage, so when applied to this program it will search relentlessly for an input that reaches the target statement, i.e. for a satisfying assignment to the formula. The hypothesis behind JFS is that this technique may sometimes be able to rapidly find satisfying assignments for formulas that are challenging for general-purpose solvers, such as floating-point formulas. However, JFS is incomplete: for an unsatisfiable formula the target statement is unreachable, thus coverage-guided fuzzing will run indefinitely, never finding an error-triggering input. We envision that JFS would be run in parallel with a complete solver as part of a portfolio.

JFS requires that the input formula is presented as a conjunction of assertions. Given the conjunction, JFS generates a C++ program that takes an assignment to the free variables of the formula as input. The program evaluates the formula on the assignment by evaluating the top-level conjuncts in turn. By construction, the program crashes if and only if all of the conjuncts are satisfied—i.e. if the input is a satisfying assignment. JFS then uses libFuzzer to automatically search for an input that triggers a crash—i.e. for a satisfying assignment.

Example. As an illustration of this idea, consider the example constraints in Listing 1, shown in SMT-LIB format. Free variables `a` and `b` of type `Float64` are declared on lines 1 and 2 respectively. On lines 3 and 4, variables `div_rne` and `div_rtp` are defined to be the division of `a` by `b` using the rounding to nearest, ties to even (RNE) and rounding toward positive infinity (RTP) rounding modes, respectively. The satisfiability problem captured by the example is the conjunction of the constraints specified in the five `assert` statements. The first four constraints state that none of `a`, `b`, `div_rne` and `div_rtp` are NaN; the last states that `div_rne` is not equal to `div_rtp`.

A possible translation of these constraints into a C++ program is shown in Listing 2. The program is a *fuzz target* for libFuzzer (with some details omitted for brevity). The guard of each `if` statement corresponds to a constraint. The fuzzer will repeatedly call `FuzzOneInput` (line 1), each time passing an input of `size` bytes via the `data` buffer. If the `abort()` statement is reached (line 11), causing the program to crash, the input corresponds to a satisfying assignment and JFS terminates and returns SAT. Otherwise, the fuzzer proceeds to try another input.

Further Details and Results. Before transforming a formula into a program, JFS performs a number of simplification and rewriting passes to make the formula more amenable to coverage-guided fuzzing. One such rewrite step involves splitting a top-level and constraint (appearing directly under assert) into two separate constraints (via two distinct assert commands). This leads to the resulting C++ program having a distinct location associated with the satisfaction of each conjunct, so that the coverage-guided fuzzer is rewarded separately for finding inputs that satisfy each conjunct.

```
1   int FuzzerTestOneInput(const uint8_t* data, size_t size) {
2     double a = makeFloatFrom(data, size, 0, 63);
3     double b = makeFloatFrom(data, size, 64, 127);
4     size_t counter = 0;
5     if (!isnan(a)) ++counter;
6     if (!isnan(b)) ++counter;
7     double a_b_rne = div_rne(a, b);
8     double a_b_rtp = div_rtp(a, b);
9     if (a_b_rne != a_b_rtp) ++counter;
10    if (!isnan(a_b_rne)) ++counter;
11    if (!isnan(a_b_rtp)) ++counter;
12    if (counter != 5)
13      return 0;
14    abort(); // TARGET REACHED
15  }
```

Fig. 3. C++ program generated by JFS for Fig. 1 using the *try-all* encoding

We experimented with two different program encodings: *fail-fast*, where the fuzz target exits as soon as it is found that the current input does *not* satisfy some constraint of the formula, illustrated by Fig. 2, and *try-all*, which evaluates all constraints of the formula even if some constraints are found not to hold, illustrated by Fig. 3. We hypothesised that *try-all* would provide a stronger coverage signal that might outweigh the additional cost associated with redundantly evaluating further constraints once a constraint has been found not to hold. However, in practice we found that *fail-fast* was significantly more efficient.

Since coverage-guided fuzzing needs an input corpus, JFS provides *smart seeds* featuring special constant values such as infinities, zeros and NaNs [26].

An evaluation over benchmarks drawn SMT-COMP suites with respect to state-of-the-art solvers at the time of the project showed JFS to be highly competitive on QF_FP and QF_BVFP benchmarks (formulas that use the floating-point theory or the combination of floating-point and bitvector theories, respectively), but uncompetitive on QF_BV benchmarks (which only use the bitvector theory). The results support the idea that JFS could be used to help unblock the search for satisfying assignments in the presence of floating-point constraints. The evaluation also confirmed that coverage guidance is an important contributor to the success of JFS: results for a version of the tool where coverage guidance is disabled show markedly worse results.

JFS only supports the combination of bitvector and floating-point theories, but the idea of encoding SMT solving as a program reachability problem to be solved using fuzzing should be straightforward to adapt to other finite-domain SMT theories.

3 JFSAMPLER: Efficient Floating-Point SMT Sampling

Traditional SMT solvers aim to find a single satisfying assignment for a satisfiable formula, but in some application domains it can be useful to obtain a diverse range of satisfying assignments for a formula; e.g. in software testing there is evidence that symbolic execution can benefit from retrieving multiple solutions from the constraints instead of exploring multiple paths from a certain program point [19].

In large, configurable systems, such as operating systems or web development frameworks, it is often useful to generate a small but representative set of valid build or deployment configurations for testing [30]. In hardware design, it is often useful to generate multiple stimuli that meet the preconditions of a functional specification, and then compare the resulting outputs with those of the hardware's logic design before the design becomes silicon [13,29].

In response to this, recent work has focused on SAT and SMT sampling techniques that find large sets of satisfying assignments for a formula, attempting to provide reasonable coverage of the formula's solution space [11–13].

The difficulty of scaling SMT solvers in the floating point domain (see Sect. 2) means that SMT sampling techniques such as SMTSAMPLER [12], which rely on using an existing SMT solver as a source of initial satisfying assignments, are also limited in this domain. In response to this we designed JFSAMPLER, an SMT sampling technique based on JFS geared towards efficient sampling for floating-point formulas [8].

Overview of JFSampler. The basic idea behind JFSAMPLER is to take the JFS approach of encoding the search for a satisfying assignment as the problem of finding an input that reaches a special *target* statement in a program. However, instead of using coverage-guided fuzzing to search for a *single* input that reaches the target location, JFSAMPLER runs fuzzing continuously for a given time budget, collecting all distinct inputs that reach the target, which correspond to distinct satisfying assignments.

However, the basic idea of reusing JFS for this task, without modification, suffers from the problem that once a solution-inducing input has been found, further solution-inducing inputs will exercise the same path through the program: the unique path that satisfies all conjuncts and hence reaches the target location. With this setup the fuzzer is not rewarded (via coverage feedback) for finding diverse solutions.

To overcome this, JFSAMPLER uses a more sophisticated encoding based on an SMT-level coverage metric proposed in the SMTSAMPLER project [12]. The new encoding, which we call the *diversity encoding*, features additional code that, if covered, will correlate with an increase in the SMT-level coverage metric used for measuring diversity.

Example. Recall again the SMT formula of Fig. 1 and its associated reachability problem encoding of Fig. 2. The diversity encoding employed by JFSAMPLER involves the addition of extra conditional code right before the `abort()` call that corresponds to the target location. At this point, an input corresponding to a satisfying assignment has been found. The additional code is designed to test the value of every bit in every non-root and non-leaf subexpression of the formula under the current satisfying assignment.

For our running example this would involve adding 128 conditional statements between lines 10 and 11 of Fig. 2 that test each of the 64 bits of each of the `double` variables `a_b_rne` and `a_b_rtp`. Each such test introduces a new program point that the fuzzer will be rewarded for reaching. This rewards the fuzzer for synthesising assignments that make the formula true in myriad different ways. It also allows the fuzzer to distinguish between them and keep them in the corpus for further refinement.

Further Details and Results. The SMTSAMPLER project proposed an effective method that takes existing satisfying assignments for a formula and combines them using a heuristic that generates candidate follow-on satisfying assignments. In JFSAMPLER we used this heuristic as the basis for a *custom mutator* for libFuzzer, the underlying fuzzer used by JFS. Our custom mutator has access to all satisfying assignments that have been shown so far. On encountering an input that achieves new coverage, the custom mutator combines the input with two randomly-selected previous satisfying assignments (if available), using a combination strategy based on the technique used by SMTSAMPLER.

We evaluated JFSAMPLER empirically, comparing it with SMTSAMPLER (which is based on the Z3 solver and its support for MaxSMT problems [6]), over the QF_FP and QF_BVFP formulas that were used in the evaluation of JFS. We found that JFSAMPLER significantly outperformed SMTSAMPLER on the QF_FP benchmarks. An ablation study confirmed that the diversity encoding and custom mutator make a substantial contribution to the performance of JFSAMPLER on these benchmarks when compared to a naïve version of JFSAMPLER that simply searches for multiple crashing inputs with respect to the program encoding emitted by JFS with no further improvements.

On the QF_BVFP benchmarks we found that SMTSAMPLER significantly outperformed the naive version of JFSAMPLER. The diversity encoding and custom mutator, when used individually in isolation, improved the performance of JFSAMPLER, but performance remained below that of SMTSAMPLER. However, these two features in combination ultimately led to a slight performance improvement over SMTSAMPLER.

4 Simulating Concurrency Memory Models

Our final case study considers the use of coverage-guided fuzzing and other reachability analysis techniques to aid in the simulation of *operational memory models* [21]. This project is distinct from the JFS and JFSAMPLER projects described in Sects. 2 and 3, but was in part inspired by the success of JFS.

Overview. The memory model of a shared memory concurrent system formally describes the potential interactions between threads that can arise through communication via shared memory locations. For reasons of efficiency, modern multi-core CPUs, accelerators and heterogeneous systems feature memory models that are *weaker* than the appealingly simple *sequentially consistent* memory model [22], as demonstrated by vendor documentation and various academic studies [1,2,20,32,34].

An operational memory model characterises the allowed memory behaviours of a system using a state machine, where *states* represent components such as store buffers, caches and queues, and *transitions* define the legal changes between these states, triggered by memory operations such as reads, writes and flushes. The memory model can then be applied to *litmus tests*—small programs that capture potential interactions between threads, where a litmus test is *allowed* if the behaviour that it captures is permitted by the memory model.

Constructing operational memory models and applying them to litmus tests is facilitated by *simulators* that reveal which behaviours of a given program are allowed. While extensive work has been done on simulating *axiomatic* memory models [2,4], there has been less work on simulation of operational models [32, 34], despite the fact that operational models are arguably more intuitive than their axiomatic counterparts.

Part of the reason for this is the overhead associated with engineering and maintaining a full-blown simulator for an operational memory model. For example the RMEM [32] state-of-the-art memory model simulator, which supports the ARM, Power, RISC-V and x86 memory models, comprises more than 60k lines of OCaml code in part because the developers had to build custom efficient reachability analyses.

An appealing idea to reduce this engineering overhead is to implement the logic of the memory model as a program that takes a particular test scenario as input. Determining whether the test scenario is allowed would then boil down to determining whether a particular state of the program that encodes the memory model is reachable when executed on an input describing the scenario of interest, and off-the-shelf reachability analysis tools for the language of interest could be leveraged to answer this question. Subsequent detailed examination of traces would then be possible by stepping through the simulator code using a standard debugger.

In recent work [21] we put this idea into practice for two different operational memory models: the x86 memory model, which allows for a comparison with the RMEM simulator, and the X+F memory model [20] associated with a system that combines an Intel Xeon CPU with a field-programmable gate array (FPGA), for which no bespoke memory model simulator exists.

Example. Figure 4 gives a flavour of our encoding—full details are in our full article about the project [21]. The program makes several nondeterministic choices – how many simulation steps to run (line 2), which thread to activate for each step (line 5), and whether each step corresponds to the thread's next

```
1   // Choose a number of steps of the simulation to run
2   int sim_steps = choose(SIMULATION_STEPS);
3   for (int i = 0; i < sim_steps; i++) {
4     // Choose a thread to take a step
5     int thread = choose(NUM_THREADS);
6     // Choose whether the CPU or the environment takes a step
7     Action action = choose(NUM_ACTIONS);
8     switch (action) {
9       case CPU_THREAD: // the CPU is to take a step
10        // Check that the thread still has work to do
11        if (!thread_ops[thread].empty()) {
12          // Pop the next instruction from the thread's list
13          Operation op = thread_ops[thread].pop();
14          // Carry out a write by appending to the CPU's store buffer
15          if (op.type == WRITE) {
16            write_to_buffer(thread, op.var, op.val);
17          }
18          // Carry out a read from the CPU's store buffer or from memory
19          if (op.type == READ) {
20            read_buffer_or_memory(thread, op.var);
21          }
22          break;
23        }
24      case FLUSH_BUFFER: // the environment is to take a step
25        // Check that the CPU's store buffer is not empty
26        if (!buffer[thread].empty()) {
27          // Flush an entry from the CPU's store buffer into memory
28          flush_buffer(thread);
29          break;
30        }
31    }
32  }
33  // Check whether the litmus test's postcondition has been reached
34  check_litmus_test();
```

Fig. 4. The pseudocode of the mechanised x86 memory model.

instruction being executed (line 9) or to the 'environment' making a transition by flushing an x86 store buffer (line 24). If executed directly, this program would not be very useful because it would be unlikely to make the right sequence of decisions to get interesting behaviours. However, it becomes useful when presented for reachability analysis, because then the question becomes: is it *possible* to resolve all of these choices so as to make the interesting behaviour emerge.

Further Details and Results. We investigated the effectiveness of three different C-analysis tools for analysis of: CBMC [10], which encodes the reachability problem as a monolithic SAT query, KLEE [7], which uses dynamic symbolic execution to explore the program in a path-by-path manner, generating an SMT query per path, and using three different coverage-guided fuzzers including lib-Fuzzer (the fuzzer behind JFS and JFSAMPLER). A common feature of all these tools is that they avoid false positives: if the tool reports 'reachable' then the error-state really is reachable. This is because, unlike many static analysis tools, they do not employ any abstraction.

The CBMC and KLEE tools can, in principle, prove that error states are *not* reachable, assuming the litmus test provided as input is loop-free. In the case of CBMC this is via the use of *unwinding assertions*, while for KLEE it involves exhaustive exploration of all program paths. While the fuzzers cannot prove unreachability of error states, when the error states *are* reachable, the fuzzers tend to discover this much more quickly than the other two tools. However, large litmus tests feature error states that can only be reached via lengthy paths that depend on an intricate schedule of thread and memory subsystem events. The fuzzers struggled on these larger litmus tests, while CBMC/KLEE performed somewhat better due to their more systematic approaches.

Our results also show that the *coverage-guidedness* of the fuzzers is valuable: when coverage guidance is disabled, the fuzzers did not perform at all well for the task of memory model simulation.

The X+F memory model to which we applied this technique is rather complicated, so it is a testimony to the generality of our approach that we were able to obtain a simulator for it with little additional effort, by encoding its logic as a C program.

5 Discussion

We conclude with a discussion of the strengths and weaknesses of coverage-guided fuzzing as a reachability analysis technique in the context of these three case studies, and how the dynamic vs. symbolic and under-approximating vs. over-approximating dimensions of a reachability analysis affect its suitability in these domains.

Effectiveness of Coverage-Guided Fuzzing. Our experience with JFS is that a fuzzing-based approach to constraint solving often works better than traditional approaches for formulas involving floating-point constraints, but that the performance of JFS was poor when applied to formulas involving only bitvector constraints, where traditional methods excel. When applied to the problem of memory model simulation, fuzzing excelled in comparison to symbolic analysis techniques for simpler litmus tests, but fared less well on larger tests where the behaviour under consideration relied on a very specific interleaving of threads and memory subsystem events. Our hypothesis is that fuzzing has the potential to outperform symbolic techniques for reachability analysis on problem instances where (a) many solutions exist, so that the probability of finding a solution via guided random search is reasonably high, and (b) the constraints that a solution must satisfy are nevertheless complex enough that symbolic solving algorithms have a difficult time navigating the underlying search tree. Investigating this hypothesis in more detail, e.g. by using model counting to see whether there is a correlation between the number of solutions to a floating-point SMT query and the ease with which JFS can solve this query, would be an interesting avenue for future work.

In both the JFS work and the memory model simulation project, experimental results confirmed that the "coverage-guided" part of coverage-guided fuzzing is an important contributor: disabling coverage guidance led to less effective solving of floating-point formulas by JFS (and thus would also negatively impact JFSAMPLER), and to markedly worse results for memory model analysis.

The JFSAMPLER approach is a particularly good match for fuzzing because in SMT sampling one is always concerned with satisfiable formulas, and typically formulas that have many different solutions.

Dynamic vs. Symbolic Reachability Analyses. In the JFS and JFSAMPLER projects we focused on a purely-dynamic reachability analysis—coverage-guided fuzzing—to analyse the program associated with an SMT formula, while in the memory model simulation work we also considered the use of techniques that perform full or partial symbolic reasoning (CBMC and KLEE).

In principle, any under-approximating analysis could be used to find satisfying assignments through analysis of the program emitted by JFS, and it would certainly be possible to try applying CBMC or KLEE to a JFS-generated program. However, given that the purpose of JFS is to provide an alternative means for solving a problem that is known to be hard for symbolic methods (namely reasoning about floating-point arithmetic), applying SAT or SMT-based analysis techniques such as CBMC or KLEE to the programs generated by JFS would make little sense. In contrast, memory model simulation involves the exploration of the possible behaviours of a nondeterministic system, something at which symbolic program analysis tools excel.

Under-approximating vs. Over-approximating Analyses. In the projects discussed here, we have focused almost entirely on the use of under-approximating program analyses, geared towards bug finding, except that in the memory model simulation work we considered the use of CBMC and KLEE for "brute force" verification, by fully unrolling program loops (CBMC) or exhaustively exploring program paths (KLEE).

In the context of JFS it is necessary to use an under-approximating analysis to find solutions to a formula from its associated program with confidence. Furthermore, the under-approximating analysis must yield inputs that triggers the bug found by the analysis if one wishes to obtain a satisfying assignment to the formula of interest rather than merely knowing that it is satisfiable. An over-approximating analysis *might* still be useful for finding solutions to a formula if there is a way to obtain a candidate input from an alarm raised by the analysis, because one could simply run the program on the candidate input to check whether it indeed reaches the program's target location.

In principle it would be possible to use a JFS-generated program to prove unsatisfiability of a given formula by using an over-approximating static verification technique to prove that the program is correct. However, similar to the discussion above regarding the limited value of applying CBMC or KLEE to JFS-generated programs, many static verification techniques are based on symbolic analysis so using them for this task would again seem somewhat circular.

For JFSAMPLER, bug finding is the *only* meaningful way to analyse the program associated with a formula, because SMT sampling involves mining a formula for a diverse range of satisfying assignments and is not concerned with proving unsatisfiability.

In the context of memory model analysis, applying an under-approximating analysis to a program obtained from a (memory model, litmus test) pair facilitates establishing that the behaviour characterised by a litmus test is *allowed*: finding a bug in the resulting program equates to confirming that a behaviour is possible. A bug report from an over-approximating analysis would merely indicate that the behaviour *might* be allowed, because the bug report could be a false alarm. In contrast, showing that a memory model behaviour is *disallowed* would require verifying that the associated program is correct. We have only investigated performing such verification via brute force methods, as discussed above, and it would be interesting to assess the utility of other verification techniques for this task.

References

1. Alglave, J., et al.: GPU concurrency: weak behaviours and programming assumptions. In: ASPLOS 2015. ACM (2015). https://doi.org/10.1145/2694344.2694391
2. Alglave, J., Maranget, L., Tautschnig, M.: Herding cats: modelling, simulation, testing, and data mining for weak memory. ACM Trans. Program. Lang. Syst. **36**(2), 1–74 (2014). https://doi.org/10.1145/2627752
3. Ali, S., Briand, L.C., Hemmati, H., Panesar-Walawege, R.K.: A systematic review of the application and empirical investigation of search-based test case generation. IEEE Trans. Softw. Eng. **36**(6), 742–762 (2010). https://doi.org/10.1109/TSE.2009.52
4. Armstrong, A., Campbell, B., Simner, B., Pulte, C., Sewell, P.: Isla: integrating full-scale ISA semantics and axiomatic concurrency models (extended version). Formal Methods Syst. Des. **63**(1), 110–133 (2024). https://doi.org/10.1007/S10703-023-00409-Y
5. Ball, T., Bounimova, E., Levin, V., Kumar, R., Lichtenberg, J.: The static driver verifier research platform. In: Touili, T., Cook, B., Jackson, P. (eds.) CAV 2010. LNCS, vol. 6174, pp. 119–122. Springer, Heidelberg (2010). https://doi.org/10.1007/978-3-642-14295-6_11
6. Bjørner, N., Phan, A.-D., Fleckenstein, L.: νz - an optimizing SMT solver. In: Baier, C., Tinelli, C. (eds.) TACAS 2015. LNCS, vol. 9035, pp. 194–199. Springer, Heidelberg (2015). https://doi.org/10.1007/978-3-662-46681-0_14
7. Cadar, C., Dunbar, D., Engler, D.: KLEE: unassisted and automatic generation of high-coverage tests for complex systems programs. In: OSDI 2008. USENIX (2008)
8. Carrasco, M., Cadar, C., Donaldson, A.F.: Scalable SMT sampling for floating-point formulas via coverage-guided fuzzing. In: ICST 2025. IEEE (2025). https://doi.org/10.1109/ICST62969.2025.10989031
9. Carter, M., He, S., Whitaker, J., Rakamaric, Z., Emmi, M.: SMACK software verification toolchain. In: Dillon, L.K., Visser, W., Williams, L.A. (eds.) ICSE 2016 Companion Volume. ACM (2016). https://doi.org/10.1145/2889160.2889163
10. Clarke, E., Kroening, D., Lerda, F.: A tool for checking ANSI-C programs. In: Jensen, K., Podelski, A. (eds.) TACAS 2004. LNCS, vol. 2988, pp. 168–176. Springer, Heidelberg (2004). https://doi.org/10.1007/978-3-540-24730-2_15
11. Delannoy, R., Meel, K.S.: On almost-uniform generation of SAT solutions: the power of 3-wise independent hashing. In: LICS 2022. ACM (2022). https://doi.org/10.1145/3531130.3533338

12. Dutra, R., Bachrach, J., Sen, K.: SMTSampler: efficient stimulus generation from complex SMT constraints. In: ICCAD 2018. ACM (2018). https://doi.org/10.1145/3240765.3240848
13. Dutra, R., Laeufer, K., Bachrach, J., Sen, K.: Efficient sampling of SAT solutions for testing. In: ICSE 2018. ACM (2018). https://doi.org/10.1145/3180155.3180248
14. Fioraldi, A., Maier, D., Eißfeldt, H., Heuse, M.: AFL++: combining incremental steps of fuzzing research. In: WOOT 2020. USENIX (2020)
15. Godefroid, P., Klarlund, N., Sen, K.: DART: directed automated random testing. In: PLDI 2005. ACM (2005). https://doi.org/10.1145/1065010.1065036
16. Godefroid, P., Levin, M.Y., Molnar, D.A.: Automated whitebox fuzz testing. In: NDSS 2008. The Internet Society (2008)
17. Google: ClusterFuzz (2025). https://github.com/google/clusterfuzz
18. Holland, J.: Adaptation in natural and artificial systems: an introductory analysis with applications to biology, control, and artificial intelligence. University of Michigan Press (1975)
19. Huang, H., Yao, P., Wu, R., Shi, Q., Zhang, C.: Pangolin: incremental hybrid fuzzing with polyhedral path abstraction. In: S&P'20. IEEE (2020). https://doi.org/10.1109/SP40000.2020.00063
20. Iorga, D., Donaldson, A.F., Sorensen, T., Wickerson, J.: The semantics of shared memory in Intel CPU/FPGA systems. Proc. ACM Program. Lang. **5**(OOPSLA), 1–28 (2021). https://doi.org/10.1145/3485497
21. Iorga, D., Wickerson, J., Donaldson, A.F.: Simulating operational memory models using off-the-shelf program analysis tools. IEEE Trans. Softw. Eng. **49**(12), 5084–5102 (2023). https://doi.org/10.1109/TSE.2023.3326056
22. Lamport, L.: How to make a multiprocessor computer that correctly executes multiprocess programs. IEEE Trans. Comput. **28**(9), 690–691 (1979). https://doi.org/10.1109/TC.1979.1675439
23. Leino, K.R.M.: Dafny: an automatic program verifier for functional correctness. In: LPAR 2010. Springer, Heidelberg (2010). https://doi.org/10.1007/978-3-642-17511-4_20
24. Leino, K.R.M., Rümmer, P.: A polymorphic intermediate verification language: design and logical encoding. In: Esparza, J., Majumdar, R. (eds.) TACAS 2010. LNCS, vol. 6015, pp. 312–327. Springer, Heidelberg (2010). https://doi.org/10.1007/978-3-642-12002-2_26
25. LibFuzzer website (2025). http://llvm.org/docs/LibFuzzer.html
26. Liew, D., Cadar, C., Donaldson, A., Stinnett, J.R.: Just fuzz it: solving floating-point constraints using coverage-guided fuzzing. In: ESEC/FSE 2019. ACM (2019). https://doi.org/10.1145/3338906.3338921
27. Liew, D., Schemmel, D., Cadar, C., Donaldson, A., Zähl, R., Wehrle, K.: Floating-point symbolic execution: a case study in N-version programming. In: ASE 2017. IEEE (2017). https://doi.org/10.1109/ASE.2017.8115670
28. Miller, B.P., Fredriksen, L., So, B.: An empirical study of the reliability of UNIX utilities. Commun. Assoc. Comput. Mach. (CACM) **33**(12), 32–44 (1990). https://doi.org/10.1145/96267.96279
29. Naveh, Y., et al.: Constraint-based random stimuli generation for hardware verification. AI Mag. **28**(3), 13–30 (2007). https://doi.org/10.1609/AIMAG.V28I3.2052
30. Plazar, Q., Acher, M., Perrouin, G., Devroey, X., Cordy, M.: Uniform sampling of SAT solutions for configurable systems: are we there yet? In: ICST 2019. IEEE (2019). https://doi.org/10.1109/ICST.2019.00032

31. Poeplau, S., Francillon, A.: Symbolic execution with SymCC: don't interpret, compile! In: USENIX Security 2020. USENIX (2020)
32. Pulte, C., Flur, S., Deacon, W., French, J., Sarkar, S., Sewell, P.: Simplifying ARM concurrency: multicopy-atomic axiomatic and operational models for ARMv8. Proc. ACM Program. Lang. **2**(POPL), 19:1–19:29 (2018). https://doi.org/10.1145/3158107
33. Rümmer, P., Wahl, T.: An SMT-LIB theory of binary floating-point arithmetic. In: SMT 2010 (2010). http://www.cprover.org/SMT-LIB-Float/smt-fpa.pdf
34. Sarkar, S., Sewell, P., Alglave, J., Maranget, L., Williams, D.: Understanding POWER multiprocessors. In: PLDI 2011. ACM (2011). https://doi.org/10.1145/1993498.1993520
35. Serebryany, K.: OSS-Fuzz – Google's continuous fuzzing service for open source software. In: Invited talk at USENIX Security 2017. USENIX (2017)
36. Zalewski, M.: Technical "whitepaper" for afl-fuzz (2025). http://lcamtuf.coredump.cx/afl/technical_details.txt

The Role of Logic and Automata in Understanding Transformers

Anthony W. Lin[1,2](✉) and Pablo Barcelo[3,4,5]

[1] Max-Planck Institute for Software Systems, Kaiserslautern, Germany
[2] University of Kaiserslautern-Landau, Kaiserslautern, Germany
awlin@mpi-sws.org
[3] Institute for Mathematical and Computational Engineering, Pontificia Universidad Católica de Chile, Santiago, Chile
[4] IMFD Chile, Santiago, Chile
[5] CENIA, Macul, Chile

Abstract. The advent of transformers has in recent years led to powerful and revolutionary Large Language Models (LLMs). Despite this, our understanding on the capability of transformers is still meager. In this invited contribution, we recount the rapid progress in the last few years to the question of what transformers can do. In particular, we will see the integral role of logic and automata (also with some help from circuit complexity) in answering this question. We also mention several open problems at the intersection of logic, automata, verification and transformers.

Keywords: Transformers · Hard Attention · LTL · Regular Languages

1 Introduction

Recent years witnessed the unprecedented emergence of Large Language Models (LLMs), which have revolutionized many aspects of our lives. LLMs are based on a new neural network model called *transformers*, which extends the classical feed-forward neural network model via *attention mechanisms* for handling texts of arbitrary lengths. Unlike Recurrent Neural Networks (RNN) [5]—which predated transformers by decades—transformers have proven to be efficiently parallelizable and able to capture long-range dependencies better in practice. Despite the rapid adoption of transformers as a mainstream ML model, some limitations of the transformer model have only been understood in recent years. One good example of such a limitation is to perform *counting* in a text, e.g., determine whether there is an even or an odd number of occurrences of a given token in a text.

In recent years, subareas of theoretical computer science—including logic, automata, and circuit complexity—have featured in the rapid development of the theory of expressivity of transformers (cf. [17]). Such a connection has organically materialized because transformers are computational models that process texts (i.e., strings) and can be studied just like formal models such as finite-state automata, Turing machines, or logics like first-order and second-order logics on strings. Multiple formal models have been developed by varying the following aspects of transformers: attention mechanisms, positional encodings, precision, and the so-called "chain of thoughts". Guided by both theory building and experimentation, a picture on the expressive power of transformers has slowly emerged. Although this picture is to date incomplete, a respectable body of works have been produced in the so-called FLaNN (Formal Languages and Neural Networks) community, consisting of logicians, automata theorists, and computational linguists.

Why this Article? This article has been written to recount *some* gems that have been discovered at the intersection of logic, automata, circuit complexity, and transformers. That is, we do not aim to be exhaustive. The choices of materials are additionally based on our subjective taste[1]. The intended audience of the article includes researchers in logic, automata, verification and programming languages. In particular, we will mention several open problems, which we believe are worth undertaking in the next years.

Highlight of Key Results. In its simplest form, a transformer can be understood as a formal model that takes an input *text* (i.e. string) and outputs a *token* (i.e. letter). More formally, a transformer gives rise to a function $f : \Sigma^* \to \Sigma$, for some finite alphabet Σ of tokens. Moreover, one could think of f as a family of formal languages $\{L_a\}_{a \in \Sigma}$, where $L_a := \{w \in \Sigma^* : f(w) = a\}$. This connection underlines the bridge between formal languages and transformers: one can simply study such formal languages L_a generated (or recognized) by transformers.

The first set of results in the paper concerns the expressivity of transformers with *unique hard attention* mechanisms (a.k.a. Unique Hard Attention Transformers, or simply UHAT). Such an attention mechanism—which finds the leftmost value that maximizes the attention score—is a simplification of *softmax attention*, which is used in practice but has proven to be tricky to analyze in theory owing to the use of such real-valued functions as e^x. The first key result that we discuss in the paper is from [2,19]. It connects formal languages definable in various fragments of first-order logic over strings extended with all numerical predicates (equivalently, subclasses of the circuit complexity class AC^0) and UHAT. In particular, the language

$$\text{PARITY} := \{w \in \{a,b\} : |w|_a \equiv 0 \pmod{2}\}$$

is well-known [1] not to be in AC^0, therefore cannot be expressed by UHAT. We cover this in Sect. 3.

The second set of results concerns the expressivity of transformers with *averaging hard attention* mechanisms (a.k.a. Average Hard Attention Transformers, or simply AHAT). Such an attention mechanism—which averages all values that

---

[1] Before working on FLaNN, the authors primarily researched in logic, automata theory, automated reasoning, finite model theory, and databases.

maximize the attention score (unlike simply taking the leftmost value)—provides another approximation of practical transformers, which use softmax attention. In particular, AHAT is tightly connected to Linear Temporal Logic extended with counting and the circuit complexity class TC^0. We cover this in Sect. 4

Finally, we discuss the limitations of both UHAT and AHAT as approximations of practical transformers. In particular, we consider a recent promising direction that restricts AHAT to uniform attention layers (i.e., each position receives the same amount of attention). The resulting model, called AHAT[U], appears to be a good approximation of softmax transformers. We also discuss the distinction between expressibility and trainability in Sect. 5.

Precision. Real-world transformers are implemented on a specific hardware that allows fixed (bit-)precision and fixed memory. Of course, one can allow more precision and more memory by upgrading the hardware. Therefore, researchers in the theory of transformers has adopted a more practical approach by specifying different precision model on a transformer $\mathcal{T}$:

1. *Fixed* precision: there is a constant c on the allowed number of bits for any computation performed by $\mathcal{T}$.
2. *Logarithmic* precision: the number of allowed bits in the computation of $\mathcal{T}$ on a string of length n is $O(\log n)$.
3. *Polynomial* precision: the number of allowed bits in the computation of $\mathcal{T}$ on a string of length n is $O(n^c)$ for some constant c.
4. *Rational* (resp. *real*) precision: this means rational (resp. real) computation is allowed with an unbounded precision.

Although the distinction is important, it overcomplicates an introductory article. For these reasons, we will assume the last precision model, and simply remark that all of the mentioned results work also for polynomial precision (and often also logarithmic precision).

Notation and Assumed Background. We assume familiarity with standard results in logic and automata, and their connections to circuit complexity. All required background could be found in the excellent book [12] by Libkin. In particular, we will consider *star-free* languages (i.e. regular languages generated by regular expressions that use concatenation, union, complementation, but no Kleene star), and their equivalent formulation using first-order logic over strings (i.e. over the embedding of strings as logical structures, e.g., aba is encoded as the structure with universe $\{1,2,3\}$, the order relation $\preceq \, \subseteq \{1,2,3\}^2$, and unary relations $U_a = \{1,3\}$ and $U_b = \{2\}$ indicating which positions labeled by a and b, respectively). By Kamp's theorem [9], the logic is equivalent to Linear Temporal Logic (LTL). First-order logic characterization of star-free languages can be extended with all numerical predicates to give us a characterization of the circuit complexity class (nonuniform) AC^0, which can be defined by a class of problems that can be solved by a family $\{C_n\}_{n \geq 0}$ of constant-depth polynomial-sized (i.e. polynomial in n) boolean circuits (with unbounded fan-ins), wherein

C_n is employed to decide input strings of length n. Note that a k-ary numerical predicate simply means a relation $R \subseteq \mathbb{N}^k$. In the sequel, we also use the fragment FO[Mon], which restricts the above use of numerical predicates only to *monadic* (i.e. unary) numerical predicates. This is a strict subset of AC^0.

The circuit complexity TC^0 extends AC^0 with majority gates, which effectively allows one to encode all standard arithmetic operations on numbers including addition, multiplication, etc. TC^0 problems are often construed in the FLaNN (Formal Languages and Neural Networks) community as *efficiently parallelizable* problems. Note that TC^0 is a subset of the circuit complexity class NC^1, which contains all problems solvable by families of polynomial-sized circuits of logarithmic depth. It is known that NC^1 contains all regular languages. [It is not known if all regular languages are contained in TC^0]. In turn, NC^1 is a subset of L, i.e., the class of problems solvable in logarithmic space.

2 Formal Models of Transformers

We define several formal models of transformers, which are based on the type of adopted attention mechanisms (i.e. hard or soft attention). We first define these semantically, and then instantiate them based on different attention mechanisms.

A transformer can be seen as a composition of several sequence-to-sequence transformations. More precisely, a *seq-to-seq transformation* is a length-preserving $f : (\mathbb{R}^l)^* \to (\mathbb{R}^h)^*$ for some positive integers l, h. That is, f maps an input sequence σ of vectors of dimension l to an output sequence $f(\sigma)$ of dimension h of the same length, i.e., $|f(\sigma)| = |\sigma|$. We write $\texttt{iDim}(f)$ (resp. $\texttt{oDim}(f)$) to denote the dimension of the input (resp. output) vectors of f, i.e., l (resp. h). A sequence $\mu := f_1, \ldots, f_k$ of seq-to-seq transformers is said to be *well-typed* if $\texttt{iDim}(f_{i+1}) = \texttt{oDim}(f_i)$ for each $i = 1, \ldots, k-1$. We assume a finite *alphabet* Σ of tokens (a.k.a. symbols or characters) not containing the *end-of-string symbol* EOS. We write $\Sigma_{\texttt{EOS}}$ to denote $\Sigma \cup \{\texttt{EOS}\}$. A transformer $\mathcal{T}$ over Σ can then be defined as a triple $(\mu, \texttt{em}, \mathbf{t})$, where μ is a well-typed sequence of seq-to-seq transformers as above, $\texttt{em} : \Sigma_{\texttt{EOS}} \to \mathbb{R}^d$ with $d = \texttt{iDim}(f_1)$ is called a *token embedding*, and $\mathbf{t} \in \mathbb{R}^s$ with $s = \texttt{oDim}(f_k)$. The token embedding $\texttt{em}$ can be extended to $\texttt{em} : \Sigma^* \to (\mathbb{R}^d)^*$ by morphism, i.e., $\texttt{em}(w_1 \cdots w_n) = \texttt{em}(w_1) \cdots \texttt{em}(w_n)$, with $w_1 \cdots w_n \in \Sigma^*$. The language $L \subseteq \Sigma^*$ accepted by $\mathcal{T}$ consists precisely of strings $w \in \Sigma^*$ such that the last vector $\mathbf{v}$ in

$$f_k(f_{k-1}(\cdots f_1(\texttt{em}(w\texttt{EOS})) \cdots)) \tag{1}$$

—that is, at position $|w|+1$ in the sequence—satisfies $\langle \mathbf{t}, \mathbf{v} \rangle > 0$, where $\langle \mathbf{t}, \mathbf{v} \rangle$ denotes the dot product of $\mathbf{t}$ and $\mathbf{v}$. That is, we first apply $f_1, \ldots, f_k$ (in this order) to the sequence $\texttt{em}(w\texttt{EOS})$ of vectors, and check if a weighted sum of the arguments in the last vector is positive.

Remark 1. The above setting of transformers does not admit *Chain of Thoughts (CoTs)*. With CoTs, a transformer $\mathcal{T}$ on input w will output symbols, which are then continuously fed back into $\mathcal{T}$ until a specific output symbol is produced.

That is, on input w, $\mathcal{T}$ produces a symbol a_1. We then run $\mathcal{T}$ on input wa_1 and produce another symbol a_2, and so on. It is known that transformers with CoTs are Turing-complete [3,14,15]. In the sequel, we do not consider transformers with CoTs. □

We have thus far defined the notion of transformers only semantically. We now discuss how to define a seq-to-seq transformation more concretely. To this end, we employ the following ideas:

1. Use *piecewise linear functions* to modify a single vector in the sequence.
2. Use *attention* to "aggregate" several vectors in the sequence.

We will discuss these in turn.

2.1 Piecewise Linear Functions

A *piecewise linear function* is a function $f : \mathbb{R}^r \to \mathbb{R}^s$ that is representable by a Feed-Forward Neural Network (FFNN). More precisely, a piecewise linear function can be defined inductively:

(Base) Each identity function $Id : \mathbb{R}^r \to \mathbb{R}^r$ is piecewise linear.
(Affine) If $f : \mathbb{R}^r \to \mathbb{R}^s$ is piecewise linear and $g : \mathbb{R}^s \to \mathbb{R}^t$ is an affine transformation[2], then the composition $f \circ g : \mathbb{R}^r \to \mathbb{R}^t$ is piecewise linear.
(ReLU) If $f : \mathbb{R}^r \to \mathbb{R}^s$ is piecewise and $i \in \{1, \ldots, s\}$, then the function $g : \mathbb{R}^r \to \mathbb{R}^s$ defined as

$$g(\mathbf{v}) = (w_1, \ldots, w_{i-1}, \max\{0, w_i\}, w_{i+1}, \ldots, w_s),$$

where $f(\mathbf{v}) = (w_1, \ldots, w_s)$, is piecewise linear.

As before, we can extend each piecewise linear function to sequences of vectors by morphisms, i.e., $f : (\mathbb{R}^r)^* \to (\mathbb{R}^s)^*$ with $f(\mathbf{v}_1, \ldots, \mathbf{v}_n) := f(\mathbf{v}_1), \ldots, f(\mathbf{v}_n)$. Notice, however, such functions can *only* modify a vector at the ith position in the sequence solely based on its values and *not* on the values of vectors at other positions. An intra-sequence aggregation of values is enabled by the so-called *attention*, which we discuss next.

2.2 Attention Layers

To define an attention layer, we assume a *weight normalizer* $\mathtt{wt} : \mathbb{R}^* \to \mathbb{R}^*$, which turns any d-sequence of weights into another d-sequence of weights. We will define some common normalizers below, which will result in hard and soft attention layers.

A seq-to-seq transformation $f : (\mathbb{R}^r)^* \to (\mathbb{R}^s)^*$ generated by an attention layer associated with $\mathtt{wt}$ is given by three piecewise linear functions A, B, C

$$A, B : \mathbb{R}^r \to \mathbb{R}^r \qquad C : \mathbb{R}^{2r} \to \mathbb{R}^s.$$

---
[2] That is, given an input vector $\mathbf{x}$, we output $A\mathbf{x}+\mathbf{c}$, where A is a linear transformation and $\mathbf{c}$ is a constant vector.

defined as follows. On input $\sigma = \mathbf{x}_1, \ldots, \mathbf{x}_n$, we have $f(\sigma) = \mathbf{y}_1, \ldots, \mathbf{y}_n$ such that
$$\mathbf{y}_i := C(\mathbf{x}_i, \mathbf{v})$$
where
$$\mathbf{v} := \sum_{j=1}^{n} \mathbf{w}(j)\mathbf{x}_j, \tag{2}$$
$$\mathbf{w} := \mathtt{wt}(\{\langle A\mathbf{x}_i, B\mathbf{x}_j\rangle\}_{j=1}^{n}). \tag{3}$$

In other words, an attention layer looks at a vector $\mathbf{x}_i$ at each position i and decides "how much attention" is to be given to vectors $\{\mathbf{x}_j\}_{j=1}^{n}$ at any position in the input sequence. To this end, one obtains a sequence of weights $\{\langle \mathbf{x}_i, \mathbf{x}_j\rangle\}_{j=1}^{n}$. After normalizing this using $\mathtt{wt}$, the result of the attention is $\mathbf{v}$, which is a weighted sum $\{\mathbf{x}_j\}_{j=1}^{n}$ over all the input vectors.

Soft Attention. Practical transformers use weight normalizers defined by the softmax function, which turns a sequence of weights into a probability distribution. In particular, given a sequence $\sigma = x_1, \ldots, x_n \in \mathbb{R}^n$, define $\mathtt{softmax}(\sigma) := y_1, \ldots, y_n$, where

$$y_i := \frac{e^{x_i}}{\sum_{j=1}^{n} e^{x_j}}.$$

A *SoftMax Attention Transformer (SMAT)* consists of seq-to-seq transformations that are defined using the softmax weight normalizer.

Hard Attention. As previously mentioned, softmax attention is sometimes rather difficult to analyze, owing to the usage of exponential functions. This led researchers to use other weight normalizers that led to the so-called *hard attention layers*. More precise, there are two common flavors: *unique hard attention* and *average hard attention*. A unique hard attention uses the weight normalizer $\mathtt{uha}$ that selects the leftmost maximum weight, i.e., $\mathtt{uha}(x_1, \ldots, x_n) = (y_1, \ldots, y_n)$, where $y_i := 1$ if i is the leftmost position in $\mathbf{x} := x_1, \ldots, x_n$ with $x_i = \max(\mathbf{x})$; or else $y_i := 0$. An average hard attention uses the weight normalizer $\mathtt{aha}$ that selects *all* positions with maximum weight, i.e., $\mathtt{aha}(x_1, \ldots, x_n) = (y_1, \ldots, y_n)$, where $y_i := 1$ if $x_i = \max(\mathbf{x})$; or else $y_i := 0$.

2.3 Positional Information

Thus far, we have actually defined a rather weak class of transformers (called *NoPE-transformers*) that cannot distinguish different positions in the input sequence. They recognize *permutation-invariant* languages, i.e., a string s is in the language L iff all of the reorderings of s are in L. There are two common ways to recover ordering: (1) *masking* and (2) *Position Embeddings (PEs)*. We will go through these in turn.

Masking. Masking is used to "hide" some positions in an input sequence to a layer with respect to a certain "anchor" position. The most commonly used type of masking is called *strict future masking*, which we will focus on in the remainder of the paper.

Intuitively, when attention is applied with respect to the position i, we looked at *all* positions and computed a normalized weight sequence accordingly. The version with strict future masking modifies this by considering only positions j *strictly before* i, i.e., $j < i$. Formally, one simply modifies Eq. 2 and Eq. 3 by the masked version:

$$\mathbf{v} = \sum_{j=1}^{i-1} \mathbf{w}(j)\mathbf{x}_j, \qquad \mathbf{w} = \mathtt{wt}(\{\langle A\mathbf{x}_i, B\mathbf{x}_j\rangle\}_{j=1}^{i-1}).$$

Position Embeddings (PEs). A *Position Embedding* is an *arbitrary* function of the form $p : \mathbb{N} \times \mathbb{N} \to \mathbb{R}^d$. The idea is that $p(i,n)$ indicates the position information of the vector at position i for a sequence of length n. Thus, to extend transformers by PEs, we first apply both the token embedding and the PE p to the input string $w = w_1 \cdots w_n$ before processing the resulting sequence of vectors in the usual way. More formally, we modify the above acceptance condition in the definition of transformers by using

$$f_k(f_{k-1}(\cdots f_1(\sigma))\cdots))$$

where, instead of Eq. 1, we use

$$\sigma := \mathtt{em}(w_1) + p(0, n+1), \cdots, \mathtt{em}(w_n) + p(n, n+1), \mathtt{em}(\mathtt{EOS}) + p(n+1, n+1).$$

At this point, it is appropriate to ask what types of PEs are reasonable. In practice, PEs may use trigonometric functions (e.g. sin) and various other information about the position in the sequence (e.g. the "absolute" position i, the length n of the sequence, etc.). Thus, researchers have studied transformers with PEs *without* any restriction whatsoever on the PEs. Remarkably, some interesting results can already be proven in this setting. We will mention some restricted classes later. We end this section with an easy result:

Proposition 1. *Each Masked UHAT (resp. AHAT) with PEs can be simulated by UHAT (resp. AHAT) with PEs with no masking.*

3 Unique Hard Attention Transformers

The first fundamental result concerning UHAT comes from [6,7], which show that their class of languages is contained in the well-studied circuit complexity class AC^0, consisting of problems solvable by constant-depth, polynomial-size Boolean circuits. More recently, this containment was proven strict [2].

Theorem 1 *([2,6,7]). UHAT with PEs is strictly subsumed in AC^0.*

Proof idea. Let us quickly discuss the proof idea behind the containment in AC^0. Fix an UHAT $\mathcal{T}$ with, say, h layers. For simplicity, let us assume the alphabet $\Sigma = \{a, b\}$. The key idea is that there is a polynomial function $p(n)$, for any possible string length n, such that the set V_n of vectors—as well as the set S_n of possible attention scores—that can be generated in the computation of the UHAT has size $|V_n| = O(p(n))$. More precisely, in the input layer after application of em and position encoding, we can generate $O(n)$ many vectors. In the next layer, there are $O(n^2)$ many vectors. In the kth layer, there are $O(n^{2^{k-1}})$ possible vectors. Therefore, we may set $p(n)$ to be $O(n^{2^h})$.

Thus, we may represent each vector in V_n and each attention score in S_n using $O(\log n)$ bits. Therefore, using a constant depth polynomial-sized boolean circuit (by a simple enumeration), we can represent the relation $\preceq \subseteq S_n \times S_n$ containing pairs (s, s') such that s has a smaller attention score as s'. Similarly, using a constant depth polynomial-sized boolean circuit, we can represent the relation $R \subseteq V_n \times V_n \times S_n$ such that $R(\mathbf{v}, \mathbf{v}', s)$ iff $\langle \mathbf{v}, \mathbf{v}' \rangle = s$. Together, this allows us to represent—using constant depth polynomial-sized boolean circuits— the function $f_\ell : V_n^n \times \{1, \ldots, n\} \to V_n$ such that $f_\ell((\mathbf{v}_1 \cdots \mathbf{v}_n), i)) = \mathbf{v}$ iff, whenever the ℓth layer has input sequence $\mathbf{v}_1 \cdots \mathbf{v}_n$, the vector at position i at layer #($\ell+1$) is $\mathbf{v}$. All in all, this gives rise to a constant-depth polynomial-sized boolean circuit C_n for input strings of length n.

To conclude the theorem, we simply use the (non-uniform) family $\{C_n\}_{n \geq 0}$ of circuits to represent $\mathcal{T}$.

Combined with well-known limitations of AC^0 (e.g. see [1,12]), the above result shows that some languages are not expressible by UHAT, including PARITY and MAJ, where the latter is defined as:

$$\text{MAJ} := \{w \in \{a,b\}^* : |w|_a \geq |w|_b\}.$$

While this provides us a ceiling of what languages are expressible as UHATs, the following two results show what UHATs are capable of. To this end, we write FO[Mon] to denote first-order logic over strings extended only by *monadic* numerical predicates (i.e. sets of numbers); recall that this would have yielded AC^0 if extended with all k-ary ($k \geq 1$) numerical predicates; see [12]. An example of monadic numerical predicates is Mod_2^3 containing all numbers that are 2 (mod 3).

Theorem 2 ([2,19]). *FO[Mon] is expressible by UHAT with PEs, as well as by masked UHATs with finite-image PEs. In addition, masked NoPE-UHAT coincides with FO, which in turn coincides with star-free languages.*

Proof idea. To prove the containment of FO[Mon] in UHAT—either with PEs or masked attention with finite image PEs—we use Kamp's theorem [9]: FO[Mon] coincides in expressivity with LTL[Mon], i.e., LTL formulas that also use monadic numerical predicates as atomic propositions. Unlike FO formulas, which have multiple variables, LTL formulas are *unary*, meaning that their semantics is a set of positions over a string. This simpler structure of LTL aligns well with the expressive power of UHATs, allowing for a proof using structural induction. In particular, we inductively show that for every LTL formula ϕ with unary numerical predicates, there exists a UHAT $\mathcal{T}_\phi$ such that on input $\sigma = \mathbf{x}_1, \ldots, \mathbf{x}_n$, corresponding to the embedding of a word $w = a_1, \ldots, a_n \in \Sigma^+$, it outputs a sequence $\mathcal{T}_\phi(\sigma) = \mathbf{y}_1, \ldots, \mathbf{y}_n$ over $\{0,1\}$ that contains a 1 precisely in those positions of w that satisfy ϕ.

Let us give some intuition on how to do the aforementioned induction proof. For the base case, we deal with only Q_a (saying that the current position has letter a) or a monadic numerical predicate $U \subseteq \mathbb{N}$. We need to set up the token embedding function em and position embedding p with a large enough dimension so that information on truth/falsehood of each atomic proposition in the given LTL formula can be read off directly. For example, for a string $w := abaa$ with the LTL formula $\mathbf{G}(\text{Mod}_2^2 \to Q_a)$, we would map w to the following sequence of vectors:

$$(1,0,0), (0,1,1), (1,0,0), (1,0,1)$$

where the vector at position i corresponds to $(Q_a(i), Q_b(i), \text{Mod}_2^2(i))$. Note, we omitted EOS and potentially other "information" in the PEs for readability.

For the inductive case, one introduces new arguments at each position (i.e. increases the dimension) to encode truth/falsehood of the formulas higher up in the parse tree. Note, we keep the information stored in the previous layer.

For boolean combinations, one can handle this with piecewise linear functions. That is, $\neg \varphi$ can be implemented by the function $1 - x_\varphi$, where x_φ encodes the value of φ at the same position in the string. For $\varphi \lor \psi$, we can implement it as $x_\varphi + \text{ReLU}(x_\psi - x_\varphi) = x_\varphi + \max(0, x_\psi - x_\varphi)$.

We next give an intuition how to do $\mathbf{F}\varphi$ and show how to do this with PEs (with no masking). For other temporal operators, the reader is referred to [2]. To this end, we assume by induction that the value x_φ and $x_{\neg\varphi}$ are available at every position in the sequence. The first step is to "nullify" the value $x_{\neg\varphi}$ at the last position n, i.e., $x_{\neg\varphi}[n] := 0$. See the proof of Lemma 1 in [2]. We then assume the use the following information at position i:

$$\mathbf{v}_i := \langle \cos(\pi(1 - 2^{-i})/10), \sin(\pi(1 - 2^{-i})/10), 1, x_{\neg\varphi} \rangle.$$

With an appropriate affine transformation B, we have

$$B\mathbf{v}_i := \langle \cos(\pi(1 - 2^{-i})/10), \sin(\pi(1 - 2^{-i})/10), -10.x_{\neg\varphi}, 0 \rangle.$$

Thus, we have

$$\langle \mathbf{v}_i, B\mathbf{v}_j \rangle := \cos(\pi(2^{-i} - 2^{-j})/10) - 10.x_{\neg\varphi}.$$

The value $\cos(\pi(2^{-i} - 2^{-j})/10)$ is maximized at position $j \geq i$ and not at $j < i$. In addition, the value $-10.x_{\neg\varphi}$ is maximized at $j = n$ (possibly also at $j < i$). Thus, it follows that $\langle \mathbf{v}_i, B\mathbf{v}_j \rangle$ is maximized at position $j \geq i$. Furthermore, it can be verified that among the value $j \geq i$ the value $\cos(\pi(2^{-i} - 2^{-j})/10)$ monotonically decreases in j. All in all, unique hard attention picks the vector $\mathbf{v}$ at the leftmost position $j \geq i$ such that $w, j \models \varphi$ (otherwise, it picks the vector $\mathbf{v}$ at position n), with which we can forward the truth/falsehood of $\mathbf{F}\varphi$. □

Corollary 1. *UHAT with PEs contain all regular languages expressible in AC^0.*

Proof. Regular languages in AC^0 are expressible in FO with unary numerical predicates [16] (more precisely, Mod_r^d containing all numbers that are r (mod d)). The corollary then follows from Theorem 2. □

Theorem 2 turns out to be powerful enough to show the following interesting "non-regular" capability of UHAT with PEs.

Corollary 2 ([2]). *Palindrome is in UHAT with PEs.*

Proof idea. Using PEs, it is possible to extend Theorem 2 with any desired family $\{\preceq_n\}_{n\geq 0}$, where $\preceq_n$ deals with strings of length n. Therefore, on strings of length n, we could use the ordering

$$1, n, 2, n-2, 3, n-3, \ldots$$

of the set $\{1, \ldots, n\}$. This essentially turns the string $abccba$ into $aabbcc$, for example. Therefore, using the unary numerical predicate Mod_1^2, we can write an LTL[Mon] (or equivalent FO[Mon]) formula that says that at each odd position i the next position $i+1$ has to have the same label as that at position i. □

We conclude our discussion of UHAT by the problem of verifying Masked UHAT with no PEs. By verifying, this could mean checking the emptiness, universality of the language, or its equivalence to (or containment in) another Masked UHAT. By Theorem 2, each Masked UHAT can be effectively turned into a finite-state automaton recognizing the same language. Owing to decidability of emptiness, universality, equivalence, and containment for finite automata, we obtain the same decidability results for Masked UHAT with no PEs.

Corollary 3 ([19]). *The problem of verifying Masked UHAT with no PEs is decidable.*

4 Logical Languages for Average Hard Attention

It is easy to construct an AHAT that recognizes MAJ. This takes AHAT beyond AC^0. The following result shows that TC^0 still upper-bounds the capability of AHAT.

Theorem 3 ([7]). *Languages recognized by AHAT are in TC^0*

The main reason behind the TC^0 upper bound of AHAT is the ability of TC^0-circuits to simulate arithmetic, which is needed in the computation of average hard attention.

For the time being no complete characterization for neither AHAT with PEs nor masked NoPE-AHAT exists. That is, we do not have an extension of Theorem 2 to AHAT. However, it is still possible to specify a logic that expresses languages that can be expressed by AHATs. The logic is called *Counting LTL*, as first defined in [2]. Intuitively, Counting LTL extends LTL with linear counting terms of the form:

$$C, C' := c\ (c \in \mathbb{Z})\ |\ \overleftarrow{\#}[\varphi]\ |\ \overrightarrow{\#}[\varphi]\ |\ C + C'\ |\ C - C',$$

and formulas of the form $C \leq C'$, where C and C' are linear counting terms. The term $\overleftarrow{\#}[\varphi]$ (resp. $\overrightarrow{\#}[\varphi]$) counts the number of times φ holds at positions before (resp. after) the one where we are evaluating the formula. The remaining terms and formulas have an intuitive meaning.

We define the fragment $\mathrm{K}_t[\#]$ of the Counting LTL, which removes all temporal operators of LTL, as well as terms of the form $\overrightarrow{\#}[\psi]$. That is, only terms of the form $\overleftarrow{\#}[\varphi]$ is allowed. For instance, if Q_a and Q_b are formulas that check whether a position in a word holds symbol a or b, respectively, then the $\mathrm{K}_t[\#]$ formula $\overleftarrow{\#}[Q_b] \leq \overleftarrow{\#}[Q_a]$ checks whether the word belongs to MAJ (if evaluated on the last position of the word). Similarly, we can define Dyck-1, the language of well-matched parenthesis words over the alphabet consisting of tokens (and). The $\mathrm{K}_t[\#]$ that checks for this language over the last position of a word in this alphabet is:

$$\overleftarrow{\#}[Q_(] = \overleftarrow{\#}[Q_)] \wedge \overleftarrow{\#}[\overleftarrow{\#}[Q_)] > \overleftarrow{\#}[Q_(]] = 0,$$

where we have used some standard logical abbreviations. It is possible to show that the Counting LTL can express PARITY, whereas $\mathrm{K}_t[\#]$ cannot express PARITY [8].

Theorem 4 ([2, 18]). *Counting LTL extended with unary numerical predicates is in AHAT with PEs. The fragment $\mathrm{K}_t[\#]$ is expressible by masked NoPE-AHAT.*

The basic idea behind the proofs of these results is that AHATs allow to compute the uniform average value among all positions that maximize the attention. This averaging mechanism allows to express many counting properties of interest. The proof is, again, by structural induction on Counting LTL formulas.

We showed that UHAT contains all regular languages in AC^0. We do not know if this is true for AHAT. That said, Theorem 4 can be used to show the following slightly weaker result:

Corollary 4 ([2]). *If TC^0 is strictly contained in NC^1, then AHAT with PEs contains all regular languages in TC^0.*

It turns out that, for the subclass of AHATs with no masking and no PEs, the following complete characterization can be proven:

Theorem 5 ([10]). *NoPE-AHAT recognizes precisely all permutation-invariant languages with semi-algebraic Parikh images.*

To explain this theorem, recall (see [11]) that the Parikh image $\mathcal{P}$ of a language is a mapping from all strings in the language to their letter counts. For example, $\mathcal{P}((ab)^*) = \{(n,n) : n \in \mathbb{N}\}$. Here, the tuple $(3,3)$ simply denotes that there are 3 occurrences of as and 3 occurrences of b's. Parikh's Theorem shows that context-free languages have semilinear Parikh images, i.e., they are definable in Presburger Arithmetic. In contrast, a relation $R \subseteq \mathbb{N}^k$ is *semi-algebraic* if it can be expressed as a finite union of nonnegative integer solutions to systems of multivariate polynomial inequalities. That is, the above theorem implies (among others) that languages L_k of the form $\{w \in \{a,b\}^* : |w|_a \geq (|w|_b)^k\}$, are expressible by NoPE-AHAT; note that L_k has no semilinear Parikh images for $k \geq 2$. Interestingly, this also shows that Counting LTL does not subsume NoPE-AHAT, since the former can only express permutation-invariant languages with semilinear Parikh images [10].

Theorem 5 yields immediately undecidability of verification of NoPE-AHAT since solvability of Diophantine equations is well-known to be undecidable [13].

Corollary 5 ([10]). *Checking whether a NoPE-AHAT recognizes a nonempty language is undecidable.*

5 Limitations of UHATs and AHATs

Having gone through some body of results in the literature, we now discuss two main limitations of these results.

Limitation 1: Soft attention vs. Hard attention. As we remarked, practical transformers are based on soft attention. It is still unclear whether the theory of expressivity of UHATs and AHATs provides a good approximation of the theory of expressivity of softmax transformers. For example, we do not know where the expressivity of softmax transformers exactly lies (e.g. do they subsume UHATs?). That said, it is known that PARITY can be captured by a softmax transformer with PEs. Thus, softmax transformers are not subsumed by UHATs [4]. Furthermore, the relationship between AHAT and softmax transformers has also not been fully delineated (for more on this, see [18,20]).

One subclass of AHAT that seems to be a promising approximation of SMAT restricts all layers to apply only *uniform* attention. More precisely, an AHAT layer is uniform if the piecewise linear functions $A, B : \mathbb{R}^r \to \mathbb{R}^r$ ensure that there exists a constant c such that $\langle A\mathbf{x}, B\mathbf{y}\rangle = c$ for all $\mathbf{x}, \mathbf{y} \in \mathbb{R}^r$. This can happen esp. when the linear transformation components of A and B map $\mathbf{x}$ and $\mathbf{y}$ to 0. The subclass is denoted by AHAT[U]. The following result is folklore and can easily be shown by noting that $\mathtt{softmax}(s_1, \ldots, s_n) = \mathtt{aha}(s_1, \ldots, s_n) = 1/n$, whenever $s_1 = \cdots = s_n$, which can be guaranteed for uniform AHAT layers.

Proposition 2. *Language recognized by AHAT[U] are also recognized by SMAT.*

The above observation was already used in [10,18] to show the power of SMAT:

Proposition 3 ([18]). *$K_t[\#]$ languages are recognizable by SMAT.*

Proposition 4 ([10]). *Permutation-invariant languages with semialgebraic Parikh images are recognizable by SMAT.*

Limitation 2: Trainability vs. expressibility. Not all expressible languages are efficiently trainable on transformers, i.e., by means of Stochastic Gradient Descent (SGD). This applies particularly to PARITY [4], which seems to be extremely difficult to train on transformers with any high enough level of accuracy, although it is expressible by a softmax transformer. This phenomenon was very recently shown to be caused by *sensitivity*. Loosely speaking, PARITY is sensitive since flipping one letter (i.e. a to b and vice versa) changes the parity of any string. Contrast this with MAJ, where there are not so many strings that change their memberships in MAJ, after flipping a letter. This was hypothesized to be the reason why MAJ is efficiently trainable, whereas PARITY is not.

One interesting upshot of the research effort in understanding trainability is the so-called *RASP-L conjecture* [21], which states that a concept is likely to length generalize (i.e. when trained on shorter strings, generalize to longer strings) precisely whenever it is expressible as a short RASP-L program. However, as noted by Huang et al. [8], this is not a precisely formulated conjecture. The authors postulated a formal version of RASP-L conjecture by replacing RASP-L with limit transformers and the logic $K_t[\#]$, for which they could successfully prove and empirically verify a length generalization theorem. In particular, this ruled out PARITY (as it is not in $K_t[\#]$), but admits MAJ. It remains to be seen if $K_t[\#]$ subsumes all concepts that admit length generalization on transformers.

6 Conclusions

We have discussed several key results employing logic and automata for understanding what is expressible in/efficiently trainable for transformers. It must be emphasized that these are only a handful of results in this rapidly growing field of FLaNN (Formal Languages and Neural Networks); for a more comprehensive (though less detailed) account of FLaNN, see the excellent survey [17]. It is our sincere hope that this article could motivate more researchers in logic and automata, as well as verification and programming languages, to take up some of the many pressing challenges in FLaNN.

Acknowledgments. We thank David Chiang, Michael Hahn, Alexander Kozachinskiy, Andy Yang, and Georg Zetzsche for the fruitful discussion. Lin is supported by the European Research Council (https://doi.org/10.13039/100010663.) under Grant No. 101089343 (LASD). Barceló is funded by ANID - Millennium Science Initiative Program - Code ICN17002, and by CENIA FB210017, Basal ANID.

References

1. Ajtai, M.: $\sum_{1}^{1}$-formulae on finite structures. Ann. Pure Appl. Log. **24**(1), 1–48 (1983)
2. Barceló, P., Kozachinskiy, A., Lin, A.W., Podolskii, V.V.: Logical languages accepted by transformer encoders with hard attention. In: ICLR (2024)

3. Bhattamishra, S., Patel, A., Goyal, N.: On the computational power of transformers and its implications in sequence modeling. In: CoNLL, pp. 455–475 (2020)
4. Chiang, D., Cholak, P.: Overcoming a theoretical limitation of self-attention. In: Muresan, S., Nakov, P., Villavicencio, A. (eds.) ACL, pp. 7654–7664 (2022)
5. Elman, J.L.: Finding structure in time. Cogn. Sci. **14**(2), 179–211 (1990)
6. Hahn, M.: Theoretical limitations of self-attention in neural sequence models. Trans. Assoc. Comput. Linguistics **8**, 156–171 (2020)
7. Hao, Y., Angluin, D., Frank, R.: Formal language recognition by hard attention transformers: perspectives from circuit complexity. Trans. Assoc. Comput. Linguistics **10**, 800–810 (2022)
8. Huang, X., et al.: A formal framework for understanding length generalization in transformers. In: ICLR (2025)
9. Kamp, H.W.: Tense Logic and the Theory of Linear Order. Ph.D. thesis, University of California, Los Angeles (1968)
10. Köcher, C., Kozachinskiy, A., Lin, A.W., Sälzer, M., Zetzsche, G.: Nope: the counting power of transformers with no positional encodings. CoRR arxiv:2505.11199 (2025)
11. Kopczynski, E., To, A.W.: Parikh images of grammars: complexity and applications. In: Proceedings of the 25th Annual IEEE Symposium on Logic in Computer Science, LICS 2010, Edinburgh, United Kingdom, 11–14 July 2010, pp. 80–89 (2010). https://doi.org/10.1109/LICS.2010.21
12. Libkin, L.: Elements of Finite Model Theory. Springer, Heidelberg (2004)
13. Matiyasevich, Y.V.: Hilbert's Tenth Problem. MIT Press, Cambridge (1993)
14. Merrill, W., Sabharwal, A.: The expressive power of transformers with chain of thought. In: ICLR (2024)
15. Pérez, J., Barceló, P., Marinkovic, J.: Attention is turing-complete. J. Mach. Learn. Res. **22**, 75:1–75:35 (2021)
16. Straubing, H.: Finite Automata, Formal Logic, and Circuit Complexity. Progress in Theoretical Computer Science, 1 edn. Birkhäuser Boston (1994). https://doi.org/10.1007/978-1-4612-0289-9
17. Strobl, L., Merrill, W., Weiss, G., Chiang, D., Angluin, D.: What formal languages can transformers express? A survey. Trans. Assoc. Comput. Linguist. **12**, 543–561 (2024)
18. Yang, A., Chiang, D.: Counting like transformers: compiling temporal counting logic into softmax transformers. CoRR arxiv:2404.04393 (2024)
19. Yang, A., Chiang, D., Angluin, D.: Masked hard-attention transformers recognize exactly the star-free languages. In: NeurIPS (2024)
20. Yang, A., Strobl, L., Chiang, D., Angluin, D.: Simulating hard attention using soft attention. CoRR arxiv:2412.09925 (2024)
21. Zhou, H., et al.: What algorithms can transformers learn? A study in length generalization. In: ICLR (2024)

Simplicity Lies in the Eye of the Beholder: A Strategic Perspective on Controllers in Reactive Synthesis

Mickael Randour(✉)

F.R.S.-FNRS and UMONS – Université de Mons, Mons, Belgium
mickael.randour@umons.ac.be

Abstract. In the game-theoretic approach to controller synthesis, we model the interaction between a system to be controlled and its environment as a game between these entities, and we seek an appropriate (e.g., winning or optimal) strategy for the system. This strategy then serves as a formal blueprint for a real-world controller. A common belief is that *simple* (e.g., using limited memory) *strategies are better*: corresponding controllers are easier to conceive and understand, and cheaper to produce and maintain.

This invited contribution focuses on the complexity of strategies in a variety of synthesis contexts. We discuss recent results concerning memory and randomness, and take a brief look at what lies beyond our traditional notions of complexity for strategies.

1 Introduction

A Story of Control. Many different fields—such as control theory, AI, and formal methods—are concerned with a common problem: how to control an agent embedded within an uncontrollable environment (possibly involving other agents) in order to achieve a given goal. This challenge was already envisioned by Church in the 1950s [29], and is now known as the *controller synthesis* problem.

My motherland is formal methods, and I have been shaped by the teachings of game theory. It is therefore no surprise that I focus on the *game-theoretic approach to controller synthesis*. The rich history linking Church's problem, logic, automata theory, and games on graphs is expertly narrated by Thomas in [50].

The Game-Theoretic Metaphor. The approach can be sketched as follows. Given a *specification* that formalizes what the system should and should not do, along with a formal model of the system, we aim to automatically construct a *controller* that ensures correct behavior, regardless of how the environment behaves.

In the simplest setting, the controller is viewed as a player, and the uncontrollable environment as its adversary. Their interaction is modeled as a *two-player*

Mickael Randour is an F.R.S.-FNRS Senior Research Associate and member of the TRAIL Institute. His work has been supported by the F.R.S.-FNRS under Grants n° F.4520.18 (MIS ManySynth) and n° T.0188.23 (PDR ControlleRS).

zero-sum game on a graph, where vertices represent *states* of the system and environment, and edges represent their possible *actions*. The players take turns moving a pebble along this graph according to *strategies*, generating a sequence of vertices called a *play*, which represents a possible behavior of the system. The controller's goal is to ensure that the play satisfies the specification, encoded as a *winning objective* (a play is winning if it belongs to the specified set). The adversarial environment's goal is to prevent this—we are effectively modeling a worst-case scenario.

Establishing winning strategies (i.e., guaranteeing victory regardless of the environment's choices) *corresponds to synthesizing implementable models of provably correct controllers*: these strategies are formal blueprints for acceptable controllers [6,45].

The state of the art extends far beyond this basic setting. Richer models incorporate, e.g., quantitative payoffs, trade-offs between objectives, or stochastic transitions. In what follows, we use the term "games" as an umbrella for: two-player *antagonistic* games as described above; *Markov decision processes* (MDPs, or $1\frac{1}{2}$-player games, where the system faces a fully probabilistic environment); and *stochastic games* (SGs, or $2\frac{1}{2}$-player games, which combine stochastic transitions with an adversarial opponent). A comprehensive introduction to games on graphs, covering classical results and recent developments, can be found in [32].

Where it All Began. Every hero has an origin story. So does every scientific endeavor. I began mine around 2010, at a time when games on graphs were already a well-established area of research, particularly for controller synthesis [36].

Historically, most game settings were studied considering a single objective for the system, either qualitative (e.g., reachability, Büchi, parity) or quantitative (e.g., mean payoff, shortest path, discounted sum). Remarkably, *pure memoryless strategies*—which we will discuss shortly—suffice to play optimally (i.e., winning whenever it is possible to win, or optimizing a given payoff function) in two-player games for virtually all classical objectives from the literature, and for both players (see, e.g., Gimbert and Zielonka's characterization [35]). The situation is similar for MDPs and SGs.

But what is a strategy? Mathematically, it is most often viewed as a *function* that maps histories (the sequence of vertices and actions up to the point of decision) to actions (typically the next vertex to visit). In general, there is no limit to how much a strategy can "remember," though infinite-memory strategies are of limited practical interest. Strategies can also rely on randomness in various ways (we will compare them in Sect. 4), the most classical one mapping histories to probability distributions over actions (so-called *behavioral* strategies). Pure memoryless strategies are often seen as the *simplest* kind of strategy: they use neither memory, nor randomness. In essence, they are simply functions from vertices to actions. Therefore, their sufficiency in most (single-objective) games can be seen as a blessing. Indeed, in the context of controller synthesis, the general consensus is that *simple strategies are always better* (e.g., [31]): corresponding controllers will be easier to conceive, and cheaper to produce and maintain. Simplicity is also strongly correlated with explainability, another desirable feature.

The First Steps. My early work contributed to a blooming effort to develop *many-sided synthesis* frameworks, which take into account the interplay between different quantitative (or qualitative) aspects and the resulting trade-offs that naturally arise in applications (e.g., decreasing the response time of a system may require additional computing power and energy consumption). I worked in several directions, such as multi-dimension games (e.g., [28]), Boolean combinations of objectives (e.g., [40]), games with heterogeneous objectives (e.g., [10]), combination of hard worst-case constraints with expected value optimization (e.g., [20]), conjunctions of percentile constraints (e.g., [48]), and efficient synthesis under simplicity constraints (e.g., [31]).

Moving from single-objective games to many-sided ones takes a heavy toll. First, there is no total order on the performance of strategies, which forces us to consider complex *Pareto frontiers* (and corresponding Pareto-optimal strategies) instead of a single optimal strategy (see an example in Sect. 4). Second, the *computational complexity* of solving a game or synthesizing a strategy increases rapidly. Finally, and most importantly in the context of this note, *strategies (at least for the system) typically need to be much more complex* to play Pareto-optimally: they might require memory, either small [16], huge [20, 26, 28], or even infinite [19, 51], randomness [28], or both [48]. I wrote an introduction to multi-objective games, discussing these costs, in [46].

Welcome to the Machine. The attentive reader may have noticed that, until now, strategies were treated as abstract mathematical objects; yet we now suggest that their (somewhat undefined) memory can be measured. This shift warrants discussion. Many papers on games adopt a purely theoretical view of strategies, often without considering their concrete representation; this is especially common when pure memoryless strategies suffice, as these are frequently viewed as "trivial to implement." When strategies involving memory are considered, they are typically represented using finite-state machines with outputs—either *Mealy* or *Moore machines*. From this perspective, the memory of a strategy corresponds to the number of states in its associated machine, and its "complexity" is often quantified by this number alone. We will return to this point shortly. For an introduction to Mealy machines and their various forms, see the survey [12].

One Meta-Theorem to Rule them All. Delving into the realm of multi-objective games opens a Pandora's box of endless specific combinations. A flurry of work has explored these combinations—often seemingly closely related and yielding similar results, yet always differing just enough to leave an independent observer wondering: What is common? What differs? Are similar results obtained for similar reasons? Can we gain transversal insight into multi-objective games and their strategies via a theoretical cross-section?

These were the questions I began to ponder after several years of working on many-sided synthesis. I was particularly inspired by Gimbert and Zielonka's characterization of memoryless-determined two-player games [35], an elegant result obtained in an elegant way. I was especially drawn to the "one-to-two-player lift" established as a corollary: it essentially states that if memoryless strategies suffice in all one-player games (which are much easier to study), for both players, then they also suffice in two-player games.

I started working with Youssouf Oualhadj, aiming to prove a similar result for finite-memory strategies. The project stayed on the back burner for a while but came back to the forefront during Pierre Vandenhove's PhD thesis, which I co-supervised with Patricia Bouyer. In collaboration with Stéphane Le Roux, we were able to prove that such a lift does not hold in full generality. However, we did provide a *comparable characterization and lift for an appropriate class of finite-memory strategies* [9]. This opened new avenues of research for me and triggered a gradual shift toward establishing *meta-theorems*—aimed at understanding the nature of strategy complexity across diverse settings.

The bulk of this note is dedicated to some of the results obtained in that direction, along the two main dimensions of strategy complexity in the classical model: *memory* [9,11,13] and *randomness* [41,42]. These results were made possible thanks to wonderful collaborators (see below), and progress along each axis was driven by outstanding PhD students: Pierre Vandenhove for memory, and James C. A. Main for randomness.

A Rabbit Hole. Let us take a step back. Simplicity is the paragon virtue of strategies, and we now have a clearer understanding of complexity requirements across a wide range of game models. Is that the end of the story? Not quite.

Indeed, all these results are deeply tied to the *Mealy machine representation* of strategies. This is the predominant model used in the literature (e.g., [32]) and it is often treated as canonical. Its use is natural, given the central role of logic and automata in synthesis approaches, and it is a versatile tool from a theoretical standpoint. Nonetheless, I will argue in Sect. 5 that our shared understanding of "simplicity" is, in part, a *side-effect of this representational choice*—one that is somewhat disconnected from the intuitive notion of simplicity one might expect in the context of controller synthesis.

While recent work has begun to explore *alternative representations*—strategy machines [33], decision trees [14,15], neural networks [21], programs [49], or structurally-enriched Mealy machines [1,5]—these efforts remain largely motivated by practical considerations. We are only scratching the surface: a thorough understanding of the *ecosystem of representation models*—and their interrelations—is still lacking, as is a *representation-agnostic theory* of complexity.

Outline. A word of warning: given the breadth of models and results covered in this survey, we adopt a somewhat informal approach. There will be occasional hand-waving and approximations to keep the discussion at a high level. Pointers to the relevant papers are provided throughout for full formal details.

In Sect. 2, we recall the main concepts of games on graphs, including the Mealy machine representation of strategies. Sections 3 and 4 are devoted to results on memory and randomness complexity, respectively, within this model. Section 5 questions our traditional (Mealy-based) notions of complexity and opens a window onto what lies beyond. We conclude briefly in Sect. 6.

2 Games for Controller Synthesis

We assume that the reader is familiar with basic concepts in games on graphs. While we recall and illustrate the main notions throughout, we often omit formal definitions. For a comprehensive treatment of the underlying theory, we refer to the recent textbook [32], which provides all necessary background.

Two-Player Games. We consider two players, $\mathcal{P}_\bigcirc$, the controller, and $\mathcal{P}_\square$, its adversarial environment. A two-player turn-based *arena* is a tuple $\mathcal{A} = (V_\bigcirc, V_\square, E)$, where $V = V_\bigcirc \uplus V_\square$ is the set of vertices, partitioned into vertices of $\mathcal{P}_\bigcirc$ and $\mathcal{P}_\square$, and E is the set of edges. We assume there are no deadlocks. Each edge e connects a source vertex $\mathsf{In}(e)$ to a target vertex $\mathsf{Out}(e)$. An example is given in Fig. 1a. We assume arenas are finite unless otherwise specified.

We use a *coloring function* $\mathfrak{c}\colon E \to C$. Colors may represent, e.g., symbols, priorities (in parity objectives), or weights (in quantitative objectives). They serve as a general abstraction for defining objectives.

A *play* is an infinite sequence $\pi = v_0 e_1 v_1 e_2 v_2 \ldots \in (VE)^\omega$ such that v_0 is a given initial vertex, and for all i, $\mathsf{In}(e_i) = v_{i-1}$ and $\mathsf{Out}(e_i) = v_i$—we allow multiple edges (with potentially different colors) between two given states; hence, it is necessary to consider edges explicitly. We sometimes identify a play with its projection onto colors, i.e., the sequence $\mathfrak{c}(e_1)\mathfrak{c}(e_2)\ldots$. For instance, in Fig. 1a, we would consider $\pi = abc(abbd)^\omega$ as a play.

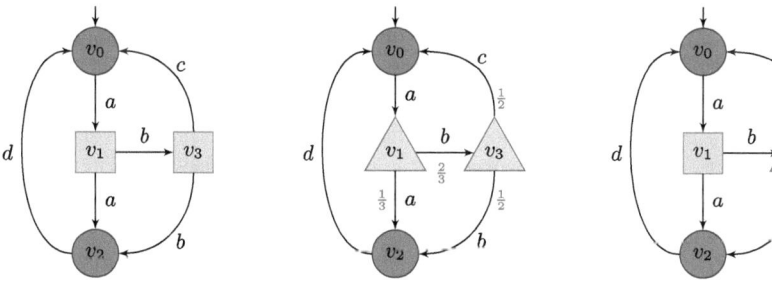

(a) Two-player game: $\mathcal{P}_\bigcirc$ and $\mathcal{P}_\square$.
(b) Markov decision process: $\mathcal{P}_\bigcirc$ and $\mathcal{P}_\triangle$.
(c) Stochastic game: $\mathcal{P}_\bigcirc$, $\mathcal{P}_\square$, and $\mathcal{P}_\triangle$.

Fig. 1. Three types of games. Vertices are partitioned between those of $\mathcal{P}_\bigcirc$, the system, $\mathcal{P}_\square$, an antagonistic adversary, and $\mathcal{P}_\triangle$, a random adversary. Letters are the colors of the edges, whereas fractions represent probability distributions in random vertices. The initial vertex is highlighted via a short arrow.

Objectives. As discussed in Sect. 1, $\mathcal{P}_\bigcirc$ aims to enforce a *specification*, while $\mathcal{P}_\square$ attempts to prevent this. The three main ways to formalize a specification using colors are as follows:

1. A *winning condition* is a set of plays that $\mathcal{P}_\bigcirc$ aims to realize. E.g., $\mathsf{Reach}(t) = \{\pi = c_0 c_1 c_2 \ldots \mid t \in \pi\}$, for $t \in C$ a given color, a *reachability* objective.
2. A *payoff function* represents a quantity to optimize, assuming $C \subset \mathbb{Q}$. E.g., the *discounted sum* function, defined as $\mathsf{DS}(\pi) = \sum_{i=0}^{\infty} \gamma^i c_i$ for $\gamma \in (0, 1)$.
3. A *preference relation* defines a total preorder over sequences of colors, thus generalizing both previous concepts.

Strategies and Optimality. Player $\mathcal{P}_\triangledown$, with $\triangledown \in \{\bigcirc, \square\}$, selects outgoing edges (from vertices in $V_\triangledown$) according to a *strategy* $\sigma_\triangledown \colon (VE)^* V_\triangledown \to E$, which must be consistent with the underlying graph. Observe that there is no bound on the length of the input history. For now, we consider only strategies that do not involve randomness—these are called *pure* strategies. We will return to randomized strategies in Sect. 4.

A particularly important subclass is that of *memoryless* strategies, i.e., functions $\sigma_\triangledown \colon V_\triangledown \to E$, where the choice depends only on the current vertex.

In the context of synthesis, we are interested in the complexity of *optimal* strategies. Recall that we currently focus on single-objective games. Suppose the objective is described by a preference relation $\sqsubseteq$ for $\mathcal{P}_\bigcirc$. Then a strategy $\sigma_\bigcirc$ is optimal if its worst-case outcome (i.e., considering all strategies of the adversary $\mathcal{P}_\square$) is at least as good, with respect to $\sqsubseteq$, as that of any other strategy $\sigma'_\bigcirc$. That is, we assume a rational adversary playing optimally and evaluate strategies based on the quality of their worst-case outcomes. See [9] for a thorough discussion of optimality under preference relations.

Stochastic Environments. The environment might not be fully antagonistic but exhibit stochastic behavior: we introduce two other types of games that incorporate a random player, denoted $\mathcal{P}_\triangle$, representing probabilistic choices.

Given a finite set S, we write $\mathfrak{D}(S)$ for the set of probability distributions over S. A *Markov decision process (MDP)* is a tuple $\mathcal{D} = (V_\bigcirc, V_\triangle, E, \delta)$, where $V = V_\bigcirc \uplus V_\triangle$ is the set of vertices, partitioned into $\mathcal{P}_\bigcirc$- and $\mathcal{P}_\triangle$-vertices. The edge set is E, and δ is the probabilistic transition function $\delta \colon V_\triangle \to \mathfrak{D}(E)$, assigning probability distributions to the outgoing edges of random vertices. An example MDP is shown in Fig. 2b. While MDPs are often defined using actions (see, e.g., [48]), we adopt the *random vertex* formalism for uniformity.

Fixing a strategy $\sigma_\bigcirc$ and an initial vertex v induces a fully stochastic process: a *Markov chain (MC)*. The objective of $\mathcal{P}_\bigcirc$ can be to maximize either the probability $\mathbb{P}_v^{\sigma_\bigcirc}[W]$ of a winning condition W, or the expected value $\mathbb{E}_v^{\sigma_\bigcirc}[f]$, for a payoff function f. For the construction of the probability measure, see [32].

Stochastic games (SGs) combine the adversarial and stochastic models of the environment. An SG is a tuple $\mathcal{S} = (V_\bigcirc, V_\square, V_\triangle, E, \delta)$, extending the structures above. An example appears in Fig. 1c. In SGs, strategies are required for both $\mathcal{P}_\bigcirc$ and $\mathcal{P}_\square$ to induce an MC. The goal of $\mathcal{P}_\bigcirc$ is to maximize, against an optimal adversary $\mathcal{P}_\square$ (i.e., worst-case scenario), the quantity $\mathbb{P}_v^{\sigma_\bigcirc, \sigma_\square}[W]$ or $\mathbb{E}_v^{\sigma_\bigcirc, \sigma_\square}[f]$.

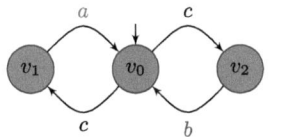
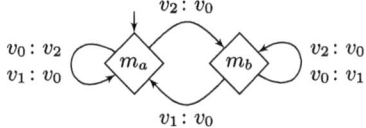

(a) Generalized Büchi game. (b) Mealy machine.

Fig. 2. Memory is needed to see a and b infinitely often. A winning strategy for $\mathcal{P}_\bigcirc$ is given as a two-state Mealy machine.

Multiple Objectives. Complex objectives arise when *combining* simple ones, and they usually require more complex strategies to play optimally. Consider the simple (one-player) game depicted in Fig. 2a: the objective of $\mathcal{P}_\bigcirc$ is to see both colors a and b infinitely often. This *generalized Büchi objective* requires memory (in this case, a single bit to alternate between v_1 and v_2), whereas single-objective Büchi games admit memoryless optimal strategies.

The classical notion of optimality is lost when moving to multi-objective games, as there is usually no total order on plays. Instead, one must consider *Pareto-optimal* strategies and the *Pareto frontier*—the set of all payoff vectors not dominated by another. We will revisit this in Sect. 4.

Mealy Machines. The classical representation of a strategy $\sigma_\triangledown$ is via a *Mealy machine*, i.e., a tuple $\mathcal{M} = (M, m_{\text{init}}, \alpha_{\text{nxt}}, \alpha_{\text{up}})$, where M is the set of memory states, m_{init} is the initial one, $\alpha_{\text{nxt}} \colon M \times V \to E$ is the next-action function, and $\alpha_{\text{up}} \colon M \times E \to M$ is the update function. A Mealy machine defines a finite-memory strategy when $|M| < \infty$, and a memoryless strategy when $|M| = 1$.

We illustrate this in Fig. 2b, where we simplify notation by identifying edges with their destination vertices. The memory state m_a (resp. m_b) encodes that the last "interesting" color observed was a (resp. b); we ignore occurrences of c, as they are neutral with respect to the objective. Transitions in the diagram represent the combined effect of α_{nxt} and α_{up}. For example, at the first step of the game, the memory is in state m_a. Since we are in vertex v_0, we follow the only corresponding transition: the next action is chosen to be v_2 and we update the memory to m_a. Following this strategy, the resulting play is $\pi = (cacb^\omega)$: i.e., $\mathcal{P}_\bigcirc$ will alternate between v_1 and v_2, and satisfy the objective.

The Ice-Cream Conundrum. Mealy machines are like ice cream: everyone has a favorite flavor and assumes it should be universally regarded as the best and canonical one. The machine above uses *chaotic* memory: its updates depend on the actual edges taken. But many other "flavors" exist: *chromatic* memory (which only depends on the color of edges), memory with or without ε-*transitions*, variants incorporating different forms of *randomness*, and more. We will encounter some of these, but refer the interested reader to [12,32] for a broader survey.

Importantly, results about strategy complexity can vary significantly depending on the chosen model—a nuance that is often overlooked in the literature.

3 Memory

Memoryless Strategies. Our starting point was Gimbert and Zielonka's characterization of *memoryless-determined two-player deterministic games* [35].

Theorem 1 (Characterization [35]). *Given a preference relation $\sqsubseteq$, memoryless strategies suffice to play optimally for both players in all finite arenas if and only if $\sqsubseteq$ and its inverse $\sqsubseteq^{-1}$ are monotone and selective.*

Intuitively, monotony corresponds to stability under prefix addition, while selectivity captures stability under cycle mixing. For example, if π and π' are two plays, and we construct a third play π'' by interleaving them, then selectivity ensures that π'' cannot be strictly better than both π and π' with respect to $\sqsubseteq$.

Beyond this characterization, Gimbert and Zielonka also derived a corollary of particular interest.

Corollary 1 (One-to-two-player lift [35]). *If $\sqsubseteq$ is such that (a) in all finite $\mathcal{P}_\bigcirc$-arenas, $\mathcal{P}_\bigcirc$ has optimal memoryless strategies, and (b) in all finite $\mathcal{P}_\square$-arenas, $\mathcal{P}_\square$ has optimal memoryless strategies, then both players have optimal memoryless strategies in all finite two-player arenas.*

One-player games (i.e., graphs) are significantly simpler. The corollary also highlights that no additional memory is needed to deal with an adversary.

Handling Finite Memory. While memoryless strategies suffice for most single-objective games, multi-objective ones may require *finite* (as in the generalized Büchi case in Sect. 2) or *infinite* memory (e.g., for multi-dimension mean-payoff games [51]). It is natural to ask whether a finite-memory analogue of Gimbert and Zielonka's result exists. Unfortunately, this hope does not hold: together with Bouyer, Le Roux, Oualhadj, and Vandenhove, we showed that there are objectives for which both players admit finite-memory optimal strategies in all one-player games, yet infinite memory is required in two-player games [9].

This led us to define a new frontier: understanding the limits of one-to-two-player lifts. We introduced *arena-independent chromatic memory structures* as suitable tools for this task. Let us revisit the generalized Büchi example. Assume $C = \{a, b, c\}$ and consider the winning condition $W = \mathsf{Buchi}(a) \cap \mathsf{Buchi}(b)$, which requires both colors to appear infinitely often. Figure 3 shows such a memory structure: it is *chromatic*, meaning that it depends only on the colors of the edges, not the edges themselves, and thus can be

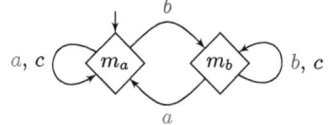

Fig. 3. A chromatic memory that suffices for $W = \mathsf{Buchi}(a) \cap \mathsf{Buchi}(b)$ in all finite arenas.

made *arena-independent*. That the structure suffices in all arenas means that it is always possible to define a suitable α_{nxt} to build an optimal Mealy machine atop it. In other words, it is possible to play optimally with a memoryless strategy when considering the product arena between the original game and this memory structure.

Using this core concept, we obtained a finite-memory analogue of Gimbert and Zielonka's [9]. Our approach relies on generalizations of monotony and selectivity interpreted *modulo a memory structure*.

Theorem 2 (Characterization [9]). *Let $\mathcal{M}_s$ be a memory structure and $\sqsubseteq$ a preference relation. Strategies based on $\mathcal{M}_s$ suffice to play optimally for both players in all finite arenas if and only if $\sqsubseteq$ and $\sqsubseteq^{-1}$ are $\mathcal{M}_s$-monotone and $\mathcal{M}_s$-selective.*

Corollary 2 (One-to-two-player lift [9]). *Let $\mathcal{M}_s$ be a memory structure. If $\sqsubseteq$ is such that (a) in all finite $\mathcal{P}_\bigcirc$-arenas, $\mathcal{P}_\bigcirc$ has optimal strategies based on $\mathcal{M}_s$, and (b) in all finite $\mathcal{P}_\square$-arenas, $\mathcal{P}_\square$ has optimal strategies based on $\mathcal{M}_s$, then both players have optimal strategies based on $\mathcal{M}_s$ in all finite two-player arenas.*

Stochastic Games. In joint work with Bouyer, Oualhadj, and Vandenhove, we further extended this result to *pure* arena-independent finite-memory strategies in stochastic games [11], notably establishing a *lift from MDPs to SGs*.

Infinite Deterministic Arenas. With Bouyer and Vandenhove, we also investigated arenas of arbitrary cardinality, allowing infinite branching [13]. Memory needs can drastically increase: e.g., (one-dimension) mean-payoff objectives require infinite memory whereas memoryless strategies suffice in finite arenas.

It had long been known that all ω-regular winning conditions admit finite-memory optimal strategies in every infinite arena [44,52]. We recently established the converse, thereby providing a *complete game-theoretic characterization of ω-regularity*. Chromatic memory structures again played a central role: given a memory structure $\mathcal{M}_s$ sufficient for a winning condition W, we constructed a parity automaton recognizing W, thus proving its ω-regularity. This construction also relies on an auxiliary structure—the *prefix-classifier* of W—which distinguishes equivalence classes of histories in the Myhill-Nerode sense.

Theorem 3 ([13]). *Chromatic-finite-memory strategies suffice for a winning condition W in all infinite arenas if and only if W is ω-regular.*

We also obtain a one-to-two-player lift, now for infinite (deterministic) arenas.

Other Criteria and Characterizations. There is a plethora of results related to memory requirements, focusing on a variety of game models. We mention a few here. Casares and Ohlmann provide a characterization through universal graphs [23]; they notably use the aforementioned ε-transitions. In [22], they also characterize the ω-regular objectives that admit *memoryless* optimal strategies.

Another active line of work focuses on establishing tight memory bounds for particular classes of objectives. We refer in particular to joint work with Bouyer, Casares, Fijalkow, and Vandenhove [7,8]. While this note has emphasized results that apply to *both* players, some works examine asymmetric requirements: e.g., one player might only require memoryless strategies even if the other one needs (potentially infinite) memory. This situation arises, for instance, in window objective games [18,26], and is further explored in [4,7,34,38].

Other lifting principles have also been studied. Le Roux and Pauly established a *two-to-multi-player lift* [39], and in joint work with them we proposed a *one-to-multi-objective lift* [40], identifying conditions under which finite-memory determinacy is preserved under combination of objectives.

Additional perspectives can be found in a survey on chromatic memory [12], co-authored with Bouyer and Vandenhove, and in a recent survey on memory in reachability problems [17], with Brihaye, Goeminne, and Main.

4 Randomness

Introducing Randomness. We may need randomness to deal with, e.g., multiple objectives, concurrent games, or imperfect information [32]. Consider the example in Fig. 4, where the goal of $\mathcal{P}_\bigcirc$ is to satisfy $\mathbb{P}^\sigma\circ[\mathrm{Reach}(a)] \geq \frac{1}{2} \wedge \mathbb{P}^\sigma\circ[\mathrm{Reach}(b)] \geq \frac{1}{2}$.) Any pure strategy can only lead to two outcomes: either v_1 is *surely* visited, and v_2 never is, or the opposite. Yet, the objective can easily be satisfied by tossing a fair coin in v_0 to pick either v_1 or v_2.

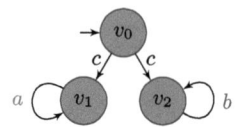

Fig. 4. Randomness is necessary to see a and b with non-zero probability.

Recall that a *pure* strategy is a function $\sigma_\triangledown\colon (VE)^*V_\triangledown \to E$. There are three classical ways to define *randomized* strategies:

- *behavioral* strategies $\sigma_\triangledown\colon (VE)^*V_\triangledown \to \mathfrak{D}(E)$, where a random choice is made *at every decision point*;
- *mixed* strategies $\mathfrak{D}(\sigma_\triangledown\colon (VE)^*V_\triangledown \to E)$, where a random choice is made *once at the beginning* of the game to pick a pure strategy to follow; and
- *general* strategies $\mathfrak{D}(\sigma_\triangledown\colon (VE)^*V_\triangledown \to \mathfrak{D}(E))$, which mix both types of randomness.

A celebrated result, known as Kuhn's theorem, states that behavioral and mixed strategies are equivalent in games with *perfect recall*—a property satisfied by all games considered in this note. This result was first established by Kuhn for games on finite trees, and generalized by Aumann [2] for infinite games. This equivalence was later extended to general strategies by Bertrand, Genest, and Gimbert [3]. However, these equivalences rely on *infinite memory* and *probability distributions with infinite support*, which may not be realistic in the context of controller synthesis.

Finite-Memory Strategies. Based on the Mealy machine representation $\mathcal{M} = (M, m_{\mathsf{init}}, \alpha_{\mathsf{nxt}}, \alpha_{\mathsf{up}})$, randomization can be implemented in different ways: via an initial distribution $\mu_{\mathsf{init}} \in \mathfrak{D}(M)$, replacing m_{init}; via the next-action function $\alpha_{\mathsf{nxt}}\colon M \times V \to \mathfrak{D}(E)$; or via the update function $\alpha_{\mathsf{up}}\colon M \times E \to \mathfrak{D}(M)$. We classify the resulting strategies using a three-letter code XYZ, where each of X, Y, Z $\in \{\mathrm{D}, \mathrm{R}\}$ denotes whether the initialization, next-action, and update functions are deterministic (D) or randomized (R). For example, DRD refers to strategies with deterministic initialization and update, but randomized action selection—the natural finite-memory analogue to *behavioral* strategies.

In joint work with Main [41], we established a *complete taxonomy* of these strategy classes, illustrated in Fig. 5. Each edge indicates a strict increase in expressive power under the notion of *outcome-equivalence*, i.e., the ability to induce the same probability distributions over plays. We directly see that Kuhn's theorem crumbles under finite memory.

Our taxonomy holds from one-player deterministic games (no collapse) up to concurrent partial-information multi-player games (equivalences still hold). While we consider here a *specification-agnostic equivalence relation*, collapses may arise for restricted classes of objectives (if the additional expressive power is not needed).

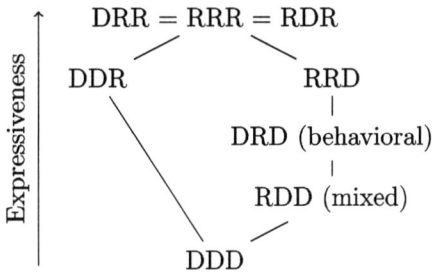

Fig. 5. Taxonomy of randomized finite-memory strategy classes.

Payoff Sets. Consider the multi-objective MDP in Fig. 6, modeling an everyday dilemma: $\mathcal{P}_\bigcirc$ starts at home (h) and aims to reach work (w). The random vertices model the available modes of transportation: the train (t), which is delayed with probability $\frac{3}{4}$, and the bike (b), which is deterministic. The time required for each transition is indicated on the edges. The objective is twofold: (i) reaching work under 40 min with high probability, and (ii) minimizing the expected time to work. Balancing these two aspects calls for reasoning about *Pareto-optimal strategies*, or more generally, characterizing the full *set of achievable payoffs*.

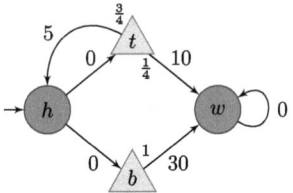

Fig. 6. From *home*, take the *train* or *bike* to reach *work*.

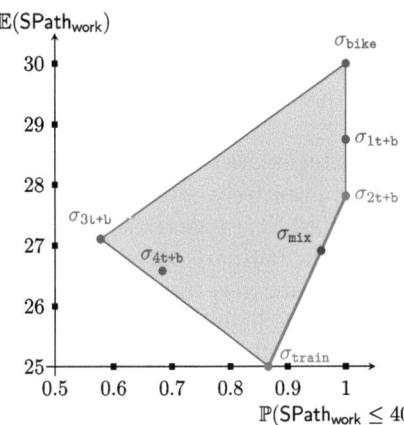

Fig. 7. Achievable payoff set.

We depict this set in Fig. 7, using the shortest-path payoff function [47]. A payoff is *Pareto-optimal* if it is not dominated by any other payoff; in our setting, this means that no other point lies in its south-east quadrant. The *Pareto frontier* is thus the line segment from σ_{train} to σ_{2t+b}, shown in green in the figure.

We highlight strategies of particular interest: σ_{train} (resp. σ_{bike}) is the pure memoryless strategy that always picks the train (resp. the bike). Strategies σ_{nt+b} try to take the train n times before switching to the bike; they are pure but use memory. Finally, σ_{mix} represents a *randomized* strategy that *mixes* σ_{2t+b} and σ_{train} to achieve a payoff that lies outside the reach of any pure strategy. In this particular example, all achievable payoffs can be obtained via mixed strategies—the lowest level of the taxonomy above.

Together with Main [42], we studied the *structure of payoff sets* in multi-objective MDPs—they are not necessarily simple polytopes as above. We notably proved that for all payoff functions with finite expectation, the set of achievable payoffs coincides with the convex hull of pure payoffs: from a strategic standpoint, this implies that *mixing a bounded number of pure strategies is sufficient to obtain any achievable payoff* (as in our example). Even when expectations are well-defined but potentially infinite, mixed strategies can still approximate any target payoff arbitrarily closely.

Trading Memory for Randomness. We close this section by building a bridge between memory and randomness. Recall the generalized Büchi game in Fig. 2: we previously saw that *pure* strategies require memory to win. Observe that a *behavioral* randomized *memoryless* strategy suffices to win with probability one: playing both v_1 and v_2 with non-zero probability ensures it. This illustrates how, in certain classes of games and objectives, randomness can compensate for the absence of memory. We notably explored this phenomenon for combinations of energy, mean-payoff and parity objectives, in a joint work with Chatterjee and Raskin [28]. Similar investigations can be found in [24,27,37,43].

Related Questions. While memory requirements are widely studied, much less is known about randomness. Cristau, David and Horn gave a first comparison of the three classical types of randomized strategies in [30]. Chatterjee et al. also studied when randomness is useful in strategies or games [25].

5 Beyond Mealy Machines

An Incomplete Story. The leitmotiv in controller synthesis is that *simpler strategies are better*. But what does "simple" really mean? The prevailing answer is: small memory and no randomness (e.g., [31])—with the underlying assumption that strategies *are* Mealy machines. We close by questioning this statement.

The Limits of Our Motto. Not all memoryless strategies are created equal. Consider Fig. 8a and let us interpret colors as actions: intuitively, strategy σ_1 (solid violet transitions) seems simpler than strategy σ_2 (dashed orange transitions)—it would be easier to explain, verify, or implement. Yet, both are encoded as trivial Mealy machines with a single memory state and are thus treated as equally simple under the standard theoretical framework. This highlights a blind spot: the *representation of the next-action function* is typically reduced to a lookup table, with little consideration for its internal structure or interpretability. In large state spaces, even such memoryless strategies may become unwieldy.

Conversely, quantitative games often require strategies with (at least) exponential memory—typically seen as prohibitively complex. Yet this is not always a barrier in practice. Consider the game in Fig. 8b, where the objective combines an

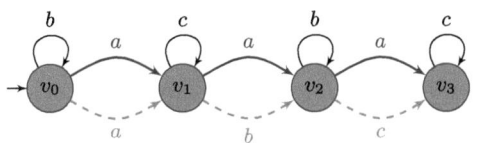
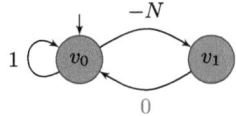

(a) Different memoryless strategies. (b) Energy-Büchi game.

Fig. 8. Strategy complexity is representation-dependent.

energy condition (ensuring a non-negative running sum) and a Büchi condition on the 0-edge. Winning requires cycling in v_0 for N steps before transitioning to v_1, and repeating. Encoding this strategy as a Mealy machine demands $N+1$ distinct memory states—one per counter value—making it pseudo-polynomial in size. But from an engineering perspective, such a strategy is easily implementable with a simple counter: the *lack of data structures* in the Mealy machine model is what is inflating the apparent complexity.

These examples suggest that our classical measure of strategy complexity, based on Mealy machines, is deeply *model-dependent*. While theoretically convenient, it may misrepresent practical simplicity. There is value in exploring *alternative representations of strategies* that reflect the true cost of implementation and allow for richer notions of simplicity.

Alternative Representations. We briefly survey several recent alternatives. Most of these approaches are motivated by practical concerns; their theoretical underpinnings remain relatively underdeveloped compared to the rich literature on Mealy machines.

A natural first idea is to augment Mealy machines with data structures to preserve internal structure rather than "flatten" it. These *structurally-enriched Mealy machines* have notably been explored for counter-based strategies: see Blahoudek et al. [5], or our recent work with Ajdarów, Main and Novotný [1]. Practically, such models yield *more succinct* and *more explainable* strategies, thanks to their transparent internal logic. They also shine a new light on a theoretical level, changing our view of which strategies are complex or not.

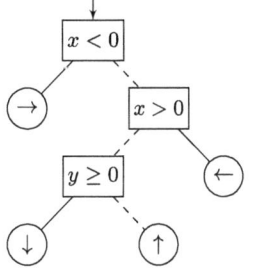

A well-studied alternative is the use of *decision trees (DTs)*, particularly suited to highly structured state and action spaces [14,15]. DT's have primarily been used to compactly represent *memoryless* strategies, serving as a compressed substitute for *large next-action tables*. For instance, consider navigation on a 2D grid with integer coordinates—say a square $[-d,d] \times [-d,d]$—with the objective of reaching the origin. A naïve lookup table would require $\mathcal{O}(d^2)$ entries, while a simple DT (such as that in Fig. 9) can represent the strategy compactly for *any* grid size—

Fig. 9. Decision tree to reach $(0,0)$ in a 2D grid.

even infinite. One key design challenge is balancing the complexity of tests at internal nodes (which affects interpretability) against the overall size of the tree.

Another family of models are *strategy machines*, inspired by Turing machines: they are powerful, yet difficult to synthesize and analyze [33]. Interestingly, this model has led to a tentative notion of *decision speed*—a rarely explored but potentially meaningful complexity measure. *Neural networks* have become prevalent in reinforcement learning and are increasingly combined with finite-state machines to aid synthesis [21]; their performance is strong, but interpretability and verifiability remain major obstacles. Finally, *programmatic representations* are closer to realistic code, thus inherently understandable. They are however strongly linked to the input format of the synthesis problem (e.g., PRISM code) and hard to generalize [49].

6 Conclusion

In controller synthesis, *simplicity is the name of the game*. But what does it truly mean? We provided a high-level overview of the two classical axes of complexity under the Mealy machine model: *memory* and *randomness*. We then pointed out critical blind spots in this framework, showing that *strategy complexity is often distinct from representation complexity*.

Looking ahead, we argue for the development of a *toolbox of diverse strategy representations*, grounded in a deeper understanding of their expressiveness, succinctness, and interpretability. Ultimately, our goal should be a principled, *representation-agnostic theory of strategy complexity*, capable of guiding both theory and practice in synthesis.

Acknowledgments. Most of the results presented in this survey [9,11,13,41,42] were obtained in the context of the F.R.S.-FNRS projects *ManySynth* and *ControlleRS*. I want to highlight the pivotal role of my outstanding PhD students, James C. A. Main and Pierre Vandenhove, whose work has been instrumental throughout. I also express my deepest gratitude to my wonderful co-authors: Patricia Bouyer, Stéphane Le Roux, and Youssouf Oualhadj.

References

1. Ajdarów, M., Main, J.C.A., Novotný, P., Randour, M.: Taming infinity one chunk at a time: concisely represented strategies in one-counter MDPs. In: Censor-Hillel, K., Grandoni, F., Ouaknine, J., Puppis, G. (eds.) 52nd International Colloquium on Automata, Languages, and Programming, ICALP 2025, Aarhus, Denmark, 8–11 July 2025. LIPIcs, vol. 334, pp. 138:1–138:19. Schloss Dagstuhl - Leibniz-Zentrum für Informatik (2025). https://doi.org/10.4230/LIPICS.ICALP.2025.138
2. Aumann, R.J..: Mixed and behavior strategies in infinite extensive games. In: Dresher, M., Shapley, L.S., Tucker, A.W. (eds.) Advances in Game Theory. (AM-52), vol. 52, pp. 627–650. Princeton University Press (1964). https://doi.org/10.1515/9781400882014-029

3. Bertrand, N., Genest, B., Gimbert, H.: Qualitative determinacy and decidability of stochastic games with signals. J. ACM **64**(5), 33:1–33:48 (2017). https://doi.org/10.1145/3107926
4. Bianco, A., Faella, M., Mogavero, F., Murano, A.: Exploring the boundary of half-positionality. Ann. Math. Artif. Intell. **62**(1–2), 55–77 (2011). https://doi.org/10.1007/S10472-011-9250-1
5. Blahoudek, F., Brázdil, T., Novotný, P., Ornik, M., Thangeda, P., Topcu, U.: Qualitative controller synthesis for consumption markov decision processes. In: Lahiri, S.K., Wang, C. (eds.) CAV 2020. LNCS, vol. 12225, pp. 421–447. Springer, Cham (2020). https://doi.org/10.1007/978-3-030-53291-8_22
6. Bloem, R., Chatterjee, K., Jobstmann, B.: Graph games and reactive synthesis. In: Handbook of Model Checking, pp. 921–962. Springer, Cham (2018). https://doi.org/10.1007/978-3-319-10575-8_27
7. Bouyer, P., Casares, A., Randour, M., Vandenhove, P.: Half-positional objectives recognized by deterministic Büchi automata. Log. Methods Comput. Sci. **20**(3) (2024). https://doi.org/10.46298/LMCS-20(3:19)2024
8. Bouyer, P., Fijalkow, N., Randour, M., Vandenhove, P.: How to play optimally for regular objectives? In: Etessami, K., Feige, U., Puppis, G. (eds.) 50th International Colloquium on Automata, Languages, and Programming, ICALP 2023, Paderborn, Germany, 10–14 July 2023. LIPIcs, vol. 261, pp. 118:1–118:18. Schloss Dagstuhl - Leibniz-Zentrum für Informatik (2023). https://doi.org/10.4230/LIPICS.ICALP.2023.118
9. Bouyer, P., Le Roux, S., Oualhadj, Y., Randour, M., Vandenhove, P.: Games where you can play optimally with arena-independent finite memory. Log. Methods Comput. Sci. **18**(1) (2022). https://doi.org/10.46298/LMCS-18(1:11)2022
10. Bouyer, P., Markey, N., Randour, M., Larsen, K.G., Laursen, S.: Average-energy games. Acta Informatica **55**(2), 91–127 (2018). https://doi.org/10.1007/S00236-016-0274-1
11. Bouyer, P., Oualhadj, Y., Randour, M., Vandenhove, P.: Arena-independent finite-memory determinacy in stochastic games. Log. Methods Comput. Sci. **19**(4) (2023). https://doi.org/10.46298/LMCS-19(4:18)2023
12. Bouyer, P., Randour, M., Vandenhove, P.: The true colors of memory: a tour of chromatic-memory strategies in zero-sum games on graphs. In: Dawar, A., Guruswami, V. (eds.) 42nd IARCS Annual Conference on Foundations of Software Technology and Theoretical Computer Science, FSTTCS 2022, IIT Madras, Chennai, India, 18–20 December 2022. LIPIcs, vol. 250, pp. 3:1–3:18. Schloss Dagstuhl - Leibniz-Zentrum für Informatik (2022). https://doi.org/10.4230/LIPICS.FSTTCS.2022.3
13. Bouyer, P., Randour, M., Vandenhove, P.: Characterizing omega-regularity through finite-memory determinacy of games on infinite graphs. TheoretiCS **2** (2023). https://doi.org/10.46298/THEORETICS.23.1
14. Brázdil, T., Chatterjee, K., Chmelík, M., Fellner, A., Křetínský, J.: Counterexample explanation by learning small strategies in markov decision processes. In: Kroening, D., Păsăreanu, C.S. (eds.) CAV 2015. LNCS, vol. 9206, pp. 158–177. Springer, Cham (2015). https://doi.org/10.1007/978-3-319-21690-4_10
15. Brázdil, T., Chatterjee, K., Křetínský, J., Toman, V.: Strategy representation by decision trees in reactive synthesis. In: Beyer, D., Huisman, M. (eds.) TACAS 2018. LNCS, vol. 10805, pp. 385–407. Springer, Cham (2018). https://doi.org/10.1007/978-3-319-89960-2_21

16. Brihaye, T., Delgrange, F., Oualhadj, Y., Randour, M.: Life is random, time is not: Markov decision processes with window objectives. Log. Methods Comput. Sci. **16**(4) (2020). https://doi.org/10.23638/LMCS-16(4:13)2020
17. Brihaye, T., Goeminne, A., Main, J.C.A., Randour, M.: Reachability games and friends: a journey through the lens of memory and complexity. In: Bouyer, P., Srinivasan, S. (eds.) 43rd IARCS Annual Conference on Foundations of Software Technology and Theoretical Computer Science, FSTTCS 2023, 18–20 December 2023, IIIT Hyderabad, Telangana, India. LIPIcs, vol. 284, pp. 1:1–1:26. Schloss Dagstuhl - Leibniz-Zentrum für Informatik (2023). https://doi.org/10.4230/LIPICS.FSTTCS.2023.1
18. Bruyère, V., Hautem, Q., Randour, M.: Window parity games: an alternative approach toward parity games with time bounds. In: Cantone, D., Delzanno, G. (eds.) Proceedings of the Seventh International Symposium on Games, Automata, Logics and Formal Verification, GandALF 2016, Catania, Italy, 14–16 September 2016. EPTCS, vol. 226, pp. 135–148 (2016https://doi.org/10.4204/EPTCS.226.10
19. Bruyère, V., Hautem, Q., Randour, M., Raskin, J.F.: Energy mean-payoff games. In: Fokkink, W.J., van Glabbeek, R. (eds.) 30th International Conference on Concurrency Theory, CONCUR 2019, Amsterdam, the Netherlands, 27–30 August 2019. LIPIcs, vol. 140, pp. 21:1–21:17. Schloss Dagstuhl - Leibniz-Zentrum für Informatik (2019https://doi.org/10.4230/LIPICS.CONCUR.2019.21
20. Bruyère, V., Filiot, E., Randour, M., Raskin, J.F.: Meet your expectations with guarantees: beyond worst-case synthesis in quantitative games. Inf. Comput. **254**, 259–295 (2017). https://doi.org/10.1016/J.IC.2016.10.011
21. Carr, S., Jansen, N., Topcu, U.: Verifiable RNN-based policies for POMDPs under temporal logic constraints. In: Bessiere, C. (ed.) Proceedings of the Twenty-Ninth International Joint Conference on Artificial Intelligence, IJCAI 2020, pp. 4121–4127. ijcai.org (2020). https://doi.org/10.24963/IJCAI.2020/570
22. Casares, A., Ohlmann, P.: Positional ω-regular languages. In: Sobocinski, P., Lago, U.D., Esparza, J. (eds.) Proceedings of the 39th Annual ACM/IEEE Symposium on Logic in Computer Science, LICS 2024, Tallinn, Estonia, 8–11 July 2024, pp. 21:1–21:14. ACM (2024). https://doi.org/10.1145/3661814.3662087
23. Casares, A., Ohlmann, P.: Characterising memory in infinite games. Log. Methods Comput. Sci. **21**(1) (2025). https://doi.org/10.46298/LMCS-21(1:28)2025
24. Chatterjee, K., de Alfaro, L., Henzinger, T.A.: Trading memory for randomness. In: 1st International Conference on Quantitative Evaluation of Systems (QEST 2004), Enschede, The Netherlands, 27–30 September 2004, pp. 206–217. IEEE Computer Society (2004). https://doi.org/10.1109/QEST.2004.1348035
25. Chatterjee, K., Doyen, L., Gimbert, H., Henzinger, T.A.: Randomness for free. Inf. Comput. **245**, 3–16 (2015). https://doi.org/10.1016/J.IC.2015.06.003
26. Chatterjee, K., Doyen, L., Randour, M., Raskin, J.F.: Looking at mean-payoff and total-payoff through windows. Inf. Comput. **242**, 25–52 (2015). https://doi.org/10.1016/J.IC.2015.03.010
27. Chatterjee, K., Henzinger, T.A., Prabhu, V.S.: Trading infinite memory for uniform randomness in timed games. In: Egerstedt, M., Mishra, B. (eds.) HSCC 2008. LNCS, vol. 4981, pp. 87–100. Springer, Heidelberg (2008). https://doi.org/10.1007/978-3-540-78929-1_7
28. Chatterjee, K., Randour, M., Raskin, J.F.: Strategy synthesis for multi-dimensional quantitative objectives. Acta Informatica **51**(3–4), 129–163 (2014). https://doi.org/10.1007/S00236-013-0182-6
29. Church, A.: Applications of recursive arithmetic to the problem of circuit synthesis. Summ. Summer Inst. Symb. Logic **1**, 3–50 (1957)

30. Cristau, J., David, C., Horn, F.: How do we remember the past in randomised strategies? In: Montanari, A., Napoli, M., Parente, M. (eds.) Proceedings First Symposium on Games, Automata, Logic, and Formal Verification, GANDALF 2010, Minori (Amalfi Coast), Italy, 17–18 June 2010. EPTCS, vol. 25, pp. 30–39 (2010). https://doi.org/10.4204/EPTCS.25.7
31. Delgrange, F., Katoen, J.-P., Quatmann, T., Randour, M.: Simple strategies in multi-objective MDPs. In: TACAS 2020. LNCS, vol. 12078, pp. 346–364. Springer, Cham (2020). https://doi.org/10.1007/978-3-030-45190-5_19
32. Fijalkow, N., et al.: Games on Graphs: From Logic and Automata to Algorithms. Cambridge University Press, Cambridge (2025). https://doi.org/10.48550/ARXIV.2305.10546, in press
33. Gelderie, M.: Strategy machines: representation and complexity of strategies in infinite games. Ph.D. thesis, RWTH Aachen University (2014). https://publications.rwth-aachen.de/record/229827/files/5025.pdf
34. Gimbert, H., Kelmendi, E.: Submixing and shift-invariant stochastic games. Int. J. Game Theory **52**(4), 1179–1214 (2023). https://doi.org/10.1007/S00182-023-00860-5
35. Gimbert, H., Zielonka, W.: Games where you can play optimally without any memory. In: Abadi, M., de Alfaro, L. (eds.) CONCUR 2005. LNCS, vol. 3653, pp. 428–442. Springer, Heidelberg (2005). https://doi.org/10.1007/11539452_33
36. Grädel, E., Thomas, W., Wilke, T. (eds.): Automata Logics, and Infinite Games. LNCS, vol. 2500. Springer, Heidelberg (2002). https://doi.org/10.1007/3-540-36387-4
37. Horn, F.: Random fruits on the Zielonka tree. In: Albers, S., Marion, J.Y. (eds.) 26th International Symposium on Theoretical Aspects of Computer Science, STACS 2009, Freiburg, Germany, 26–28 February 2009, Proceedings. LIPIcs, vol. 3, pp. 541–552. Schloss Dagstuhl - Leibniz-Zentrum für Informatik (2009). https://doi.org/10.4230/LIPICS.STACS.2009.1848
38. Kopczyński, E.: Half-positional Determinacy of Infinite Games. Ph.D. thesis, Warsaw University (2008). https://www.mimuw.edu.pl/~erykk/papers/hpwc.pdf
39. Le Roux, S., Pauly, A.: Extending finite memory determinacy to multiplayer games. In: Lomuscio, A., Vardi, M.Y. (eds.) Proceedings of the 4th International Workshop on Strategic Reasoning, SR 2016, New York City, USA, 10 July 2016. EPTCS, vol. 218, pp. 27–40 (2016). https://doi.org/10.4204/EPTCS.218.3
40. Le Roux, S., Pauly, A., Randour, M.: Extending finite-memory determinacy by Boolean combination of winning conditions. In: Ganguly, S., Pandya, P.K. (eds.) 38th IARCS Annual Conference on Foundations of Software Technology and Theoretical Computer Science, FSTTCS 2018, Ahmedabad, India, 11–13 December 2018. LIPIcs, vol. 122, pp. 38:1–38:20. Schloss Dagstuhl - Leibniz-Zentrum für Informatik (2018). https://doi.org/10.4230/LIPICS.FSTTCS.2018.38
41. Main, J.C.A., Randour, M.: Different strokes in randomised strategies: revisiting Kuhn's theorem under finite-memory assumptions. Inf. Comput. **301** (2024). https://doi.org/10.1016/J.IC.2024.105229
42. Main, J.C.A., Randour, M.: Mixing any cocktail with limited ingredients: on the structure of payoff sets in multi-objective MDPs and its impact on randomised strategies. CoRR arxiv:2502.18296 (2025). https://doi.org/10.48550/ARXIV.2502.18296
43. Monmege, B., Parreaux, J., Reynier, P.A.: Playing stochastically in weighted timed games to emulate memory. Log. Methods Comput. Sci. **21**(1) (2025). https://doi.org/10.46298/LMCS-21(1:19)2025

44. Mostowski, A.W.: Regular expressions for infinite trees and a standard form of automata. In: Skowron, A. (ed.) SCT 1984. LNCS, vol. 208, pp. 157–168. Springer, Heidelberg (1985). https://doi.org/10.1007/3-540-16066-3_15
45. Randour, M.: Automated synthesis of reliable and efficient systems through game theory: a case study. In: Proceedings of the European Conference on Complex Systems 2012, ECCS 2012, Brussels, Belgium, 2–7 September 2012. pp. 731–738. Springer Proceedings in Complexity. Springer, Heidelberg (2013). https://doi.org/10.1007/978-3-319-00395-5_90
46. Randour, M.: Games with multiple objectives. In: Fijalkow, N. (ed.) Games on Graphs: From Logic and Automata to Algorithms, pp. 488–527. Cambridge University Press, Cambridge (2025). https://doi.org/10.48550/ARXIV.2305.10546
47. Randour, M., Raskin, J.-F., Sankur, O.: Variations on the stochastic shortest path problem. In: D'Souza, D., Lal, A., Larsen, K.G. (eds.) VMCAI 2015. LNCS, vol. 8931, pp. 1–18. Springer, Heidelberg (2015). https://doi.org/10.1007/978-3-662-46081-8_1
48. Randour, M., Raskin, J.-F., Sankur, O.: Percentile queries in multi-dimensional Markov decision processes. Formal Methods Syst. Des. (2), 207–248 (2017). https://doi.org/10.1007/s10703-016-0262-7
49. Shabadi, G., Fijalkow, N., Matricon, T.: Programmatic reinforcement learning: navigating gridworlds. In: Proceedings of the International AAAI Workshop on Generalization in Planning, GenPlan 2025 (2025). https://doi.org/10.48550/ARXIV.2402.11650
50. Thomas, W.: Facets of synthesis: revisiting church's problem. In: de Alfaro, L. (ed.) FoSSaCS 2009. LNCS, vol. 5504, pp. 1–14. Springer, Heidelberg (2009). https://doi.org/10.1007/978-3-642-00596-1_1
51. Velner, Y., Chatterjee, K., Doyen, L., Henzinger, T.A., Rabinovich, A.M., Raskin, J.F.: The complexity of multi-mean-payoff and multi-energy games. Inf. Comput. **241**, 177–196 (2015). https://doi.org/10.1016/J.IC.2015.03.001
52. Zielonka, W.: Infinite games on finitely coloured graphs with applications to automata on infinite trees. Theor. Comput. Sci. **200**(1), 135–183 (1998). https://doi.org/10.1016/S0304-3975(98)00009-7

Regular Papers

Word Equations with Length Constraints via Weak Arithmetics and Matrix Reachability Problems

Joel D. Day and Matthew Konefal$^{(\boxtimes)}$

Department of Computer Science, Loughborough University, Loughborough, UK
{j.day,m.konefal2}@lboro.ac.uk

Abstract. The problem of determining whether a word equation admits a solution whose lengths belong to some semilinear set is fundamental to the theory of string-solving. Despite this and connections to Hilbert's Tenth Problem, its decidability has been open since 1968. We prove this problem decidable for the class of quadratic word equations in at most two unknowns. We use a novel technique recasting the problem as half-space reachability in the matrix semigroup $\mathrm{SA}(n, \mathbb{N})$. We also look at word equations' length abstractions: the sets of their solutions' length-vectors. It is known that these need not be semilinear; with the above problem in mind we seek to understand their expressivity. We show that the length abstractions of quadratic word equations differ finitely from unions of arithmetic progressions. Tools for showing inexpressibility by regular and quadratic word equations result. Each relation definable in existential Presburger arithmetic with divisibility (**EPAD**) can be built from word equations' length abstractions. Conversely, the length abstraction of $Xw_1Y \doteq Yw_2X$ is shown **EPAD**-equidefinable with a certain vector reachability problem.

Keywords: Word Equations · Matrix Reachability Problems · EPAD · Length Abstractions

1 Introduction

A *word equation* (WE) is a formal equality between two *sides*: words made of terminals $\mathsf{a} \in \Sigma$ and unknowns $X \in \Xi$. It is solved by those assignments to the unknowns making the sides graphically equal. For instance, $(X,Y) \mapsto (\mathsf{a}^5, \mathsf{b})$ is a solution of the WE $E_1 : X\mathsf{ab} \doteq \mathsf{a}XY$. A popular use-case for WEs is *string-solving* [3], which usually requires WEs to be solved modulo certain constraints. E_1, for instance, becomes unsatisfiable if we require that the length of Y exceeds 1. String-solving has been applied to cybersecurity [36], code repair [50], and formal verification [54]. WEs themselves are central to formal language theory [2, 8,9]. Results on WEs have found applications in combinatorial group theory [40], database theory [21], and information extraction [22].

Makanin [39] famously showed that *satisfiability* is decidable for WEs. (Jeż reproves this elegantly in [30]). In 1968, Matiyasevich formulated [42] the satisfiability problem *modulo length constraints* (SatLC) for WEs. He also reduced it to Hilbert's 10th problem, hoping that undecidability would carry from the former problem to the latter. SatLC has received great attention since; it was adapted to groups in [11], while [3,5,7,54] demonstrate how it arises and is tackled in practice. However, its decidability remains open in general [26]. Algorithms are known only if the length constraints are "inequations" [12] (*i.e.* assertions that unknowns' lengths must differ) or if the class of WEs is heavily restricted, as in [33] and the following:

In [23], SatLC is shown decidable for WEs having a "solved form". Such WEs are (effectively) *parameterisable*, *i.e.* their solution sets are describable by finitely many word and integer parameters. Already $Xw_1Y \doteq Yw_2X$ is non-parameterisable [27]. Similarly, [1] handles the "acyclic" case, in which unknowns do not have mutual dependencies. Most recently, SatLC was proved decidable for regular-ordered WEs [35] (ROWEs), which are those where each side contains each unknown at most once, and there is an order over Ξ which, left-to-right, the unknowns of both sides respect. Structurally the solution sets of ROWEs are still simple: they are low in the hierarchy of the well-studied classes, each higher tier begetting new complexity. Above ROWEs lie *regular* WEs (RWEs) [41], which permit one occurrence of each unknown per side, and *quadratic* WEs (QWEs), which permit two occurrences of each unknown in total. Thus, E_1 is a ROWE, while $XX \doteq Y$ is quadratic but not regular. As well as being the frontier of SatLC, RWEs are the largest class of WEs for which the conjecture "satisfiability is in NP" has been verified [15].

Progressing to larger classes of WEs means understanding these WEs' expressivity. If SatLC is undecidable, the culprit will be conspiracy between what WEs and length constraints - by themselves - can do. Thus, we speak of *WE-expressible* relations on words: those described by (the solutions of) some WE. For instance, $XY \doteq YX$ expresses the binary relation of commuting, while $X \doteq Y$a expresses - via the unknown X - the unary relation of "ending with a". WE-expressibility was defined in [31], but the idea dates to at least [6], wherein the equivalence relation of length-equality was proved WE-inexpressible.

In [35], the term *length abstraction* was introduced to mean the set of vectors describing a WE's solutions' lengths. The length abstraction of E_1, for instance, is $\text{Len}(E_1) = \{(n,1) : n \in \mathbb{N}\}$. Determining when $\text{Len}(E)$ has a suitable finite representation has been called the "crux of the difficulty" of SatLC [23]. Restrictions on the solution sets of WEs from a certain class often transfer to the length abstractions of this class. For example, the structure of $\text{Sol}(E)$ when $|\Xi| = 1$ is understood [32]; it follows that in this case $\text{Len}(E)$ is Presburger-definable. For the case $|\Xi| = 2$, [29] describes how $\text{Sol}(E)$ can grow, but the analysis is too macroscopic for our purposes.

Let $\mathcal{C}$ be a class of WEs, and let $\mathcal{T}$ be a decidable arithmetic such as PA of Presburger [48], EPAD of Lipshitz & Bel'tyukov [4,37], BA of Büchi [24], or SA of Semënov [19]. If, for each $E \in \mathcal{C}$, $\text{Len}(E)$ is effectively $\mathcal{T}$-definable, we can decide

SatLC for $\mathcal{C}$. (The unsatisfiability of $E_1 \wedge |Y| > 1$, for example, follows from the unsatisfiability of the formula $(x,y) \in \text{LEN}(E_1) \wedge y > 1$, which can be rewritten in PA). This approach was used, *e.g.*, in [23] with $\mathcal{T} = \text{PA}$. However, we need more, for even ROWEs E can have PA-undefinable $\text{LEN}(E)$ (Proposition 2). As in [35], we choose $\mathcal{T} = \text{EPAD}$; our reason is as follows: WEs' length abstractions capture $+$ and $|$ (Proposition 3). We take $\mathcal{T}$ existential because the first-order theory PAD of $\langle \mathbb{N}; +, | \rangle$ equals that of $\langle \mathbb{N}; +, \times \rangle$ [49]. The MRDP Theorem (*i.e.* the solution to Hilbert's Tenth Problem) [43] then gives that PAD is undecidable (and defines all recursively enumerable sets). EPAD is thus our minimal choice, but it is also at the apogée, in the sense that the $\exists\forall$-theory of $\langle \mathbb{N}; +, | \rangle$ is already undecidable [38]. The positional and powering predicates, respectively, of BA and SA seem less pertinent.

Another way to express sets of vectors is via *reachability problems* in matrix semigroups. Where $\mathcal{R}$ is (a rational subset of) such a semigroup, the *vector reachability problem* (VRP) for $\mathcal{R}$ asks whether some $M \in \mathcal{R}$ transforms one input vector $\boldsymbol{u}$ into another $\boldsymbol{v}$, while the *halfspace reachability problem* (HRP) for $\mathcal{R}$ asks, given vectors $\boldsymbol{u},\boldsymbol{v}$ and scalar k, if any $M \in \mathcal{R}$ verifies $\boldsymbol{u}^T M \boldsymbol{v} \geq k$. Though these problems are undecidable in general [13], restricting the form or dimension of the matrices involved can make them decidable [13,47]. Under this condition, a partially-specified instance of either problem wherein $\boldsymbol{v}$ is not given defines a recursive set of vectors: the $\boldsymbol{v}$ for which the resulting instance is positive. Though more specialised than the arithmetics above, these "decidable theories" will be shown to be a good fit for WEs, as the problems we study reformulate naturally as VRPs or HRPs.

In brief, this paper aims to chart the boundaries of PA-definability, EPAD-definability, and decidability. Our Sect. 3 gives the first systematic study of length abstractions. Its main result is that the length abstractions of QWEs differ finitely from infinite unions of nonconstant arithmetic progressions. This is necessary for EPAD-definability. It yields also tools for showing inexpressibility by QWEs (and RWEs), which are among the first of their type. In Sect. 4, we introduce a technique for SatLC based on a reduction to a HRP. Using this, we establish that SatLC is decidable for the class of QWEs in ≤ 2 unknowns. Linking everything, Sect. 5 gives a VRP which must be captured in EPAD (yet seems not to be) if RWEs' length abstractions are to be proved EPAD-definable. Due to space constraints, we omit our proofs. Many of them utilise the famous *Nielsen graph algorithm* for QWEs, which we introduce informally in Sect. 4.

2 Preliminaries

Let Σ, Ξ be disjoint alphabets of *terminals* and *unknowns*, respectively. A *word equation* (*WE*) is a pair $E : (\alpha, \beta)$, usually denoted $\alpha \doteq \beta$, where $\alpha, \beta \in (\Sigma \cup \Xi)^*$. Its *e-solutions* are the morphisms $h : (\Sigma \cup \Xi)^* \to \Sigma^*$ acting as the identity on Σ, and such that $h(\alpha) = h(\beta)$. Its *solutions* are those e-solutions which are nonerasing; these form a set $\text{SOL}(E)$, and - for expositional convenience - shall be our main focus. A WE (α, β) is called *quadratic* (or a QWE) if each $X \in \Xi$ occurs at most twice in $\alpha\beta$. It is called *regular* (or a RWE) if each $X \in \Xi$ occurs at most once in α and at most once in β. It is called *regular-ordered* (or a $ROWE$) if it is a RWE, and there is a total order $\prec$ on Ξ, such that $X_1 \prec X_2$ whenever X_1 occurs before X_2 in α or in β. It is called *strictly regular-ordered* (or a $SROWE$) if it is a ROWE and each $X \in \Xi$ occurs twice.

By $\mathbb{N}_0$ (resp. $\mathbb{N}_1$) we mean the nonnegative (resp. positive) integers. A relation $R \subseteq (\Sigma^*)^n$ is *expressed by* $\mathcal{X}$ *in* E if E is a WE, $\mathcal{X}$ is an n-tuple of unknowns of E, and $R = \mathcal{L}(E, \mathcal{X}) := \{h(\mathcal{X}) : h \in \text{SOL}(E)\}$. In this case, we call R *WE-expressible*. Similarly, we speak of *RWE-expressible* relations, etc. The *length* $|w|$ of a word w is the number of its letters. The *empty word* ε is the word of length 0. The *length abstraction* $\text{LEN}(R)$ of a relation $R \subseteq (\Sigma^*)^n$ is the subset of $\mathbb{N}_0^n$ obtained by mapping each $(w_1, ..., w_n) \in R$ to $(|w_1|, ..., |w_n|)$. For a WE E where there is an obvious way to order the set Ξ of unknowns, we use $\text{LEN}(E)$ as shorthand for $\text{LEN}(\mathcal{L}(E, \mathcal{X}))$, where $\mathcal{X}$ is a $|\Xi|$-tuple implementing the obvious ordering. This notation will also be used in the absence of an obvious ordering if our argument is independent of the ordering chosen. e-Solution sets generate e-length abstractions.

Presburger arithmetic (PA) is the first-order theory of $\langle \mathbb{N}_0; +, \leq \rangle$. *Existential Presburger arithmetic with divisibility* (EPAD) is the $\exists$-theory of $\langle \mathbb{N}_0; +, \leq, | \rangle$. $R \subseteq \mathbb{N}_0^n$ is called PA- (resp. EPAD-) *definable* if there is a PA- (resp. EPAD-)formula $\varphi(x_1, ..., x_n)$ whose satisfying assignments are exactly those mapping $(x_1, ..., x_n)$ into R. The quantifier-elimination procedure from [48] establishes PA-definability as implying EPAD-definability.

The *satisfiability problem* is to decide whether $\text{SOL}(E) \neq \emptyset$, given a WE E. The satisfiability problem *modulo length constraints* (SatLC) is to decide whether $\text{LEN}(E) \cap R \neq \emptyset$, given a WE E and a *length constraint*: a quantifier-free PA-formula φ defining some $R \subseteq \mathbb{N}_0^{|\Xi|}$. We also use *Boolean combinations* of WEs and length constraints; their (e-)solution sets and the semantics of their satisfiability problem are defined in the natural way. One can transform [31, §2] a Boolean combination C of WEs into a single WE E with the property that $\text{SOL}(C)$ is a projection of $\text{SOL}(E)$. This transformation fails, in general, to preserve properties such as quadracity and regularity.

Remark 1 (Robustness of Length Constraints). It is not expressively restrictive to define length constraints as positive Boolean combinations of length-equalities $|X| = |Y|$: the rest of PA can be modelled using additional WEs, (which can be absorbed into the original WE, as above). In fact, one can show (*q.v.* Proposition 3) that the length abstraction of $((Za \doteq aZ \wedge X \doteq Zb) \vee X \doteq b) \wedge XY \doteq YX$, projected onto its first two components, is $\{(x, y) \in \mathbb{N}_1^2 : x \,|\, y\}$. This and the normal form of EPAD-formulas given in [38] show that - once length-equalities are allowed - additional WEs can model all of EPAD. So no expressivity is gained using EPAD-formulas as length constraints for WEs. The same cannot necessarily be said for subclasses of WEs (*e.g.* the QWEs and RWEs).

Let $\$ \in \{\mathbb{Z}, \mathbb{N}_0\}$. $\mathrm{GL}(n, \$)$ (resp. $\mathrm{SL}(n, \$)$) is the set of matrices in $\$^{n \times n}$ having determinant ± 1 (resp. $+1$). $\mathrm{SA}(n, \$)$ is the set

$$\left\{ \begin{pmatrix} M & \boldsymbol{v} \\ \boldsymbol{0}^T & 1 \end{pmatrix} : M \in \mathrm{SL}(n, \$), \boldsymbol{v} \in \$^n \right\}.$$

These sets are multiplicative monoids, and groups if $\$ = \mathbb{Z}$. Famously [25], $\mathrm{SL}(2, \mathbb{N}_0)$ is freely generated by $\left(\begin{smallmatrix}1 & 1\\ 0 & 1\end{smallmatrix}\right)$ and $\left(\begin{smallmatrix}1 & 0\\ 1 & 1\end{smallmatrix}\right)$. Where $S_1, ..., S_n$ are subsets of some (common) multiplicative semigroup, we will call $\{s_1 \times \cdots \times s_n : s_i \in S_i\}$ the *product-set* of the S_i. Given a multiplicative monoid, its class of *rational subsets* contains all its finite subsets, and as few infinite subsets as possible to achieve closure under the unary operation of Kleene star, and the binary operations of union and taking product-sets.

Sets S_1, S_2 will be said to *differ finitely* if their symmetric difference is finite. An *arithmetic progression* in $\mathbb{N}_0^n$ is a set $\boldsymbol{a} + \boldsymbol{d}\mathbb{N}_0$ ($\boldsymbol{a}, \boldsymbol{d} \in \mathbb{N}_0^n$). It is *nonconstant* if $\boldsymbol{d} \neq \boldsymbol{0}$.

3 Word Equations' Length Abstractions

In this section we study the expressivity of the length abstractions of various WE-expressible relations. Of interest are three thresholds: those of PA-definability, EPAD-definability, and decidability. Our methods are sometimes indirect; Proposition 1, *e.g.*, establishes PA-definability for certain WE-expressible languages by proving them regular. Our main results (Theorems 1 and 2) say that the length abstractions of RWE-expressible relations and QWEs differ finitely from infinite unions of nonconstant arithmetic progressions. This overapproximates EPAD-definability: it is necessary for it [38], but easily shown insufficient. These results cannot be strengthened to speak of *finite* unions, since then we would guarantee PA-definability, contradicting [35, Thm. 3.5]. Table 1 collects results of this ilk, both known and from the present section.

Table 1. Results on length abstractions of WE-expressible relations. Citations are provided for known results; ours are given with specific references. The columns correspond to classes of WEs, their generality increasing left-to-right. The row labels indicate by which unknowns the relation is expressed. Top-to-bottom then, the rows refer to sets $\mathrm{LEN}(\mathcal{L}(E, X)))$, $\mathrm{LEN}(\mathcal{L}(E, (X, ..., Y)))$ and $\mathrm{LEN}(E)$. "∪APs" means "unions of nonconstant arithmetic progressions".

	SROWEs	ROWEs	RWEs	QWEs	WEs
1 Unknown	PA [14]	PA (Prp. 1)	∪APs (Thm. 1)		Not PA (Cor. 3)
Arbitrary Unknowns	PA [14]	EPAD [35]	∪APs (Thm. 1)		Not PA [35]
		Not PA (Prp. 2)	Not PA [35]		
All Unknowns	PA [14]	EPAD [35]	∪APs (Thm. 2)		Not PA [35]
		Not PA (Prp. 2)	Not PA [35]		

3.1 Regular Word Equations

We begin by looking at RWEs and their subclasses. First, we give a necessary condition on the length abstractions of RWE-expressible relations:

Theorem 1. *Let E be a RWE, and let $\mathcal{X}$ be a tuple of its unknowns. Then $\text{LEN}(\mathcal{L}(E, \mathcal{X}))$ differs finitely from a (possibly empty or infinite) union of non-constant arithmetic progressions.*

It follows that not every recursively enumerable subset of $\mathbb{N}_0$ is obtainable as $\text{LEN}(L) \cap R$, where L is a RWE-expressible language and $R \subseteq \mathbb{N}_0$ is as defined by a length constraint. Indeed, the class of subsets of $\mathbb{N}_0$ differing finitely from unions of arithmetic progressions is closed under intersection with PA-definable (*i.e.* eventually periodic) subsets of $\mathbb{N}_0$. Structure such as this should be expected if there is any hope for SatLC to be proved decidable for RWEs.

Theorem 1 can also be used for showing RWE-inexpressibility. For instance, let $L_{\text{fib}} \subseteq \{\mathsf{a},\mathsf{b}\}^*$ be the language containing $w_{-2} := \mathsf{a}$, $w_{-1} := \mathsf{b}$, and $w_i := w_{i-1}w_{i-2}$ for all $i \in \mathbb{N}_0$. The WE-expressibility of L_{fib} was posed as an open problem in [31]. In [28], the techniques of [31] were shown insufficient to resolve the problem. Since then there has been no news, yet since $\text{LEN}(L_{\text{fib}})$ is the set of Fibonacci numbers, which get further apart as they increase, Theorem 1 yields the following:

Corollary 1. L_{fib} *is not RWE-expressible.*

Next we recall [35, Lem. 3.6] that $\text{LEN}(X\mathsf{ab}Y \doteq Y\mathsf{ab}X)$ is not PA-definable. In light of this, we investigate for which WEs PA-definability arises. The length abstractions of SROWEs (and thus SROWE-expressible relations) are known to be PA-definable [14]. For general ROWEs, one can prove the following:

Proposition 1. *ROWE-expressible languages are regular, and thus have PA-definable length abstractions.*

This prompts the question (asked[1] also in [35, §5]) as to whether PA-definability extends to the length abstractions of ROWE-expressible *relations*. In fact, it does not:

Proposition 2. $\text{LEN}(X\mathsf{a}Y \doteq Z\mathsf{b}X) = \{(x,y,y) \in \mathbb{N}_1^3 : y+1 \nmid x+1\}$. *Thus, there exists a ROWE E such that $\text{LEN}(E)$ is not PA-definable.*

One can intuit Proposition 2 using [44, Prp. 1.3.4], which (roughly) says $X\mathsf{a}Y \doteq Z\mathsf{b}X$ forces $\mathsf{a}Y$ and $Z\mathsf{b}$ to conjugate. Further, it implies that - once a length $y \in \mathbb{N}_1$ has been fixed for Y (and thus Z) in a solution h - the possible $|h(X)|$ lie in at most $y+1$ arithmetic progressions, the gaps of which are all $y+1$, and the offsets of which depend on the number n of cyclic shifts needed to

---
[1] Actually, [35] asked this about e-length abstractions. However, where E is a given WE, PA-undefinability clearly carries from $\text{LEN}(E)$ to the corresponding e-length abstraction.

take aY to Zb. For almost every $n \in \{0, 1, ..., y\}$, one can choose $h(Y)$ and $h(Z)$ so that a whole arithmetic progression of $|h(X)|$-values does occur alongside our $|h(Y)|$-value. There is one exception: a certain n-value will cause the explicit a of aY and b of Zb to line up, meaning that one arithmetic progression will be missing. The "does not divide" results from this phenomenon.

3.2 Beyond Regular Word Equations

We now stop requiring regularity, studying what WEs' length abstractions - in totality - can do. Our first result is a weaker version of Theorem 1 for a general QWE E; we establish the same conditions, but only for $\text{Len}(E)$ and not its projections. (Taking projections need not preserve these conditions: consider $\{(2^m, n) : m, n \in \mathbb{N}_0\}$). To prove it the lemmas of Theorem 1 are reused; these tell that $\text{Len}(E)$ comprises arithmetic progressions as soon as the QWE E has any unknown occurring "regularly" (*i.e.* at most once per side). Thus, the only case left is that in which all unknowns occur twice on one side; this is handled inductively. We suspect Theorem 1 holds fully for QWEs, but we leave this open.

Theorem 2. *Let E be a QWE. Then $\text{Len}(E)$ differs finitely from a (possibly empty or infinite) union of nonconstant arithmetic progressions.*

We know no WE that violates Theorem 1 or 2; EPAD may suffice to capture all WEs' length abstractions. As [35] alludes to without proof, a kind of converse *can* be established:

Proposition 3. *Let $S \subseteq \mathbb{N}_0^n$ be EPAD-definable. Then S may be obtained*

(i) from a WE-length-constraint pair by projection of its e-length abstraction, or
(ii) from finitely many WEs' e-length abstractions by projection, union, and intersection.

During the proof of *(i)*, the following becomes apparent:

Corollary 2. *EPAD-satisfiability reduces in polynomial time to e-satisfiability for positive Boolean combinations of WEs and length constraints.*

This is intriguing: the latter problem may yet be in NP, which would improve the current NEXP upper bound [34] for EPAD-satisfiability. Another consequence, using an observation of [34], is the following:

Proposition 4. *There are WE-length-constraint pairs whose minimal e-solution size is doubly exponential in the length of the pair.*

For WEs alone, an upper bound is conjectured on the minimal e-solution size which is *singly* exponential in the length of the WE [46]. We leave open whether, for each EPAD-definable $S \subseteq \mathbb{N}_0^n$, there exists a WE E, such that S is a projection of its e-length abstraction.

Until now, we have not met a WE-expressible *language* with a PA-undefinable length abstraction. For instance, $E : X\text{ab}Y \doteq Y\text{ab}X$ has a PA-undefinable

LEN(E) [35], but the languages expressed by X and Y (individually) in E have length abstractions which equal $\mathbb{N}_1$ (and are thus PA-definable). We end §3 by showing this is not true in general.

Proposition 5. *Let $n \in \mathbb{N}_1$, and - for each $i \in \{1, ..., n\}$ - let $p_i \in \mathbb{N}_1$, $q_i \in \mathbb{N}_1$, $S_i := p_i + q_i \mathbb{N}_0$. Let S be the product-set of the S_i. Then $S = \mathrm{LEN}(L)$ for some WE-expressible language L.*

The product-set S of $2 + \mathbb{N}_0$ with itself is the (PA-undefinable) set of composite numbers. From Proposition 5 we may thus conclude the following:

Corollary 3. *There exists a WE-expressible language L such that $\mathrm{LEN}(L)$ is not PA-definable.*

Note that the product-sets of arithmetic progressions from Proposition 5 can all be proved EPAD-definable.

4 A Decision Procedure via Reachability in $\mathrm{SA}(n, \mathbb{N})$

Let QWE_n be the class of QWEs having $\leq n$ unknowns. In this section, we show that SatLC for QWE_n can be reduced to a halfspace-type reachability problem for rational subsets of $\mathrm{SA}(n, \mathbb{N}_0)$. This is a key step towards our main result: that SatLC is decidable for QWE_2. Note that QWE_2 contains WEs which are non-parameterisable and whose length abstractions may yet be EPAD-undefinable (*q.v.* Sect. 5).

In the sequel, E is a QWE over $\Sigma \cup \Xi$, and $\Xi = \{X_1, ..., X_{k-1}\}$ WLOG. For each $C \in \{\square, \triangledown\}$, $X_s \in \Xi$, and $\ell \in \Sigma \cup \Xi \setminus \{X_s\}$, let $\psi^C_{(\ell, X_s)}$ be the endomorphism of $(\Sigma \cup \Xi)^*$ acting as the identity on $\Sigma \cup \Xi \setminus \{X_s\}$, and mapping X_s to

$$\begin{cases} \ell X_s & \text{if } C = \square, \\ \ell & \text{if } C = \triangledown. \end{cases}$$

Let $\mathfrak{P}$ be the set of these endomorphisms, and let $\mathfrak{P}^*$ be the (free) concatenative monoid generated thereby[2]. The *Nielsen transformations* $\psi \in \mathfrak{P}$ are the operations underlying the Nielsen graph *algorithm* for WEs' satisfiability problem. Intuitively, this algorithm iteratively guesses which of the symbols at the left end of a WE's sides is mapped to a longer word in its solution, reflecting this guess by rewriting the longer symbol X as ℓX (*i.e.* $\psi^\square_{(\ell, X)}$), where ℓ is the shorter symbol, and X is reinterpreted as "what remains of the old X". Since unknowns never represent ε, we also allow the guess that X and ℓ line up exactly, reflecting this with the rewriting $X \mapsto \ell$ (*i.e.* $\psi^\triangledown_{(\ell, X)}$). Each rewriting sets up cancellation, and for QWEs this cancellation suffices to preserve quadracity, the net change in the length of the QWE being non-positive. Thus a finite set of QWEs

---
[2] $\mathfrak{P}$ also generates a (non-free) compositional monoid; the coexistence of these monoids will be used in, *e.g.*, the statement of our Proposition 6.

is reachable, and [9] the original QWE is satisfiable iff $\varepsilon \doteq \varepsilon$ is among them. The working of the algorithm is naturally illustrated as a graph $\mathcal{G}(E)$. Figure 1 gives an example. It can be shown that the paths through $\mathcal{G}(E)$ correspond to the solutions of E.

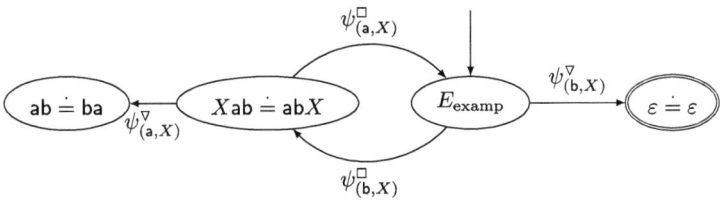

Fig. 1. On $E_{\text{examp}} := X\mathrm{ab} \doteq \mathrm{ba}X$, the Nielsen graph algorithm generates the above graph $\mathcal{G}(E_{\text{examp}})$. Since $\varepsilon \doteq \varepsilon$ appears, E_{examp} is satisfiable.

Let v_{init} be the k-dimensional column vector having zeros in all but its kth row entry, which is a 1. For each $s \in \{1, ..., k-1\}$ and each $t \in \{1, ..., k\} \setminus \{s\}$, let $M^{(s,t)} \in \mathbb{N}_0^{k \times k}$ be such that

$$M^{(s,t)}_{i,j} = \begin{cases} 1 & \text{if } i = j \text{ or } (i,j) = (s,t), \\ 0 & \text{otherwise.} \end{cases}$$

Ostensibly, the set $\mathfrak{M}$ of these matrices is included in $\mathrm{SA}(|\Xi|, \mathbb{N}_0)$. Let $\sigma : \mathfrak{P} \to \mathfrak{M}$ be the function such that

$$\sigma\left(\psi^C_{(\ell, X_s)}\right) = \begin{cases} M^{(s,t)} & \text{if } \ell = X_t \in \Xi, \\ M^{(s,k)} & \text{if } \ell \in \Sigma. \end{cases}$$

The motivation for associating $\sigma(\psi^C_{(\ell, X_s)})$ with $\psi^C_{(\ell, X_s)}$ is as follows:

Lemma 1. Set $g := h \circ \psi^C_{(\ell, X_s)}$, where h is an endomorphism of $(\Sigma \cup \Xi)^*$ acting as the identity on Σ, and such that $h(X_s) = \varepsilon$ if $C = \nabla$. Then

$$\begin{pmatrix} |g(X_1)| \\ \vdots \\ |g(X_{k-1})| \\ 1 \end{pmatrix} = \sigma(\psi^C_{(\ell, X_s)}) \begin{pmatrix} |h(X_1)| \\ \vdots \\ |h(X_{k-1})| \\ 1 \end{pmatrix}.$$

Due to Lemma 1, if $\mathcal{G}(E)$ is treated as a deterministic finite automaton (DFA) with alphabet $\mathfrak{P}$, and σ is applied letter-wise to the words of its language, one obtains an encoding of $\mathrm{LEN}(E)$ in a set $\mathcal{R}$ of matrices. By construction, $\mathcal{R}$ is a rational subset of $\mathrm{SA}(|\Xi|, \mathbb{N}_0)$. Figure 2 gives an example. Formally,

Proposition 6. Let E be a QWE and $\mathfrak{P}$ be as above. Then there exists a regular language $L \subseteq \mathfrak{P}^*$ such that

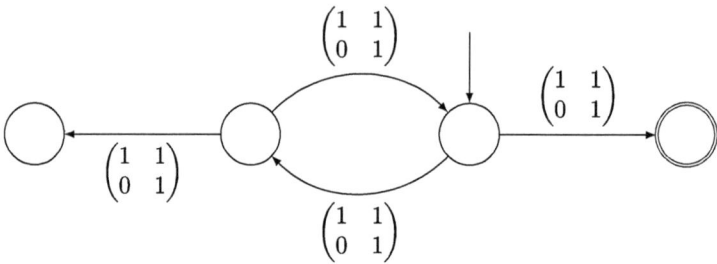

Fig. 2. Applying σ to the edge labels of the graph from Fig. 1 yields what we interpret as a DFA. The language $\mathcal{R}_{\text{examp}}$ of matrices it recognises is a rational subset of $\text{SA}(1, \mathbb{N}_0) \cong (\mathbb{N}_0, +)$. One verifies that $\mathcal{R}_{\text{examp}} v_{\text{init}} = (\text{Len}(E_{\text{examp}}), 1)$; Proposition 6 says that this is true more generally.

- $\text{Sol}(E) = \{\psi_n \circ \cdots \circ \psi_2 \circ \psi_1 : \psi_1 \psi_2 \cdots \psi_n \in L\}$, and
- the rational subset $\mathcal{R} := \sigma(L) \subseteq \text{SA}(|\Xi|, \mathbb{N}_0)$ satisfies $\mathcal{R} v_{\text{init}} = (\text{Len}(E), 1)$.

Now let x be the k-dimensional column vector having the variables $|X_1|$, ..., $|X_{k-1}|$ as its first $k - 1$ entries, and the number 1 as its kth. Ostensibly, any length constraint for E can be converted into a finite disjunction of relations of the form $D x \geq 0$, where $D \in \mathbb{Z}^{m \times k}$, $m \in \mathbb{N}_1$. (All our inequalities between vectors are element-wise). In this way, SatLC for E reduces to taking each D in turn, and checking - for each - if there exists $M \in \mathcal{R}$ such that $DM v_{\text{init}} \geq 0$. §4 rests on this reduction to reachability, and the following related results:

Theorem 3. *([13] [Thm. 16]) Let $t \in \mathbb{Z}$, and let $u, v \in \mathbb{Z}^2$. Then the set $\{M \in \text{GL}(2, \mathbb{Z}) : u^T M v \geq t\}$ is an effectively rational subset of $\text{GL}(2, \mathbb{Z})$.*

Proposition 7. *([13] [Cor. 6]) Rational subsets of $\text{GL}(2, \mathbb{Z})$ are effectively closed under Boolean operations. That is, given rational subsets L, L' of $\text{GL}(2, \mathbb{Z})$, one can compute regular expressions for $L \cap L'$, $L \cup L'$, and $\text{GL}(2, \mathbb{Z}) \backslash L$.*

Now we have the ingredients for our decision procedure. Let E be a QWE_{k-1}, and let $Dx \geq 0$ ($D \in \mathbb{Z}^{m \times k}$) represent a length constraint for it. First, the Nielsen graph algorithm should be used to compute the rational subset $\mathcal{R}$ (from Proposition 6) representing $\text{Len}(E)$. Deciding this instance of SatLC then amounts to checking the subset $\mathcal{R}' := \{M \in \mathcal{R} : DM v_{\text{init}} \geq 0\}$ of $\mathcal{R}$ for its emptiness. We can write $\mathcal{R}'$ as the intersection of $m + 1$ subsets of $\text{GL}(k, \mathbb{Z})$, viz. $\mathcal{R}$ and m of the form from Theorem 3 (with $t = 0$, $v = v_{\text{init}}$). Thus, we require both that the latter sets are computable, and that they (and $\mathcal{R}$) have computable intersections. In the case $k = 2$ (i.e. $|\Xi| = 1$), we get this immediately from Theorem 3 and Proposition 7. Handling QWE_1 is rather trivial, though.

Instead, we consider the case $|\Xi| = 2$ (i.e. $k = 3$). Thus we deal with 3×3 matrices; the main difficulty in extending our technique is reducing that dimensionality. We proceed with a casewise analysis of the possible arrangements of the unknowns of E. Known results (such as that for ROWEs [35]) deal with many arrangements, and we are able to concentrate on just one, viz. $\cdots X \cdots Y \cdots \dot= \cdots Y \cdots X \cdots$. In this case, we examine the structure of $\mathcal{G}(E)$ using the results of [15]. This permits the question to be reformulated as a HRP involving 2×2 matrices. Hence the above results become applicable, and we are able to prove our result:

Theorem 4. *SatLC is decidable for* QWE_2.

Whether these techniques can be adapted to QWEs with 3 or more unknowns is most intriguing. The HRP and VRP become undecidable in higher dimensions [18]. In fact, there is a fixed $S \subseteq \mathbb{Z}^{10 \times 10}$ of cardinality 5, such that - given $\boldsymbol{u}, \boldsymbol{v} \in \mathbb{Z}^{10}$ - it is undecidable whether $\exists M \in S^* : \boldsymbol{u}^T M \boldsymbol{v} > 1$ [16]. In this may lie a method to establish undecidability for SatLC; alternately, the structure of $\mathcal{G}(E)$ may continue to allow simplification.

5 $Xw_1Y = Yw_2X$ and Its Length Abstraction

Harbinger of PA-undefinability, unparameterisable, base of the remarkable [29, Thm. 36], and our main preoccupation in proving Theorem 4, the WE E : $Xw_1Y \dot= Yw_2X$ $(w_1, w_2 \in \Sigma^+)$ appears as the present step on the path to SatLC. As the simplest non-ROWE, E is the first hurdle any result on ROWEs must pass before being generalised. For instance, E illustrates that Proposition 1 is tight: $\mathcal{L}(Xba Y \dot= YabX, X)$ is so closely related [29, Exmp. 13] to the (non-regular [20, Thm. 5.5]) language of finite Sturmian words that it inherits its non-regularity. Similarly, we are keen that the EPAD-definability of [35] be taken beyond ROWEs. Thus, we generalise [35, Lem. 3.6] as far as we can to E. We first allow an arbitrary (but common) terminal word w on the two sides. Recall that the *periods* of a word w are those $d \in \mathbb{N}_1$ such that positions d letters apart in w always hold the same letter. For instance, every $d > 1$ is a period of aba.

Proposition 8. *Let* $w \in \Sigma^*$, $\boldsymbol{v} := (x, y) \in \mathbb{N}_1^2$. *Then* $\boldsymbol{v} \in \mathrm{LEN}(XwY \dot= YwX)$ *iff* $\gcd(x + |w|, y + |w|)$ *is a period of* w.

Our proof is combinatorial; an alternate one might use the known parameterisation [27, 1.23] of $\mathrm{SOL}(XWY \dot= YWX)$.

Next, we stop requiring the two w to coincide. In doing so we leave the class of parameterisable WEs [45]. The technical-looking premise below ensures that the WE "acts the same as" $XabY \dot= YbaX$, which we can study directly.

Proposition 9. *Let* $u, v \in \Sigma^+$ *be equal-length words such that the only triples* (t_1, t_2, t_3) *satisfying* $uv = t_1t_2t_3$, $vu = t_3t_2t_1$ *are* (u, v, ε), (u, ε, v), *and* (ε, u, v). *Then* $\mathrm{LEN}(XuvY \dot= YvuX) = \{(x, y) \in \mathbb{N}_1^2 : \gcd(x + |uv|, y + |uv|) = |u|\}$.

Propositions 8 and 9 both establish effective EPAD-definability for the length abstractions of some small class of WEs. Thus they each imply decidability for small subcases of SatLC.

We will call the VRP for $\mathcal{R} \subseteq \mathbb{N}_0^{n \times n}$ EPAD-*definable* if all the sets $\mathcal{R}v$ ($v \in \mathbb{N}_0^n$) are EPAD-definable. Our final results connect all our titular elements, and suggest it may be hard to go beyond Proposition 9:

Proposition 10. *The following assertions are equivalent:*

(1) $\text{LEN}(E)$ is EPAD-definable for all WEs $E : Xw_1Y \doteq Yw_2X$ ($w_1, w_2 \in \Sigma^+$).
(2) The VRP for $\text{SL}(2, \mathbb{N}_0)$ is EPAD-definable.

The proof of Proposition 10 uses the description of $\text{SOL}(XZY \doteq YVX)$ from [53]. Evidence against *(2)* above is that the *joint spectral radius* (JSR) of the basis of $\text{SL}(2, \mathbb{N}_0)$ is $\frac{1}{2}(1+\sqrt{5}) > 1$ [52]. In monoids generated by a single matrix, this condition turns out to preclude EPAD-definability of the VRP. In fact, we can characterise the cyclic matrix monoids having EPAD-definable VRPs:

Proposition 11. *Let $S := \{M\}$, where $M \in \mathbb{N}_0^{n \times n}$, and let $\rho(S)$ be the JSR of S. Then the VRP for S^* is EPAD-definable iff*

- *$\rho(S) < 1$, or*
- *$\rho(S) = 1$, and every Jordan block of M coming from an eigenvalue of norm 1 has dimension ≤ 2.*

The difficulty in extending Proposition 11 is that, though $\rho(S) > 1$ always signifies exponential growth, if the matrices of S^* are "generated sequentially" there is a clear way to extract - using EPAD - an exponentially-growing *sequence*.

6 Conclusions and Further Work

We must mention [10, Thm. 4], which says that, for *any* WE E, there is an alphabet $\Delta \supseteq \Sigma \cup \Xi$ and a regular subset L of the (free) concatenative monoid generated by the set $\mathfrak{P}$ of endomorphisms of Δ^*, such that

$$\text{SOL}(E) = \{(\psi_n \circ \cdots \circ \psi_1)|_{(\Sigma \cup \Xi)^*} : \psi_1 \cdots \psi_n \in L\}.$$

Hence, using the incidence matrices of the $\psi_i \in \mathfrak{P}$ (similarly to our Proposition 6), $\text{LEN}(E)$ can be encoded in a rational subset of $\mathbb{N}_0^{|\Delta| \times |\Delta|}$. It follows that the reduction described in Sect. 4 is always possible in some form. The six-hundredfold blowup from $|\Sigma| + |E|$ to $|\Delta|$ required by [10] prevents results such as Theorem 3 from being applicable here.

A *regular constraint* for a WE E is a Cartesian product R of $|\Xi|$ regular languages over Σ. The satisfiability problem *modulo a regular constraint* is to decide whether $\text{SOL}(E) \cap R \neq \emptyset$, given WE E and a regular constraint R for E. (Here $\text{SOL}(E)$ should be interpreted as a set of tuples of words). This problem is decidable [51]. One can adapt the Nielsen graph algorithm so it takes a QWE E

and regular constraint R, and generates a graph $\mathcal{G}(E, R)$ describing $\text{Sol}(E) \cap R$ by way of the labels $h \in \mathfrak{P}^*$ of the paths linking two particular states [17]. Thus, Proposition 6 withstands the addition of a regular constraint. That said, even for trivial E, one can vary R so as to endow $\mathcal{G}(E, R)$ with arbitrarily complex structure; managing this seems challenging.

Another goal would be to characterise the class of WEs having PA- (resp. EPAD-)definable length abstractions. If EPAD is insufficient in general, other weak arithmetics might be interesting lenses onto this topic. A first step may be to determine if the VRP for $\text{SL}(2, \mathbb{N}_0)$ is EPAD-definable. Conceivably, something resembling Proposition 11 could be formulated for monoids with multiple generators. Per Proposition 10, any results here will have consequences for WEs.

Lastly, one could try to retread the path of [29] to generalise our work, particularly on $Xw_1Y \doteq Yw_2X$, to all WEs in two unknowns.

Acknowledgments. We are grateful to Benjamin Przybocki, who discussed Proposition 3 with us. He noticed both Corollary 2 and Proposition 4. We would also like to thank the anonymous reviewers for their valuable feedback.

Disclosure of Interests. The authors have no competing interests to declare that are relevant to the content of this article.

References

1. Abdulla, P.A., et al.: String constraints for verification. In: Biere, A., Bloem, R. (eds.) CAV 2014. LNCS, vol. 8559, pp. 150–166. Springer, Cham (2014). https://doi.org/10.1007/978-3-319-08867-9_10
2. Albert, M.H., Lawrence, J.: A proof of Ehrenfeucht's conjecture. Theor. Comput. Sci. **41**, 121–123 (1985)https://doi.org/10.1016/0304-3975(85)90066-0
3. Amadini, R.: A survey on string constraint solving. ACM Comput. Surv. **55**(2), 16:1–16:38 (2023). https://doi.org/10.1145/3484198
4. Bel'tyukov, A.P.: Decidability of the universal theory of natural numbers with addition and divisibility. Zapiski Nauchnykh Seminarov POMI **60**, 15–28 (1976), https://www.mathnet.ru/eng/znsl2066
5. Berzish, M., et al.: An SMT solver for regular expressions and linear arithmetic over string length. In: Silva, A., Leino, K.R.M. (eds.) CAV 2021. LNCS, vol. 12760, pp. 289–312. Springer, Cham (2021). https://doi.org/10.1007/978-3-030-81688-9_14
6. Büchi, J.R., Senger, S.: Definability in the existential theory of concatenation and undecidable extensions of this theory. Math. Log. Q. **34**(4), 337–342 (1988). https://doi.org/10.1002/MALQ.19880340410
7. Chen, Y., Chocholatý, D., Havlena, V., Holík, L., Lengál, O., Síc, J.: Solving string constraints with lengths by stabilization. Proc. ACM Program. Lang. **7**(OOPSLA2), 2112–2141 (2023). https://doi.org/10.1145/3622872
8. Choffrut, C., Karhumäki, J.: Combinatorics of words. In: Rozenberg, G., Salomaa, A. (eds.) Handbook of Formal Languages, pp. 329–438. Springer, Heidelberg (1997). https://doi.org/10.1007/978-3-642-59136-5_6
9. Christian, C.: Equations in words. In: Lothaire, M. (ed.) Combinatorics on Words, pp. 162–183. Cambridge Mathematical Library, Cambridge University Press (1997). https://doi.org/10.1017/CBO9780511566097.012

10. Ciobanu, L., Diekert, V., Elder, M.: Solution sets for equations over free groups are EDT0L languages. Int. J. Algebra Comput. **26**(5), 843–886 (2016). https://doi.org/10.1142/S0218196716500363
11. Ciobanu, L., Evetts, A., Levine, A.: Effective equation solving, constraints, and growth in virtually abelian groups. SIAM J. Appl. Algebra Geom. **9**(1), 235–260 (2025). https://doi.org/10.1137/23M1604679
12. Ciobanu, L., Zetzsche, G.: Slice closures of indexed languages and word equations with counting constraints. In: Sobocinski, P., Lago, U.D., Esparza, J. (eds.) Proceedings of the 39th Annual ACM/IEEE Symposium on Logic in Computer Science, LICS 2024, Tallinn, Estonia, 8–11 July 2024, pp. 25:1–25:12. ACM (2024). https://doi.org/10.1145/3661814.3662134
13. Colcombet, T., Ouaknine, J., Semukhin, P., Worrell, J.: On reachability problems for low-dimensional matrix semigroups. In: Baier, C., Chatzigiannakis, I., Flocchini, P., Leonardi, S. (eds.) 46th International Colloquium on Automata, Languages, and Programming, ICALP 2019, July 9-12, 2019, Patras, Greece. LIPIcs, vol. 132, pp. 44:1–44:15. Schloss Dagstuhl - Leibniz-Zentrum für Informatik (2019). https://doi.org/10.4230/LIPICS.ICALP.2019.44
14. Day, J.D., Ganesh, V., He, P., Manea, F., Nowotka, D.: The satisfiability of extended word equations: the boundary between decidability and undecidability. CoRR **abs/1802.00523** (2018). https://doi.org/10.48550/ARXIV.1802.00523
15. Day, J.D., Manea, F.: On the structure of solution sets to regular word equations. In: Czumaj, A., Dawar, A., Merelli, E. (eds.) 47th International Colloquium on Automata, Languages, and Programming, ICALP 2020, July 8-11, 2020, Saarbrücken, Germany (Virtual Conference). LIPIcs, vol. 168, pp. 124:1–124:16. Schloss Dagstuhl - Leibniz-Zentrum für Informatik (2020). https://doi.org/10.4230/LIPICS.ICALP.2020.124
16. D'Costa, J.R.: Reachability and escape problems in linear dynamical systems. PhD thesis, University of Oxford (2024)
17. Diekert, V., Robson, J.M.: Quadratic word equations, pp. 314–326. Springer, Berlin, Heidelberg (1999). https://doi.org/10.1007/978-3-642-60207-8_28
18. Dong, R.: Recent advances in algorithmic problems for semigroups. ACM SIGLOG News **10**(4), 3–23 (2023). https://doi.org/10.1145/3636362.3636365
19. Draghici, A., Haase, C., Manea, F.: Semënov arithmetic, affine VASS, and string constraints. In: Beyersdorff, O., Kanté, M.M., Kupferman, O., Lokshtanov, D. (eds.) 41st International Symposium on Theoretical Aspects of Computer Science, STACS 2024, 12–14 March 2024, Clermont-Ferrand, France. LIPIcs, vol. 289, pp. 29:1–29:19. Schloss Dagstuhl - Leibniz-Zentrum für Informatik (2024). https://doi.org/10.4230/LIPICS.STACS.2024.29
20. Dulucq, S., Gouyou-Beauchamps, D.: Sur les facteurs des suites de Sturm. Theoret. Comput. Sci. **71**(3), 381–400 (1990). https://doi.org/10.1016/0304-3975(90)90050-R
21. Figueira, D., Jeż, A., Lin, A.W.: Data path queries over embedded graph databases. In: Libkin, L., Barceló, P. (eds.) PODS 2022: International Conference on Management of Data, Philadelphia, PA, USA, 12–17 June 2022, pp. 189–201. ACM (2022). https://doi.org/10.1145/3517804.3524159
22. Freydenberger, D.D., Peterfreund, L.: The theory of concatenation over finite models. In: Bansal, N., Merelli, E., Worrell, J. (eds.) 48th International Colloquium on Automata, Languages, and Programming, ICALP 2021, 12–16 July 2021, Glasgow, Scotland (Virtual Conference). LIPIcs, vol. 198, pp. 130:1–130:17. Schloss Dagstuhl - Leibniz-Zentrum für Informatik (2021). https://doi.org/10.4230/LIPICS.ICALP.2021.130

23. Ganesh, V., Minnes, M., Solar-Lezama, A., Rinard, M.: Word equations with length constraints: what's decidable? In: Biere, A., Nahir, A., Vos, T. (eds.) HVC 2012. LNCS, vol. 7857, pp. 209–226. Springer, Heidelberg (2013). https://doi.org/10.1007/978-3-642-39611-3_21
24. Haase, C., Różycki, J.: On the expressiveness of Büchi arithmetic. In: FOSSACS 2021. LNCS, vol. 12650, pp. 310–323. Springer, Cham (2021). https://doi.org/10.1007/978-3-030-71995-1_16
25. Harju, T., Karhumäki, J.: Morphisms. In: Rozenberg, G., Salomaa, A. (eds.) Handbook of Formal Languages, pp. 439–510. Springer, Heidelberg (1997). https://doi.org/10.1007/978-3-642-59136-5_7
26. Havlena, V., Holík, L., Lengál, O., Síč, J.: Cooking string-integer conversions with noodles. In: Chakraborty, S., Jiang, J.R. (eds.) 27th International Conference on Theory and Applications of Satisfiability Testing, SAT 2024, 21–24 August 2024, Pune, India. LIPIcs, vol. 305, pp. 14:1–14:19. Schloss Dagstuhl - Leibniz-Zentrum für Informatik (2024). https://doi.org/10.4230/LIPICS.SAT.2024.14
27. Hmelevskii, J.I., Kandall, G.A.: Equations in free semigroups. In: Proceedings of the Steklov Institute of Mathematics ; no. 107 (1971), AMS, Providence, Rhode Island (1976), https://www.mathnet.ru/eng/tm2975
28. Ilie, L.: Subwords and power-free words are not expressible by word equations. Fundam. Informaticae **38**(1–2), 109–118 (1999). https://doi.org/10.3233/FI-1999-381209
29. Ilie, L., Plandowski, W.: Two-variable word equations. In: Reichel, H., Tison, S. (eds.) STACS 2000. LNCS, vol. 1770, pp. 122–132. Springer, Heidelberg (2000). https://doi.org/10.1007/3-540-46541-3_10
30. Jeż, A.: Recompression: a simple and powerful technique for word equations. In: Portier, N., Wilke, T. (eds.) 30th International Symposium on Theoretical Aspects of Computer Science, STACS 2013, February 27 - March 2, 2013, Kiel, Germany. LIPIcs, vol. 20, pp. 233–244. Schloss Dagstuhl - Leibniz-Zentrum für Informatik (2013). https://doi.org/10.4230/LIPICS.STACS.2013.233
31. Karhumäki, J., Mignosi, F., Plandowski, W.: The expressibility of languages and relations by word equations. J. ACM **47**(3), 483–505 (2000). https://doi.org/10.1145/337244.337255
32. Laine, M., Plandowski, W.: Word equations with one unknown. Int. J. Found. Comput. Sci. **22**(2), 345–375 (2011). https://doi.org/10.1142/S0129054111008088
33. Le, Q.L., He, M.: A decision procedure for string logic with quadratic equations, regular expressions and length constraints. In: Ryu, S. (ed.) APLAS 2018. LNCS, vol. 11275, pp. 350–372. Springer, Cham (2018). https://doi.org/10.1007/978-3-030-02768-1_19
34. Lechner, A., Ouaknine, J., Worrell, J.: On the complexity of linear arithmetic with divisibility. In: 2015 30th Annual ACM/IEEE Symposium on Logic in Computer Science, pp. 667–676 (2015). https://doi.org/10.1109/LICS.2015.67
35. Lin, A.W., Majumdar, R.: Quadratic word equations with length constraints, counter systems, and Presburger arithmetic with divisibility. Log. Methods Comput. Sci. **17**(4) (2021). https://doi.org/10.46298/LMCS-17(4:4)2021
36. Lin, A.W., Barceló, P.: String solving with word equations and transducers: towards a logic for analysing mutation XSS. In: Bodík, R., Majumdar, R. (eds.) Proceedings of the 43rd Annual ACM SIGPLAN-SIGACT Symposium on Principles of Programming Languages, POPL 2016, St. Petersburg, FL, USA, January 20 - 22, 2016. pp. 123–136. ACM (2016). https://doi.org/10.1145/2837614.2837641

37. Lipshitz, L.: The diophantine problem for addition and divisibility. Trans. Am. Math. Soc. **235**, 271–283 (1978). https://doi.org/10.1090/S0002-9947-1978-0469886-1
38. Lipshitz, L.: Some remarks on the Diophantine problem for addition and divisibility. Bull. Soc. Math. Belg. Sér. B **33**(1), 41–52 (1981)
39. Makanin, G.S.: The problem of solvability of equations in a free semigroup. Math. USSR-Sbornik **32**, 129–198 (1977). https://doi.org/10.1070/SM1977v032n02ABEH002376
40. Makanin, G.S.: Equations in a free group. Math. USSR-Izvestiya **21**(3), 483–546 (1983). https://doi.org/10.1070/IM1983v021n03ABEH001803
41. Manea, F., Nowotka, D., Schmid, M.L.: On the solvability problem for restricted classes of word equations. In: Brlek, S., Reutenauer, C. (eds.) DLT 2016. LNCS, vol. 9840, pp. 306–318. Springer, Heidelberg (2016). https://doi.org/10.1007/978-3-662-53132-7_25
42. Matiyasevich, Y.V.: A connection between systems of words-and-lengths equations and Hilbert's tenth problem. Zapiski Nauchnykh Seminarov POMI **8**, 132–144 (1968), https://www.mathnet.ru/eng/znsl2256
43. Matiyasevich, Y.V.: Enumerable sets are Diophantine. In: Soviet Math. Dokl. vol. 11, pp. 354–358 (1970). https://doi.org/10.1142/9789812564894_0013
44. Perrin, D.: Words. In: Lothaire, M. (ed.) Combinatorics on Words, pp. 1–17. Cambridge Mathematical Library, Cambridge University Press (1997). https://doi.org/10.1017/CBO9780511566097.004
45. Plandowski, W.: An efficient algorithm for solving word equations. In: Kleinberg, J.M. (ed.) Proceedings of the 38th Annual ACM Symposium on Theory of Computing, Seattle, WA, USA, 21–23 May 2006, pp. 467–476. ACM (2006). https://doi.org/10.1145/1132516.1132584
46. Plandowski, W., Rytter, W.: Application of Lempel-Ziv encodings to the solution of words equations. In: Larsen, K.G., Skyum, S., Winskel, G. (eds.) Automata, Languages and Programming, 25th International Colloquium, ICALP 1998, Aalborg, Denmark, 13–17 July 1998, Proceedings. LNCS, vol. 1443, pp. 731–742. Springer (1998). https://doi.org/10.1007/BFB0055097
47. Potapov, I., Semukhin, P.: Vector and scalar reachability problems in $SL(2,\mathbb{Z})$. J. Comput. Syst. Sci. **100**, 30–43 (2019). https://doi.org/10.1016/j.jcss.2018.09.003
48. Presburger, M.: Über die Vollständigkeit eines gewissen Systems der Arithmetik ganzer Zahlen, in welchem die Addition als einzige Operation hervortritt. (On the completeness of a certain system of arithmetic of whole numbers in which addition occurs as the only operation). In: Comptes-Rendus du Ier Congres des Mathematiciens des Pays Slavs (1929)
49. Robinson, J.: Definability and decision problems in arithmetic. J. Symb. Log. **14**(2), 98–114 (1949). https://doi.org/10.2307/2266510
50. Samimi, H., Schäfer, M., Artzi, S., Millstein, T.D., Tip, F., Hendren, L.J.: Automated repair of HTML generation errors in PHP applications using string constraint solving. In: Glinz, M., Murphy, G.C., Pezzè, M. (eds.) 34th International Conference on Software Engineering, ICSE 2012, 2–9 June 2012, Zurich, Switzerland, pp. 277–287. IEEE Computer Society (2012). https://doi.org/10.1109/ICSE.2012.6227186
51. Schulz, K.U.: Makanin's algorithm for word equations - two improvements and a generalization. In: Schulz, K.U. (ed.) IWWERT 1990. LNCS, vol. 572, pp. 85–150. Springer, Heidelberg (1992). https://doi.org/10.1007/3-540-55124-7_4
52. Theys, J.: Joint spectral radius: theory and approximations. PhD thesis, Université catholique de Louvain (2005)

53. Weinbaum, C.: Word equation $ABC = CDA$, $B \neq D$. Pac. J. Math. **213**(1), 157–162 (2004). https://doi.org/10.2140/pjm.2004.213.157
54. Zheng, Y., et al.: **Z3str2**: an efficient solver for strings, regular expressions, and length constraints. Formal Methods Syst. Des. **50**(2–3), 249–288 (2017). https://doi.org/10.1007/S10703-016-0263-6

Word Chain Generators for Prefix Normal Words

Duncan Adamson[1], Moritz Dudey[2], Pamela Fleischmann[2(✉)], and Annika Huch[2]

[1] University of St Andrews, St Andrews, UK
duncan.adamson@st-andrews.ac.uk
[2] Kiel University, Kiel, Germany
stu227171@mail.uni-kiel.de, {fpa,ahu}@informatik.uni-kiel.de

Abstract. In 2011, Fici and Lipták introduced prefix normal words. A binary word is prefix normal if it has no factor (substring) that contains more occurrences of the letter 1 than the prefix of the same length. Among the open problems regarding this topic are the enumeration of prefix normal words and efficient testing methods. We show a range of characteristics of prefix normal words. These include properties of factors that are responsible for a word not being prefix normal. With word chains and generators, we introduce new ways of relating words of the same length to each other.

1 Introduction

A binary word is *prefix normal* if it contains no factor (substring) with more occurrences of the letter 1 than the prefix of the same length [11]. For example, 1101 is prefix normal, as every factor contains no more 1s than the prefix of the same length, while 100101 is not prefix normal since its factor 101 contains more 1s than the prefix of the same length, 100. Research on prefix normal words emerged from work on *binary jumbled pattern matching* (BJPM) [2,13]. This is a combinatorial problem asking, given a word over the binary alphabet $\{0,1\}$ and two natural numbers x and y, whether the word contains a factor with x 0s and y 1s. Note that the order in which these letters appear is considered irrelevant (hence the name *jumbled* pattern matching). The description of a word by the number of occurrences of each letter in the underlying alphabet is given by its *Parikh vector* [15,16]. If we look at all factors of length $x+y$ of a word (sliding window), we can answer a BJPM query in linear time. The *indexed* jumbled pattern matching problem (e.g., [14]) is concerned with answering multiple such queries on the same word. One way to construct an index is by computing the maximum amount of 1s and 0s in any factor of any length. With that, we can create a new word where each prefix contains exactly the maximum number of 1s in any factor of the same length of the original word. In [11], this new word was introduced as the *prefix normal form* of a word with respect to 1. For example, for 00101 we will find that the word 10100 fulfils this property. Thus, with the

© The Author(s), under exclusive license to Springer Nature Switzerland AG 2026
P. Ganty and A. Mansutti (Eds.): RP 2025, LNCS 16230, pp. 68–82, 2026.
https://doi.org/10.1007/978-3-032-09524-4_5

prefix normal form of a word this maximum can be looked up in linear time, or for multiple searches a look-up table can be computed in linear time such that the maximum can be obtained in $O(1)$. As shown in [7], for any number between the maximum and the minimum of 0s in a word, the word always contains a factor with that number of occurrences of 0s. That explains why prefix normal forms can be used to create an index for BJPM queries [11]. Calculating the prefix normal form of a word in such a manner requires a quadratic amount of time, and indeed provides one of the main motivations for studying prefix normal words: A more efficient way to generate prefix normal forms would lead to a faster solution for the indexed BJPM problem. As presented in [2], all BJPM queries for a word can be answered in $O(1)$ if its prefix normal form is known.

In order to obtain the prefix normal form of a word, Fici and Lipták introduced in [11] the notion of *prefix normal equivalence*. A word is *prefix normal equivalent* to another if both have the same prefix normal form. The proof that this is indeed an equivalence relation was presented by Burcsi et al. in [5]. They also showed that each equivalence class has a unique prefix normal representative. Fici and Lipták's work [11] includes a characterization of prefix normality using two functions: For a binary word w, f_w maps a natural number x to the maximum number of 1s in any factor of length x in w. The other function, p_w, maps a natural number y to the amount of 1s in the prefix of w of length y. A word w is prefix normal if f_w equals p_w. Further, Fici and Lipták proved that the language of prefix normal words is not context-free, and explored the relationship between Lyndon and prefix normal words, showing that every prefix normal word is a pre-necklace. Among the open problems stated are the unknown size and number of prefix normal equivalence classes. It was shown in [4] that prefix normal words form a *bubble language*. This means that the first occurrence of 01 in a prefix normal word can be swapped to 10, resulting in another prefix normal word. This property has been used in new algorithms for testing [3] and an algorithm generating all prefix normal words of a fixed length [6]. In [12], prefix normal palindromes and *collapsing* prefix normal words were examined. Two words v and w of the same length *collapse* if $1v$ and $1w$ are prefix normal equivalent. *Infinite* prefix normal words were studied in [8,9]. In [5], it was conjectured that the number of prefix normal words of length n is bounded by $2^{n-\Theta((\log n)^2)}$. This has since been proven in [1]. A closed form formula for the number of prefix normal words or a generating function has not been found yet.

Our Contribution. In our work, we start by presenting observations on (minimal) factors of prefix normal words. We point out several characteristics of factors that are responsible for a word not being prefix normal. Our approach to the enumeration of prefix normal words, comes with the new concept of *prefix normal word chains* and *prefix normal word chain generators*. Here, we observe graphs of binary words of length n where every node is labelled with some $w \in \Sigma^n$ and two nodes are adjacent if their two words differ in only one letter (Fig. 1). We investigate the paths starting with 1^n and reaching 0^n within those graphs that only contain prefix normal words (marked non-transparent in Fig. 1). Since every path corresponds to a permutation of $\{1, \ldots, n\}$ that indicated at which position a letter is changed from 1 to 0, we introduce those special permutations as *prefix normal word chain generators* whereas we refer to the respective sequence of prefix normal words along this path as *prefix normal word chain*. After some basic observations, we examine under which circumstances two numbers in a prefix normal generator can be swapped to construct a new such generator. Then, we connect prefix normal generators with the iterative construction of prefix normal words introduced in [3,6,12].

Structure of the Work. Section 2 is about basic definitions. Results on more efficient testing on prefix normality can be found in Sect. 2.1. Section 3 is concerned with the enumeration of prefix normal words. There, we introduce *word chains* and *word chain generators*.

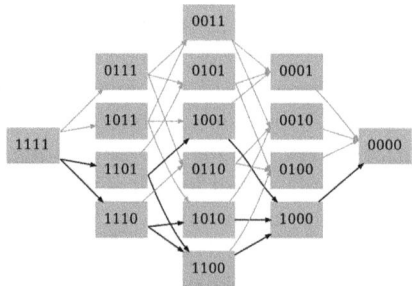

Fig. 1. Graph of words of length 4 with paths corresponding to prefix normal words in black.

2 Preliminaries

Let $\mathbb{N}$ be the set of natural numbers starting with 1. Let $\mathbb{N}_0 = \mathbb{N} \cup \{0\}$. For $n \in \mathbb{N}$ let $[n] = \{i \in \mathbb{N} \mid i \leq n\}$ and $[n]_0 = [n] \cup \{0\}$. For a sequence s, let $s[k]$ be the kth element of that sequence.

An *alphabet* is a finite set of elements called *letters*. We will often use the binary alphabet $\Sigma = \{0, 1\}$. A *word* is a sequence of letters. Let Σ^* denote the set of all finite words consisting of letters in Σ. For $w \in \Sigma^*$, $|w|$ is the *number of letters* in w and $w[i]$ is the ith letter in w for $i \in [|w|]$. Let ε denote the *empty word*, meaning $|\varepsilon| = 0$. Let $x, y, z \in \Sigma^*$ with $w = xyz$, then $y \in \Sigma^*$ is called a *factor*. Let $\text{Fact}(w)$ denote the set of factors of a word w, and for $k \in \mathbb{N}$, let $\text{Fact}_k(w)$ be the set of factors of length k of w. Further, x is a *prefix* of w, which is denoted as $x \leq_p w$, and z is a *suffix* of w, denoted as $z \leq_s w$. Let for any $k \in [|w|]$, $\text{pref}_k(w)$ denote the prefix of length k of a word w, and $\text{suf}_k(w)$ the suffix of length k of a word w. Let $w[i:k]$ denote the factor $w[i] \cdots w[k]$, for $i, k \in [|w|]$ with $i \leq k$. If $i > k$ or, for some $i, k \in \mathbb{Z}$, $i \notin [|w|]$ or $k \notin [|w|]$, then $w[i:k] = \varepsilon$. For $v \in \Sigma^*$ and $w \in \Sigma^*$, we define $|w|_v = |\{(i,j) \in [|w|]^2 \mid w[i:j] = v\}|$, i.e. $|w|_v$ denotes the number of times v occurs as a factor in w. Two factors x and y of a word $w \in \Sigma^*$ *overlap* with each other if they share at least one index of w, i.e. if there exist $i, j, k, l \in \mathbb{N}$ with $x = w[i : j], y = w[k : l]$ and $i \leq j, k \leq l$, then x and y overlap if $\{i, i + 1, ..., j\} \cap \{k, k + 1, ..., l\} \neq \emptyset$. We say that for $w \in \Sigma^*$ and a factor $v = w[k : l]$ for $k, l \in [|w|]$, v *includes* an index $a \in [|w|]$ of w if $k \leq a \leq l$. A word w is called a *palindrome* if $w = w^R$ where w^R denotes the word in reverse. For a word $w \in \Sigma^*$ and $k \in \mathbb{N}_0$, let $w^0 = \varepsilon$ and $(w)^{k+1} = ww^k$. A *block* is a maximal unary factor of a word, i.e. $u \in \text{Fact}(w)$ is a block if there exists $\mathbf{a} \in \Sigma$ such that $u = \mathbf{a}^{|u|}$ and either $u \leq_p w$ and $w[|u| + 1] \neq \mathbf{a}$ or $u \leq_s w$ and $w[|w| - |u|] \neq \mathbf{a}$ or there exist $x, y \in \Sigma^+$ with $w = xuy$ and $x[|x|] \neq \mathbf{a} \neq y[1]$. Blocks of 0s and 1s are called 0- and 1-runs, resp., in [6].

A *permutation* of a set S is a bijection σ from S to S. We call $|S|$ the *length of the permutation*. As an example, for the set $S = \{1, 2, 3, 4\}$, the bijection $\sigma : S \to S, x \mapsto (x + 1) \bmod 4$ is a permutation, and the bijection $\gamma : S \to S$, $\gamma(1) = 4, \gamma(2) = 2, \gamma(3) = 1, \gamma(4) = 3$ is also a permutation. We will write these as words $\sigma = 2341$ and $\gamma = 4213$.

The *Hamming distance* between two words of equal length is the number of indices at which the two words differ: for $u, v \in \Sigma^*$, with $|u| = |v|$, $d_H(u, v) = |\{i \mid i \in [|u|], x[i] \neq y[i]\}|$ denotes the Hamming distance between u and v.

After these basic definitions, we introduce the main object of interest - the prefix normal words.

Definition 1. *For $w \in \Sigma^*$, define the* prefix-ones function *and the* maximum-ones function *by $p_w : [|w|] \to [|w|], i \mapsto |\text{pref}_i(w)|_1, f_w : [|w|] \to [|w|], i \mapsto \max_{u \in \text{Fact}_i(w)} |u|_1$.*

Definition 2. *A word $w \in \{0, 1\}^*$ is called* prefix normal *if and only if $p_w = f_w$. If w is not prefix normal there exists an $i \in [|w|]$ and $x \in \text{Fact}_i(w)$ with $|x|_1 > \text{pref}_i(w)$. We say that x is responsible for w not being prefix normal.*

For instance, the word $w = 111001011011$ is not prefix normal, because its factor $w[8 : 12] = 11011$ contains more 1s than the prefix of length five, namely 11100. The prefix normal words up to length three are: 0, 1, 00, 10, 11, 000, 100, 101, 110, 111.

For the definition of a word chain, we need a function that takes a word and flips a single 1 to a 0. For example, $\text{flip}_1(001) = 001$ while $\text{flip}_2(11010) = 10010$. A similar operation that allows for flipping 0s to 1s was defined in [8].

Definition 3. *Let $i \in \mathbb{N}$. Define $\text{flip}_i : \Sigma^* \to \Sigma^*, w \mapsto w[1 : i-1] \cdot 0 \cdot w[i+1 : |w|]$, if $1 \leq i \leq |w|$ and $w \mapsto w$, otherwise.*

Definition 4. *For two words $p, c \in \Sigma^*$, we call p a* parent *of c and c a* child *of p if and only if $|p| = |c|$, $|p|_1 - 1 = |c|_1$ and $d_H(p, c) = 1$. Let $\text{Parents}(w)$ denote the set of all parents of a word $w \in \Sigma^*$ and let $\text{Children}(w)$ denote the set of all of its children. Let $\text{Parents}_{PN}(w) = \{x \in \Sigma^* \mid x \in \text{Parents}(w) \land x \text{ prefix is normal}\}$.*

Remark 5. Another way of describing this relation of parents and children is by using the flip function defined above. Let $w \in \Sigma^*$. Then for all $i \in [|w|]$, if $w[i] = 1$ then w is a parent of $\text{flip}_i(w)$ and $\text{flip}_i(w)$ is a child of w.

Our last definition introduces the prefix normal word chains and their generators. The idea is to start with the prefix normal word 1^k for some $k \in \mathbb{N}$ and flipping successively 1s to 0s in such an order that every word in the chain (until we reach 0^k) is prefix normal. This sequence of prefix normal words is given by a permutation of the word's indices.

Definition 6. *Let σ be a permutation on $[m]$ for some $m \in \mathbb{N}$. Let the sequence of words emerging from applying the flip function at the indices given by σ be*

$$c_\sigma = \Big(\mathrm{flip}_{\sigma(n-1)}(1^m) \circ \cdots \circ \mathrm{flip}_{\sigma(1)}(1^m) \Big)_{n \in [m+1]}.$$

Such a sequence c_σ is a word chain and σ its word chain generator. We call c_σ a prefix normal word chain and σ the corresponding prefix normal word chain generator (or for short prefix normal generator) iff for all $n \in \mathbb{N}$, $c_\sigma[n]$ is prefix normal. Here, we denote by $c_\sigma[n]$ the projection onto the nth component of c_σ.

Note that word chains can also be defined recursively, i.e., $c_\sigma[1] = 1^m$ and $c_\sigma[n+1] = \mathrm{flip}_{\sigma(n)}(c_\sigma[n])$ for a permutation σ on $[m]$ for some $m, n \in \mathbb{N}, n \leq m$. The sequence $\sigma = 231$ is a prefix normal word chain generator, because it creates the prefix normal word chain $c_\sigma = (111, 101, 100, 000)$. The permutation $\gamma = 213$ generates the sequence $s_\gamma = (111, 101, 001, 000)$ and is not a prefix normal generator as $s_\gamma[3] = 001$ is not a prefix normal word.

2.1 Observations on (Minimal) Factors of Prefix Normal Words

In this section we present observations about prefix normal words. The first idea is to characterize the minimal factor that contains more 1s than the prefix of the same length.

Observation 7. *The factor minimal in length that is responsible for a word not being prefix normal always begins and ends with a 1.*

Observation 8. *The factor minimal in length responsible for a word not being prefix normal always contains just one more 1 than the prefix of the same length.*

By Observation 7, we only have to check all factors in a word that both begin and end with a 1, which reduces the complexity compared to checking every factor. For $w \in \Sigma^*$, the number of factors that w contains is $\frac{|w| \cdot (|w|+1)}{2}$. Note that $|\mathrm{Fact}(w)| \leq \frac{|w| \cdot (|w|+1)}{2}$, because factors can occur more than once. With the following result, we give the number of factors within a word starting and ending with a 1, including the trivial factor 1.

Remark 9. *Let $w \in \Sigma^*$. Let $\mathrm{Fact1}(w)$ denote the set of factors of w both starting and ending in 1. Let $f^{1\ldots 1}(w) = \sum_{u \in \mathrm{Fact1}(w)} |w|_u$ denote the number of factors in w starting and ending with 1. Then*

$$f^{1\ldots 1}(w) = \frac{|w|_1(|w|_1 - 1)}{2}.$$

Before we determine which characteristics at least one factor of a non prefix normal word must have, we need the following observation about prefix normal words in general.

Observation 10. *A factor that includes any part of the first block of 1s of a word cannot be responsible for the word not being prefix normal. Let $w \in \Sigma^*, k \in \mathbb{N}, u \in \Sigma^*$ such that $w = 1^k u$ and a factor $x = w[i : j], i, j \in [|w|], i \leq j$. If $i \leq k$, then $|x|_1 \leq |\operatorname{pref}_{|x|}(w)|_1$.*

Informally, if a factor *is* responsible for a word not being prefix normal, meaning it *does* contain more 1s than the prefix of the same length, then it does *not* begin inside the first block of 1 s in the word. With this observation, we can now show the following proposition that extends Observation 7 that does not refer to blocks of 1s.

Proposition 11. *In any word that is not prefix normal, there is a factor that starts and ends with a block of 1 s and contains more 1s than the prefix of the same length (see Fig. 2).*

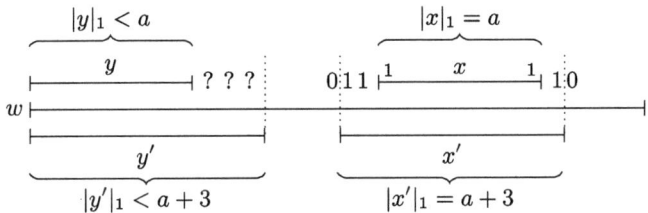

Fig. 2. Illustration of Proposition 11

Note that in Fig. 2, x is the minimal factor containing more 1s than the prefix of the same length.

3 Word Chains and Generators

In this section, we provide insights to the problem of enumerating prefix normal words. We approach this by restricting our field of view to just the prefix normal words of fixed length. This leads to *prefix normal word chains* and *prefix normal word chain generators*. First, we give some results on these chains before we connect them to the known field of *extension-critical* prefix normal words. Note that for any permutation σ, $c_\sigma[1] = 1^m$, because the flip function is applied 0 times. The following remarks both follow immediately from the definition.

Remark 12. Let $n \in \mathbb{N}$, σ a permutation of length n, c_σ the corresponding word chain and $i \in [n]$. Then, there exist $u, v \in \Sigma^*$ with $|u| = \sigma(i)-1$ and $|v| = n-\sigma(i)$ such that $c_\sigma[i] = u1v$ and $c_\sigma[i+1] = u0v$.

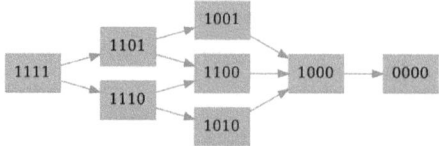

Fig. 3. Prefix normal words of length 4

Remark 13. Let $n \in \mathbb{N}$, σ a permutation of length n and c_σ the corresponding word chain. For any $i \leq n$, we have $|c_\sigma[i+1]|_0 = i$.

Every prefix normal word except those in $\{1\}^*$ has at least one prefix normal parent, since the first occurrence of 0 can always be replaced by a 1. With this in mind, the following remark is clear.

Remark 14. Every prefix normal word is part of a prefix normal word chain: every word is part of a word chain and thus, by definition every prefix normal word is part of a prefix normal word chain (Fig. 3).

Every word in a prefix normal word chain except the last one begins with 1. The index 1 can only be flipped if all other letters are 0s.

Lemma 15. *Let $n \in \mathbb{N}$ and let σ be a permutation over $[n]$. If σ is a prefix normal word chain generator, then $\sigma(n) = 1$.*

Furthermore, the first flip from 1 to 0 cannot happen in the first half of a word. Otherwise, the second half would contain an extra 1 compared to the first, which also is the prefix of the same length.

Lemma 16. *Let $n \in \mathbb{N}$ and let σ be a word chain generator of length n. If σ is a prefix normal word chain generator, then $\sigma(1) \geq \lceil \frac{n+1}{2} \rceil$.*

For each word length, there is a finite number of prefix normal word chain generators. Often, these are in a relationship with each other: two adjacent numbers in a generator can be swapped, leading to a different prefix normal word chain generator. This word chain will be equal to the first except for one word. A question arises whether all prefix normal word chain generators of a certain length are connected by such swaps. If so, which swaps are allowed, and which lead to the resulting word chain not being prefix normal?

Now, we fix the preliminaries for the following statements. let σ be a prefix normal word chain generator of length n, and c_σ the arising prefix normal word chain. Let $j \in [n-1]$, and let $x, y \in \text{Fact}(\sigma)$ such that $\sigma = x\, \sigma[j]\, \sigma[j+1]\, y$. Let $\sigma' = x\, \sigma[j+1]\, \sigma[j]\, y$ (see Fig. 4).

Lemma 17. *If $\sigma[j] < \sigma[j+1]$, then σ' is also a prefix normal word chain generator.*

```
                    x      σ[j]   σ[j+1]    y
        σ  ├──────────────┼──────┼──────────┼────────┤

                    x      σ[j+1]  σ[j]     y
        σ' ├──────────────┼──────┼──────────┼────────┤
```

Fig. 4. Prefix normal generator σ and *potentially* prefix normal generator σ'

Proof. Notice that $c_{\sigma'}$ is the new word chain arising from the generator σ'. The words $c_\sigma[|x|+1]$, $c_\sigma[|x|+2]$ and $c_\sigma[|x|+3]$ are all prefix normal, since they appear in the prefix normal word chain c_σ. We also know that all words in the first word chain c_σ are the same as in $c_{\sigma'}$ except for one: $c_\sigma[i] = c_{\sigma'}[i]$ for $i \in [n], i \neq |x|+1$. The question therefore is whether this word, $c_{\sigma'}[|x|+1]$, is prefix normal (see Fig. 5).

We know that $c_\sigma[|x|+1]$ contains 1s at its indices $\sigma[j]$ and $\sigma[j+1]$, i.e., the $\sigma[j]^{\text{th}}$ letter and the $\sigma[j+1]^{\text{th}}$ letter are both 1. That is because these two have not yet been flipped to 0, as each number can only appear once in a generator, so they do not occur in x. We also know that $c_\sigma[|x|+3]$ contains 0s at both indices $\sigma[j]$ and $\sigma[j+1]$. It is only in between these two words that the word chains differ at exactly these two indices: $c_\sigma[|x|+2]$ contains a 0 as its $\sigma[j]^{\text{th}}$ letter and a 1 as its $\sigma[j+1]^{\text{th}}$ letter, whereas $c_{\sigma'}[|x|+1]$ contains a 1 at index $\sigma[j]$ and a 0 at index $\sigma[j+1]$.

Let u be a factor of $c_{\sigma'}[|x|+1]$, which is not a prefix. We will consider several cases based on which of the two letters at index $\sigma[j]$ and $\sigma[j+1]$ are included in u. If u contains only the 0, which is the $\sigma[j+1]^{\text{th}}$ letter of $c_{\sigma'}[|x|+1]$, or none of the two letters, then it is also a factor of $c_\sigma[|x|+3]$, whose prefixes contain either the same amount or one 1 less than $c_{\sigma'}[|x|+1]$ which is prefix normal. If u contains only the 1, i.e., the $\sigma[j]^{\text{th}}$ letter of $c_{\sigma'}[|x|+1]$, then it is also a factor of $c_\sigma[|x|+1]$, which is prefix normal. There are, of course, prefixes of $c_\sigma[|x|+1]$ that contain more 1s than prefixes of the same length of $c_{\sigma'}[|x|+1]$, namely those of length at least $\sigma[j+1]$. These are not relevant here, however, as u must not contain the $\sigma[j+1]^{\text{th}}$ letter in $c_{\sigma'}[|x|+1]$ and therefore is shorter than $\sigma[j+1]$. The last remaining case is if u contains both letters, the 1 and the 0. In that case, u contains the same amount of 1s as the factor starting and ending at the same index in $c_\sigma[|x|+2]$. This word is prefix normal and its prefixes contain either the same amount or one 1 less than the prefixes of $c_{\sigma'}[|x|+1]$. So u cannot contain more 1s than the prefix of the same length. Therefore, $c_{\sigma'}[|x|+1]$ is prefix normal. □

So in this case, σ' is always a prefix normal generator. This is in line with our intuition because, starting with the prefix normal word $c_\sigma[|x|+2]$, a 1 is moved closer to the beginning of the word and a 0 closer to the word's end (see Fig. 5). Lemma 17 also implies that prefix normal words form a bubble language, which has already been shown in [4]. Now we will move on to the case of $\sigma[j] > \sigma[j+1]$ (Fig. 6).

Lemma 18. *Let $\sigma[j] > \sigma[j+1]$. Let v be a factor of $c_\sigma[|x|+1]$ where:*
- v includes the index $\sigma[j]$ and does not include the index $\sigma[j+1]$,

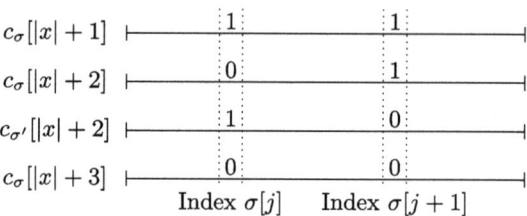

Fig. 5. Case $\sigma[j] < \sigma[j+1]$

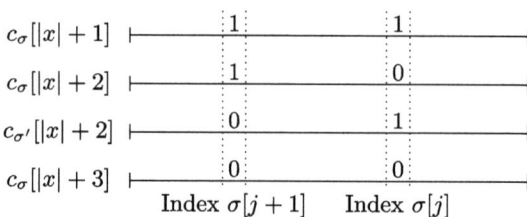

Fig. 6. Case $\sigma[j] > \sigma[j+1]$

- $\sigma[j+1] \leq |v| < \sigma[j]$, and,
- $|v|_1 = |\operatorname{pref}_{|v|}(c_\sigma[|x|+1])|_1$.

Then, σ' is a prefix normal word chain generator if and only if no such v exists in $c_\sigma[|x|+1]$.

Proof. For the reasons given in the proof of Lemma 17, we only have to look at $c_{\sigma'}[|x|+2]$ to determine whether $c_{\sigma'}$ is a prefix normal word chain and, therefore, σ' a prefix normal generator. Any factor of $c_{\sigma'}[|x|+2]$ that includes both the 1 at $\sigma[j]$ and the 0 at $\sigma[j+1]$ cannot contain more 1s than the prefix of the same length, since $c_\sigma[|x|+2]$ contains a factor of the same length with the same amount of 1s and is prefix normal. Also, any factor of $c_{\sigma'}[|x|+2]$ that does not contain the 1 at index $\sigma[j]$ cannot contain more 1s than the prefix of the same length because it is also a factor of $c_\sigma[|x|+3]$, whose prefixes contain either the same amount of 1s as $c_{\sigma'}[|x|+2]$ or one less and which is prefix normal, too. Therefore, any factor of $c_{\sigma'}[|x|+2]$ containing more 1s than the prefix of their length must contain the 1 at index $\sigma[j]$, but not the 0 at $\sigma[j+1]$.

If the length of such a factor is less than $\sigma[j+1]$, the prefix of the same length will not reach the 0 at that index. In that case, however, both factor and prefix also appear in $c_\sigma[|x|+1]$, which is prefix normal. If such a factor is at least $\sigma[j]$ letters long, then the prefix of the same length includes the 1 at index $\sigma[j]$, and both prefix and factor will have one more 1 than the corresponding prefix and factor in $c_\sigma[|x|+2]$. And since $c_\sigma[|x|+2]$ is prefix normal, adding a 1 to both factor and prefix will also result in a factor with still at most as many 1s as the prefix. This further narrows the factors potentially containing more 1s than the corresponding prefix down to factors of a length of at least $\sigma[j+1]$ and at most $\sigma[j]-1$.

We can now use these observations about $c_{\sigma'}[|x|+2]$ and apply them to $c_\sigma[|x|+1]$. Let v be a factor of $c_\sigma[|x|+1]$ that has all four properties stated in the lemma. This means that there is a factor v' in $c_{\sigma'}[|x|+2]$ at the same position as v with $|v'|_1 = |v|_1$. Because the 1 at index $\sigma[j+1]$ has now been flipped and this 1 is included in the prefix of length $|v'|$, we have $|\operatorname{pref}_{|v|}(c_\sigma[|x|+1])|_1 - 1 = |\operatorname{pref}_{|v'|}(c_{\sigma'}[|x+2|])|_1$. This means that v' contains exactly one more 1 than the prefix of the same length. So $c_{\sigma'}[|x|+1]$ is not prefix normal, and the swap results in $c_{\sigma'}$ not being a prefix normal generator.

To show that, if σ' is a prefix normal word chain generator, no such v can exist, it suffices to show the following: If a v fulfilling the properties exists, then σ' is not a prefix normal word chain generator. If such a v exists in $c_\sigma[|x|+1]$, then we know the following about $c_{\sigma'}[|x|+2] = uvw$ for some $u, w \in \Sigma^*$: $|u| > 0$, because v excludes the letter at index $\sigma[j+1]$ but includes the letter at index $\sigma[j]$, and $\sigma[j] > \sigma[j+1]$. We also know $|v|_1 > |pref_{|v|}(c_{\sigma'}[|x|+2])|_1$, because this prefix now contains a 0 at index $\sigma[j+1]$ which v does not contain, and both are otherwise unchanged when compared to the factors at the same indices in $c_\sigma(|x|+1)$, where v contained the same amount of 1s as the prefix of the same length. Together, these two observations give us that v is a factor that is not a prefix and contains more 1s than the prefix of the same length. So $c_{\sigma'}(|x|+2)$ is not a prefix normal word and σ' is not a prefix normal word chain generator. □

Remark 19. In the light of Lemma 18, the question arises how many possible factors one needs to test for deciding whether σ' is a prefix normal word chain generator. The worst case on the number of possible factors v that need to be tested on the conditions of Lemma 18 happens for $\sigma[j+1] = 2$ and $\sigma[j] = \lfloor \frac{n-2}{2} \rfloor$ ($\sigma[j+1] = 1$ is not relevant by Lemma 15). By the first condition (v includes the index $\sigma[j] = \lfloor \frac{n-2}{2} \rfloor$ and does not include the index $\sigma[j+1] = 2$) we know that there exist $\lfloor \frac{n-2}{2} \rfloor$ such factors of length $\lfloor \frac{n-2}{2} \rfloor$, $\lfloor \frac{n-2}{2} \rfloor - 1$ such factors of length $\lfloor \frac{n-2}{2} \rfloor - 1$ etc. Thus, the number of factors is upper bounded by the Gaussian sum from 1 to $\lfloor \frac{n-2}{2} \rfloor$, i.e., $\frac{\frac{n-2}{2}(\frac{n-2}{2}+1)}{2} = \frac{n^2-2n}{8}$.

Note, that the largest possible range in the condition $\sigma[j+1] \leq |v| < \sigma[j]$ of Lemma 18 is given by $\sigma[j+1] = 2$ and $\sigma[j] = n$ does not automatically lead to the worst case. By the first condition (v includes index $\sigma[j] = n$ and does not include index $\sigma[j+1] = 2$) we know that every suffix of $c_\sigma[|x|+1]$ from the 3rd position on needs to be inspected (the suffix of length 1 is not relevant since we ensured that $\sigma(n) = 1$ holds for prefix normal word chain generator), i.e., less factors than in the previously described case.

Further, one can apply the results from Sect. 2.1 in order to reduce the number of inspected factors even more (e.g., only inspecting those that start and end with the letter 1).

We will now combine these last two lemmas into one general observation about the prefix normality of generators after a swap.

Theorem 20. *A generator σ' is not a prefix normal generator if and only if*
- *$\sigma[j] > \sigma[j+1]$, and*
- *a factor $v \in \text{Fact}(c_\sigma[|x|+1])$ exists such that*
    - *v includes index $\sigma[j]$ and does not include index $\sigma[j+1]$,*
    - *$\sigma[j+1] \leq |v| < \sigma[j]$ and*
    - *$|v|_1 \geq |\text{pref}_{|v|}(c_\sigma[|x|+1])|_1$.*

So we do not have to look at $c_{\sigma'}[|x|+2]$ or indeed any part of $c_{\sigma'}$ itself to determine whether it is prefix normal. It is sufficient to check a prefix normal word chain generator by looking at pairs of entries like this: If the first of the two numbers is smaller than the second, swapping them will lead to a prefix normal generator. If the first number is larger than the second, we can test the last word in the corresponding word chain - where all flips up to the two that the number pair specifies, already happened - for factors with the specified characteristics.

Instead of swapping elements of prefix normal generators, we can also increase their length. Prefix normal word chains and generators differ from the way prefix normal words have been studied in that the words we consider are all of the same length [3,5]. Now, we combine this view with the classical approach of appending letters to words. First, we will look at complete graphs and ask the question: What is the difference between a graph showing all prefix normal words of length n and its successor showing all prefix normal words of length $n+1$? First we give some easy results we need to proceed.

Remark 21. (cf. [11]) If a word $w \in \Sigma^*$ is prefix normal, so are $1w$ and $w0$.

Lemma 22. *The number of prefix normal words of length n is equal to the number of prefix normal words of length $n+1$ ending in 0.*

So we know more than half of the prefix normal words of length $n+1$ by appending a 0 to all words of length n. But how do we find the remaining words? One way to look at it is by asking: To which words of the graph of prefix normal words of length n can we append a 1 with the resulting word being prefix normal? And what are the characteristics of words where this is not possible? In [3,6], a prefix normal word w is called *extension-critical* if $w1$ is not prefix normal.

Lemma 23. *If 1 is appended to a prefix normal palindrome containing at least one 0, the resulting word is not prefix normal.*

Going one step further, we can combine prefix normal word chain generators with the approach of appending letters like this: Take a prefix normal generator of length n, and insert the number $n+1$ at different points. This means appending one letter to each word in the word chain. In which cases will the new generator (not) be prefix normal? It follows immediately from Lemma 15 that $n+1$ cannot be inserted at the last position of the new generator. Let σ be a prefix normal generator of length n and c_σ its corresponding prefix normal word chain. Let $w = c_\sigma[i]$ for some $i \in \mathbb{N}$. Let τ be the word chain generator arising from the insertion of $n+1$ into σ such that w is the last word in c_σ to which 1 is appended, while 0 is appended to its child $c_\sigma[i+1]$ and all other words that follow in the word chain. So

$$\tau = (\sigma[1], \sigma[2], ..., \sigma[i-1], n+1, \sigma[i], ..., \sigma[n]), \text{ and}$$

$$c_\tau = (c_\sigma[1]1, c_\sigma[2]1, ..., w1, c_\sigma[i+1]0, ..., c_\sigma[n+1]0, 0^{n+1}).$$

Now, if $w1$ is not prefix normal, neither is c_τ. But even if $w1$ is prefix normal, there are cases where c_τ is not. What we want to show is that, if $w1$ is prefix normal, it is always part of a prefix normal word chain that was constructed in the same way as c_τ. We know from Remark 21 that $c_\sigma[i+1]0$ and all words following it in c_τ are prefix normal. So we only have to look at the first i words in c_τ. We do this by showing that every word like $w1$ and the words before it in c_τ have at least one prefix normal parent. If we look at all prefix normal words of length $n \in \mathbb{N}$ as a graph, this means the following: We append 1 to every word in the graph that contains less than $k \in \mathbb{N}$ 0s, and we append 0 to every word in the graph containing at least k 0s. Then, we take a prefix normal word containing exactly $k-1$ occurrences of 0 from the resulting graph. This word will be connected to the first and last word in the graph (the two words containing only 1s and 0s) via a word chain of only prefix normal words.

Proposition 24. *Let w be a prefix normal word. If $w1$ is also prefix normal, then w has at least one prefix normal parent v such that $v1$ is also prefix normal.*

Further, we know that for a word chain generator that is not prefix normal, a right shift of the entry n always results again in a word chain generator that is not prefix normal.

Proposition 25. *If σ' is a prefix normal word chain generator of length $n \in \mathbb{N}$ with $\sigma'[i] = n$ for some $i \in [2, n-1]$ then σ given by $\sigma = \sigma[1..i-2]\sigma[i]\sigma[i-1]\sigma[i+1..n]$ is a prefix normal word chain generator.*

Using the previous results, we give the idea of constructing the prefix normal generators of length $n+1$ from those of length n the following way: We take a prefix normal generator of length n and insert $n+1$ as the new first element of the sequence. This means creating the new prefix normal word 1^{n+1} as the new first element of the word chain and appending a 0 to all other words in it. We know that the resulting generator is prefix normal. Now, we move the number $n+1$ one position to the right, i.e., we swap it with the element succeeding it in the sequence. If the resulting generator is prefix normal, we add it to the list of prefix normal generators of length $n+1$ and repeat this step. If not, we stop (by contraposition of Proposition 25) and continue with the next generator of length n. From Proposition 24, we know that we are not missing prefix normal words of length $n+1$ ending in 1. That is because every prefix normal word to which a 1 can be appended is connected to a parent to which a 1 can also be appended. So every such prefix normal word will be found because it is part of a word chain where all previous words also remain prefix normal if a 1 is appended. This algorithm can also include a check whether the word at the current position of the generator of length n is a palindrome, in which case no 1 can be appended because of Lemma 23.

To find all prefix normal words of a certain length, this algorithm has to be applied recursively, with the prefix normal generator of length 1 being given as (1). It has to test almost every prefix normal word of length n on whether it is extension-critical to find those of length $n+1$. Its complexity therefore is too high for it to be a practical approach. However, as stated similarly in the conclusion of [4], the number of binary words grows much more rapidly than that of prefix normal words. So if all prefix normal words of length n are given, the algorithm performs significantly better than generating all binary words of length $n + 1$ and then filtering on the condition from the definition of prefix normality.

Since permutations and the symmetric group are well-studied objects, we investigated two basic notions of permutations. First, the closure of composition within the symmetric group and second, the parity of a permutation.

Remark 26. Note that the composition of two prefix normal generators is never a prefix normal generator again. Let σ, σ' be prefix normal generators over $[n]$ for $n \in \mathbb{N}$. By Lemma 15 we have $\sigma(n) = \sigma'(n) = 1$. Thus, we get $(\sigma \circ \sigma')(n) = \sigma(\sigma'(n)) = \sigma(1) \neq 1$, i.e., the composition cannot be a prefix normal generator.

Remark 27. There exist prefix normal generators with even and odd parity witnessed by $(3241), (3421), (4321), (4231)$ that have $4, 5, 6, 5$ inversions, so even, odd, even, odd parities, resp.

To conclude this section, we presented a new perspective on prefix normal words of a fixed length n giving a characterisation of prefix normal word chains. For now a connection to existing results in theory about permutations and the symmetric group cannot be made and stay open for further research.

4 Conclusion

Our work on prefix normal words is based on [11]. In Sect. 2.1, we presented characteristics of factors that contain more 1s than the prefix of the same length, hence exploiting the word that is responsible for them being not prefix normal.

Our work in Sect. 4 focused on relationships between words of equal length. We introduced word chains and word chain generators that describe multiple words of the same length. While these concepts enabled us to think about prefix normal words in a more abstract way, the problem of enumerating prefix normal words is open to further investigation. Research into both the relationship between generators of the same length as well as inserting one number into a generator is necessary. We have shown that testing whether a swap will lead to a new prefix normal generator can be done by testing only specific factors of a word. Is there a path from any prefix normal generator to all others, just by swapping neighbouring numbers? If so, all prefix normal words of a length could be found by checking possible swap positions. It is not yet clear whether these positions follow a pattern or are further related to the symmetric group. In [10], *weighted* prefix normal words were introduced, extending the concept of prefix normality to arbitrary finite alphabets. Word chain generators could become *word chain generating matrices* with $n \in \mathbb{N}$ rows to represent word chains of words over alphabets containing $n+1$ letters.

References

1. Balister, P., Gerke, S.: The asymptotic number of prefix normal words. Theor. Comput. Sci. **784**, 75–80 (2019)
2. Burcsi, P., Cicalese, F., Fici, G., Lipták, Z.: Algorithms for jumbled pattern matching in strings. Int. J. Found. Comput. Sci. **23**(2), 357–374 (2012)
3. Burcsi, P., Fici, G., Lipták, Z., Ruskey, F., Sawada, J.: Normal, abby normal, prefix normal. In: FUN 2014, Proceedings, volume 8496 of *LNCS*, pp. 74–88. Springer (2014)
4. Burcsi, P., Fici, G., Lipták, Z., Ruskey, F., Sawada, J.: On combinatorial generation of prefix normal words. In: Kulikov, A.S., Kuznetsov, S.O., Pevzner, P. (eds.) CPM 2014. LNCS, vol. 8486, pp. 60–69. Springer, Cham (2014). https://doi.org/10.1007/978-3-319-07566-2_7
5. Burcsi, P., Fici, G., Lipták, Z., Ruskey, F., Sawada, J.: On prefix normal words and prefix normal forms. Theor. Comput. Sci. **659**, 1–13 (2017)
6. Burcsi, P., Fici, G., Lipták, Z., Raman, R., Sawada, J.: Generating a gray code for prefix normal words in amortized polylogarithmic time per word. Theor. Comput. Sci. **842**, 86–99 (2020)
7. Cicalese, F., Fici, G., Lipták, Z.: Searching for jumbled patterns in strings. In: Proceedings of the Prague Stringology Conference 2009, pp. 105–117. Prague Stringology Club, Department of Computer Science and Engineering, Faculty of Electrical Engineering, Czech Technical University in Prague (2009)
8. Cicalese, F., Lipták, Z., Rossi, M.: Bubble-flip - a new generation algorithm for prefix normal words. Theor. Comput. Sci. **743**, 38–52 (2018)
9. Cicalese, F., Lipták, Z., Rossi, M.: On infinite prefix normal words. Theor. Comput. Sci. **859**, 134–148 (2021)
10. Eikmeier, Y., Fleischmann, P., Kulczynski, M., Nowotka, D.: Weighted prefix normal words: mind the gap. In: Moreira, N., Reis, R. (eds.) DLT 2021. LNCS, vol. 12811, pp. 143–154. Springer, Cham (2021). https://doi.org/10.1007/978-3-030-81508-0_12
11. Fici, G., Lipták, Z.: On prefix normal words. In: Mauri, G., Leporati, A. (eds.) DLT 2011. LNCS, vol. 6795, pp. 228–238. Springer, Heidelberg (2011). https://doi.org/10.1007/978-3-642-22321-1_20
12. Fleischmann, P., Kulczynski, M., Nowotka, D., Poulsen, D.B.: On collapsing prefix normal words. In: Leporati, A., Martín-Vide, C., Shapira, D., Zandron, C. (eds.) LATA 2020. LNCS, vol. 12038, pp. 412–424. Springer, Cham (2020). https://doi.org/10.1007/978-3-030-40608-0_29
13. Gagie, T., Hermelin, D., Landau, G.M., Weimann, O.: Binary jumbled pattern matching on trees and tree-like structures. Algorithmica **73**(3), 571–588 (2015)
14. Lee, L.-K., Lewenstein, M., Zhang, Q.: Parikh matching in the streaming model. In: Calderón-Benavides, L., González-Caro, C., Chávez, E., Ziviani, N. (eds.) SPIRE 2012. LNCS, vol. 7608, pp. 336–341. Springer, Heidelberg (2012). https://doi.org/10.1007/978-3-642-34109-0_35
15. Parikh, R.: On context-free languages. J. ACM **13**(4), 570–581 (1966)

16. Salomaa, A.: Counting (scattered) subwords. Bull. EATheor. Comput. Sci. **81**, 165–179 (2003)

Reachability and Mortality for Two-Dimensional RHPCD Systems Are co-NP-hard

Olga Tveretina[(✉)]

Department of Computer Science, University of Hertfordshire, Hatfield, UK
o.tveretina@herts.ac.uk

Abstract. Reachability and mortality are fundamental problems in the study of hybrid dynamical systems. Reachability investigates whether a system can evolve from an initial state to a designated target state, while mortality asks whether the system inevitably halts or reaches a deadlock state under its given dynamics. In this work, we study these problems for two-dimensional restricted hierarchical piecewise constant derivative systems (2-RHPCD), a class characterised by a hierarchical structure and piecewise-constant dynamics. We prove that both reachability and mortality are co-NP-hard for bounded 2-RHPCD systems. In particular, our result resolves the open question posed in [4] concerning the complexity of the mortality problem for 2-RHPCD systems.

1 Introduction

The reachability and mortality problems are fundamental to the analysis of hybrid dynamical systems, which exhibit both continuous and discrete dynamics. The reachability problem asks whether a system can evolve from an initial state to a target state. The mortality problem determines when and how the system may terminate or reach a failure state. Both problems are generally undecidable, with decidability established only for certain subclasses [9,10].

While much existing research has focused on the decidability of these problems, less attention has been given to their computational complexity. Nonetheless, a few studies have addressed this aspect [3–5,15,16]. Investigating the computational complexity of both problems is essential for establishing theoretical limits on the efficiency of the underlying algorithms.

This work studies the computational complexity of the reachability and mortality problems for restricted hierarchical piecewise constant derivative systems (RHPCD) [6]. Key features of this model include identity resets, elementary flows ($\{0, \pm 1\}$ derivatives), and non-comparative guards (aligned with the x- and y-axes). These restrictions are intended to make RHPCD systems more tractable and to provide a framework for analysing reachability in hybrid systems [4].

The computational complexity of RHPCD systems has predominantly been studied in the context of co-NP-hardness and PSPACE membership [4]. These complexity classes are typically characterised as follows: (1) *co-NP-hard:* Problems in this class are at least as hard as the hardest problems in co-NP. In practice, verifying a no answer generally requires exponential time, as it often involves proving that no solution exists—a task typically more difficult than finding a solution. (2) *PSPACE:* Problems in this class may require exponential time to solve in the worst case but can be solved using only polynomial space.

While the reachability problem is decidable for bounded n-RHPCD systems, and mortality is decidable for bounded 2-RHPCD systems, both reachability and mortality for bounded 3-RHPCD systems are co-NP-hard [4]. Furthermore, mortality is known to be in PSPACE for $n=2$ [4]. However, despite their theoretical significance, the computational complexity bounds for these problems in the general RHPCD framework remain incompletely understood and merit further investigation.

This study demonstrates that, for 2-RHPCD systems, both the reachability and mortality problems are co-NP-hard. Additionally, it solves the open question posed in [4] concerning the complexity of the mortality problem in two dimensions, specifically for the class of 2-RHPCD systems.

Our proofs are based on encoding the NP-complete *simultaneous incongruences problem* [8,14] into the reachability and mortality problems for 2-RHPCD systems. A key distinction between this study and the work in [4] lies in the challenge of simulating the simultaneous incongruences problem within a lower-dimensional RHPCD model. Moreover, this study differs fundamentally from [16], where the increase in the value of x—a potential solution to the simultaneous incongruences problem—relies on discontinuous resets. In contrast, this approach focuses on the class of 2-RHPCD systems, which strictly prohibit such discontinuities.

A key technical feature of our construction is a mechanism for explicitly reconstructing discrete integer values from continuous trajectories, entirely without resets or discontinuous jumps. This is achieved through a systematic arrangement of locations and guards that compel the system to traverse a piecewise-linear staircase in the continuous state space. Each step encodes a portion of the integer value, enabling precise calculation via controlled incrementation. To the best of our knowledge, this is the first use of such a *purely flow-based reconstruction method* in the context of continuous-dynamics systems.

The remainder of this paper is organised as follows. Section 2 defines the formal 2-RHPCD model. Section 3 introduces the NP-complete simultaneous incongruences problem. Section 4 presents the reduction to reachability and mortality, and Sect. 5 provides co-NP-hardness proofs for both problems. Finally, Sect. 6 offers conclusions and suggests potential directions for future research.

2 Preliminaries

In this section, we introduce the formal framework underlying two-dimensional restricted hierarchical piecewise constant derivative systems (2-RHPCD). We proceed in three stages: (i) basic geometric notions, (ii) restricted piecewise constant derivative systems (2-RPCD), and (iii) their hierarchical extension to 2-RHPCD. Our presentation follows the notation of [1,2,11–13,16].

Section 2.1 recalls fundamental definitions and terminology. Section 2.2 defines 2-RPCD systems and their hierarchical extension, 2-RHPCD. Finally, Sect. 2.3 formalises the reachability and mortality problems for 2-RHPCD systems.

2.1 Definitions

We write $\mathbb{R}^2$ to denote the two-dimensional Euclidean space. Points in $\mathbb{R}^2$ are denoted by boldface letters such as $\boldsymbol{x}$ or $\boldsymbol{y}$.

A *convex open rectangular set* $r \subseteq \mathbb{R}^2$ is a nonempty open rectangle of the form $r = (a_1, b_1) \times (a_2, b_2)$, where $a_1, b_1, a_2, b_2 \in \mathbb{Q}$ with $a_1 < b_1$ and $a_2 < b_2$.

Given a set $\mathcal{X} \subseteq \mathbb{R}^2$, we let $\mathsf{cl}(\mathcal{X})$ denote its *topological closure*, and $\mathsf{int}(\mathcal{X})$ its *interior*, i.e., the set of all points $\boldsymbol{x} \in \mathcal{X}$ for which there exists $\varepsilon > 0$ such that the open ε-neighbourhood $N_\varepsilon(\boldsymbol{x}) \subseteq \mathcal{X}$.

A *finite rectangular partition* of a set $\mathcal{X} \subseteq \mathbb{R}^2$ is a finite collection $\mathcal{R} = \{r_1, \ldots, r_k\}$ of convex open rectangular sets, called *regions*, such that:

1. Each region is nonempty: $r_i \neq \varnothing$ for all i;
2. Regions are pairwise disjoint: $r_i \cap r_j = \varnothing$ for all $i \neq j$;
3. The closures of the regions cover the domain: $\bigcup_{i=1}^{k} \mathsf{cl}(r_i) = \mathcal{X}$.

An *edge* in a collection $\mathcal{R}$ is a nonempty one-dimensional set of the form $e = \mathsf{int}\left(\mathsf{cl}(r_i) \cap \mathsf{cl}(r_j)\right)$, for some $r_i, r_j \in \mathcal{R}$ with $i \neq j$, where $\mathsf{int}(e)$ denotes the relative interior of the line segment e, i.e., the segment without its endpoints. Let $\mathsf{B}_1(\mathcal{R})$ denote the set of all such edges.

A *vertex* of $\mathcal{R}$ is a singleton set of the form $v = \mathsf{cl}(e_i) \cap \mathsf{cl}(e_j)$, for some $e_i, e_j \in \mathsf{B}_1(\mathcal{R})$ with $i \neq j$, such that this intersection is a single point. Let $\mathsf{B}_0(\mathcal{R})$ denote the set of all vertices.

The set of *border elements* of $\mathcal{R}$ is defined as $\mathsf{B}(\mathcal{R}) = \mathsf{B}_0(\mathcal{R}) \cup \mathsf{B}_1(\mathcal{R})$. Then $\mathcal{R} \cup \mathsf{B}(\mathcal{R})$ forms a disjoint cover of $\mathcal{X}$. We note that, to ensure that each border element is uniquely associated with exactly one region throughout the rest of the paper, we define regions as open, axis-aligned rectangles and treat their border elements, such as edges and vertices, separately.

Finally, for each region $r \in \mathcal{R}$, we define its boundary as $\partial(r) = \mathsf{cl}(r) \setminus \mathsf{int}(r)$. The boundaries of $\mathcal{R}$ are then defined as the set $\partial(\mathcal{R}) = \bigcup_{r \in \mathcal{R}} \partial(r)$.

2.2 Restricted Hierarchical Piecewise Constant Derivative Systems

We define a two-dimensional restricted piecewise constant derivative system (2-RPCD) as follows.

Definition 1 (2-RPCD). *We define a two-dimensional restricted piecewise constant derivative system (2-RPCD) as a triple $\mathcal{A} = (\mathcal{R}, \varphi, \psi)$, where:*

1. $\mathcal{R} = \{r_1, \ldots, r_k\}$ is a finite rectangular partition of a domain $\mathcal{X} \subseteq \mathbb{R}^2$;
2. $\varphi : \mathcal{R} \to \mathbb{R}^2$ assigns a constant vector $\varphi(r) = (x_1, x_2)$, where $x_1, x_2 \in \{-1, 0, 1\}$, to each region $r \in \mathcal{R}$;
3. $\psi : \mathsf{B}(\mathcal{R}) \to \mathcal{R}$ assigns to each border element a unique owning region.

This model exhibits deterministic continuous dynamics within each region. Consequently, reachability analysis focuses on computing discrete successors across border elements.

Definition 2 (Step, adapted from [11]). *Let $\mathcal{A} = (\mathcal{R}, \varphi, \psi)$ be a 2-RPCD, and let $\boldsymbol{x}, \boldsymbol{x}' \in \mathbb{R}^2$ be two distinct points. The pair $(\boldsymbol{x}, \boldsymbol{x}')$ is a step in $\mathcal{A}$ if there exists a region $r \in \mathcal{R}$ and a scalar $t > 0$ such that:*

1. $\boldsymbol{x}, \boldsymbol{x}' \in \partial(r)$;
2. $\boldsymbol{x}' = \boldsymbol{x} + t \cdot \varphi(r)$;
3. *For all $0 < t' < t$, the point $\boldsymbol{x} + t' \cdot \varphi(r) \in r$.*

Building on the notion of a step, we define a trajectory as a finite sequence of consecutive steps starting from a boundary point.

Definition 3 (Trajectory in 2-RPCD). *Let $\mathcal{A} = (\mathcal{R}, \varphi, \psi)$ be a 2-RPCD, and let $\boldsymbol{x}_0 \in \partial(\mathcal{R})$. A trajectory of length $k \in \mathbb{N}$ rooted at $\boldsymbol{x}_0$ is a finite sequence*

$$\tau_{\boldsymbol{x}_0}^k = \boldsymbol{x}_0, \boldsymbol{x}_1, \ldots, \boldsymbol{x}_k$$

such that for every i, $0 \leq i \leq k-1$, the pair $(\boldsymbol{x}_i, \boldsymbol{x}_{i+1})$ is a step.

We now lift this to a hierarchical model in which each location is a 2-RPCD by incorporating two key notions: non-comparative guards, where transitions are guarded by constraints that do not involve comparisons between state variables, and identity resets, where state variables preserve their current values during transitions.

Definition 4 (2-RHPCD). *A two-dimensional restricted hierarchical piecewise constant derivative system (2-RHPCD) is a pair $\mathcal{H} = (\mathsf{L}, \mathsf{T})$, where:*

1. *L is a finite set of locations, where each location $\ell \in \mathsf{L}$ is a 2-RPCD $\mathcal{A}^\ell = (\mathcal{R}^\ell, \varphi^\ell, \psi^\ell)$;*
2. *$\mathsf{T} \subseteq \mathsf{L} \times \mathsf{Guards} \times \mathsf{L}$ is a finite set of transitions, where a guard $g \in \mathsf{Guards}$ is a constraint of the form $x_i = b_i$ with $i \in \{1, 2\}$ and $b_i \in \mathbb{Q}$;*
3. *The reset relation for each transition is the identity relation on the continuous variables, i.e., all continuous variables are preserved during discrete jumps.*

The system evolves as follows: within each location $\ell \in \mathsf{L}$, the continuous state $\boldsymbol{x} \in \mathbb{R}^2$ evolves according to the location's dynamics. A discrete transition from location ℓ to ℓ' occurs whenever the current state satisfies the corresponding guard condition g; that is, $(\ell, g, \ell') \in \mathsf{T}$.

For the remainder of the paper, we restrict our analysis to bounded 2-RHPCD systems, that is, systems in which all locations/regions are bounded. This assumption avoids technical complications arising from unbounded regions and aligns with prior work on RHPCD complexity [4].

Definition 5 (Trajectory in a 2-RHPCD). *Let $\mathcal{H} = (\mathsf{L}, \mathsf{T})$ be a 2-RHPCD. A* trajectory *of length $k \in \mathbb{N}$ rooted at $\boldsymbol{x}_0 \in \partial(\ell_0)$ for some location $\ell_0 \in \mathsf{L}$ is a sequence*

$$\tau_{(\ell_0, \boldsymbol{x}_0)}^k = (\ell_0, \boldsymbol{x}_0), (\ell_1, \boldsymbol{x}_1), \ldots, (\ell_k, \boldsymbol{x}_k)$$

such that for each $0 \leq i < k$, exactly one of the following conditions holds:

1. *$\ell_i = \ell_{i+1}$ and $\boldsymbol{x}_i \neq \boldsymbol{x}_{i+1}$; moreover, $(\boldsymbol{x}_i, \boldsymbol{x}_{i+1})$ is a step;*
2. *$\ell_i \neq \ell_{i+1}$ and $\boldsymbol{x}_i = \boldsymbol{x}_{i+1}$; moreover, a discrete transition $(\ell_i, g, \ell_{i+1}) \in \mathsf{T}$ is enabled at $\boldsymbol{x}_i$.*

2.3 Reachability and Mortality Problems

We now formalise two fundamental decision problems concerning the behaviour of 2-RHPCD systems: the *reachability* problem and the *mortality* problem.

Definition 6 (Reachability Problem). *Let $\mathcal{H} = (\mathsf{L}, \mathsf{T})$ be a 2-RHPCD, and let $\boldsymbol{x}_0 \in \mathbb{R}^2$ be an initial state such that $\boldsymbol{x}_0 \in \partial(\ell_0)$ for some location $\ell_0 \in \mathsf{L}$. Given a target (final) configuration, the* reachability problem *asks whether there exists a trajectory $\tau_{(\ell_0, \boldsymbol{x}_0)}^k = (\ell_0, \boldsymbol{x}_0), (\ell_1, \boldsymbol{x}_1), \ldots, (\ell_k, \boldsymbol{x}_k)$ such that $(\ell_k, \boldsymbol{x}_k) = (\ell_f, \boldsymbol{x}_f)$, i.e., the trajectory reaches a specified final (target) state $\boldsymbol{x}_f$ in a final location ℓ_f.*

In other words, the reachability problem asks whether the system, starting from a given initial configuration, can evolve through a finite sequence of transitions to eventually reach a desired target configuration.

Definition 7 (Mortality Problem). *Let $\mathcal{H} = (\mathsf{L}, \mathsf{T})$ be a 2-RHPCD. The* mortality problem *asks whether every trajectory, starting from any initial configuration $(\ell_0, \boldsymbol{x}_0)$ with $\ell_0 \in \mathsf{L}$ and $\boldsymbol{x}_0 \in \partial(\ell_0)$, has finite length and necessarily terminates—that is, whether all executions eventually reach a point where no further transitions are possible.*

The mortality problem thus concerns whether the system is guaranteed to terminate after a finite number of steps, regardless of its starting point. Alternatively, it may be phrased as asking whether the system inevitably reaches a *deadlock state*, a configuration from which no further progress is possible.

3 The NP-Complete Problem of Simultaneous Incongruences

In this section, we introduce the *simultaneous incongruences problem*, a classical decision problem known to be NP-complete [8]. In Sect. 4, we demonstrate how this problem can be reduced to the reachability and mortality problems for 2-RHPCD systems, thereby establishing the co-NP-hardness of these problems.

3.1 Problem Definition

The simultaneous incongruences problem asks whether an integer can simultaneously avoid satisfying a given collection of congruences. Specifically, given a finite set of pairs of integers (a_i, b_i), the goal is to determine whether there exists an integer x such that $x \not\equiv a_i \pmod{b_i}$ for all i.

Definition 8 (Simultaneous Incongruences Problem). *Assume a set of ordered pairs of positive integers $S = \{(a_1, b_1), \ldots, (a_k, b_k)\}$, where $k \geq 1$ and each pair satisfies $a_i \leq b_i$ for $1 \leq i \leq k$. The simultaneous incongruences problem asks whether there exists an $x \in \mathbb{Z}^+$ such that*

$$x \not\equiv a_i \pmod{b_i} \quad \text{for } 1 \leq i \leq k.$$

That is, the objective is to find whether an integer can simultaneously fail to satisfy a given set of congruences.

3.2 Bounding the Solution Space via the Least Common Multiple

The *least common multiple* (LCM) of a finite set of integers is the smallest positive integer divisible by all the integers in the set. For integers $x_1, \ldots, x_n$, we write $\text{lcm}(x_1, \ldots, x_n)$ to denote their LCM.

The LCM plays a crucial role in bounding the solution space for the simultaneous incongruences problem. Since congruences are periodic modulo each b_i, any potential solution repeats with a period equal to the LCM of $b_1, \ldots, b_k$. Consequently, it suffices to consider the interval $[1, \Lambda]$, where $\Lambda = \text{lcm}(b_1, \ldots, b_k)$, without loss of generality.

Theorem 1 ([4]). *Let $S = \{(a_1, b_1), \ldots, (a_k, b_k)\}$ be a set of congruence pairs. Then there exists an integer $x \in \mathbb{Z}^+$ such that $x \not\equiv a_i \pmod{b_i}$ for all i if and only if there exists such an x in the range $1 \leq x \leq \Lambda$, where $\Lambda = \text{lcm}(b_1, \ldots, b_k)$.*

The LCM of two integers x and y can be computed via their greatest common divisor, $\gcd(x, y)$, as

$$\text{lcm}(x, y) = \frac{|x \cdot y|}{\gcd(x, y)}.$$

This approach extends to computing the LCM of n integers by applying the operation repeatedly: $\text{lcm}(x_1, x_2, \ldots, x_n) = \text{lcm}(\ldots (\text{lcm}(x_1, x_2), x_3), \ldots, x_n)$. A pairwise LCM computation requires computing a GCD, and the Euclidean algorithm for computing it runs in $O(\log(\min(x_i, x_j)))$ time [7]. Finally, the complexity of computing the LCM of n numbers is provided by Proposition 1 below.

Proposition 1. *The complexity of computing the least common multiple of n integers $x_1, x_2, \ldots, x_n$ is $O(n \cdot \log(\max(x_1, \ldots, x_n)))$.*

Therefore, the computational effort required to bound the solution space for $S = \{(a_1, b_1), \ldots, (a_k, b_k)\}$ depends polynomially on the number of congruences k and logarithmically on the magnitude of $b_1, \ldots, b_k$.

4 Reduction of the Simultaneous Incongruences Problem to Reachability and Mortality for 2-RHPCD

In this section, we provide technical details of our reduction from the simultaneous incongruences problem to both the reachability and mortality problems for 2-RHPCD systems. The core idea is to construct a 2-RHPCD $\mathcal{H}(\mathcal{S})$ from a given instance $\mathcal{S} = \{(a_1, b_1), \ldots, (a_k, b_k)\}$ of the simultaneous incongruences problem, so that trajectories in $\mathcal{H}(\mathcal{S})$ correspond precisely to choices of an integer $x \in [1, \Lambda]$, where $\Lambda = \operatorname{lcm}(b_1, \ldots, b_k)$. The key observations are as follows:

1. If there exists an integer x satisfying all incongruences $x \not\equiv a_i \pmod{b_i}$ then:
   (a) the designated final state is unreachable (the reachability problem evaluates to no),
   (b) at least one infinite trajectory arises (the mortality problem evaluates to no).
2. Conversely, if no such x exists, then:
   (a) the final state is reachable (the answer to the reachability problem is yes),
   (b) every trajectory terminates (the answer to the mortality problem is yes).

Throughout the paper, for any rational interval $I \subset \mathbb{Q}$ (closed, open or half-open) and any $x, y \in \mathbb{Q}$, we write $I \times \{y\} = \{(x, y) \mid x \in I\}$ and $\{x\} \times I = \{(x, y) \mid y \in I\}$ to denote the corresponding horizontal and vertical slices of the plane.

We now proceed to describe the locations, flows, and guards of $\mathcal{H}(\mathcal{S})$ in detail, and to prove that its reachability and mortality behaviours exactly mirror the existence or non-existence of a solution to the original simultaneous incongruences instance.

4.1 Reduction to the Reachability Problem for 2-RHPCD

In the rest of the paper, we will denote by $\mathcal{H}(\mathcal{S})$ the 2-RHPCD system constructed to simulate the simultaneous incongruences problem for a given set $\mathcal{S} = \{(a_1, b_1), \ldots, (a_k, b_k)\}$. By Theorem 1, we need only check whether there exists an integer x in the range $1 \leq x \leq \Lambda$, with $\Lambda = \operatorname{lcm}(b_1, \ldots, b_k)$, such that $x \not\equiv a_i \pmod{b_i}$ for all i. We encode a candidate value x as $x - \frac{x}{\Lambda+1}$.

The corresponding $\mathcal{H}(\mathcal{S})$ consists of the locations Loc^0, Loc_i^1, Loc_i^2, $\operatorname{Loc}_{i,j}^3$, $\operatorname{Loc}_{i,j}^4$, Loc_i^5, Loc_i^6, where $0 \leq i \leq k$ and $1 \leq j \leq b_i$. These locations are specifically used to (1) perform the modulo checks for each (a_i, b_i), and (2) increment the candidate value by one as necessary. Now we define the following auxiliary expressions:

- $c_l = l\left(1 - \frac{1}{\Lambda+1}\right)$, where $0 \leq l \leq \Lambda+2$;
- $c_l^{i,j} = j + l\left(1 - \frac{2}{\Lambda+1}\right)$, where $0 \leq l \leq \Lambda+1$ and $1 \leq j \leq b_i$;
- $d_l^i = a_i + \frac{1}{\Lambda+1} + l\left(1 - \frac{2}{\Lambda+1}\right)$, where $0 \leq l \leq \Lambda+2$ and $1 \leq i \leq k$.

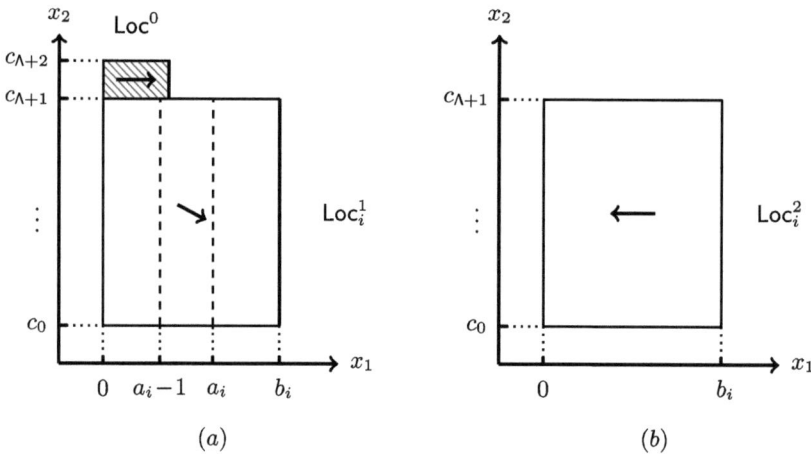

Fig. 1. Locations Loc_i^1 and Loc_i^2 (for $1 \leq i \leq k$) encode the simultaneous incongruences problem. The location Loc^0 contains the final state.

Intuitively, the constants c_l, $c_l^{i,j}$, and d_l^i define a fine-grained geometric grid in the continuous state space, serving as landmarks for encoding and manipulating integers purely via flows.

We detail the construction of locations $\mathsf{Loc}^0, \mathsf{Loc}_i^1$, and Loc_i^2 ($0 \leq i \leq k$) shown in Fig. 1:

1. $\mathsf{Loc}^0 = [0,1] \times [c_{\Lambda+1}, c_{\Lambda+2})$ with flow vector $\boldsymbol{x} = (1,0)$;
2. $\mathsf{Loc}_i^1 = [0, b_i] \times [c_0, c_{\Lambda+1}]$ with flow vector $\boldsymbol{x} = (1,-1)$;
3. $\mathsf{Loc}_i^2 = [0, b_i] \times [c_0, c_{\Lambda+1}]$ with flow vector $\boldsymbol{x} = (-1,0)$.

To verify whether a given integer $x \in [1, \Lambda]$ solves the simultaneous incongruences problem, we simulate it by a trajectory starting at $(0, x - \frac{x}{\Lambda+1}) \in \mathsf{Loc}_1^1$. Here, the value x is encoded as $x - \frac{x}{\Lambda+1}$.

Each Loc_i^1 implements the modulo check for (a_i, b_i). The flow causes the trajectory to reduce the coordinate x_2 by b_i when moving from the interval $\{0\} \times [0, c_{\Lambda+1})$ to the interval $\{b_i\} \times [0, c_{\Lambda+1})$. When the trajectory reaches $(x', 0) \in (j-1, j) \times \{0\}$ for some $1 \leq j \leq b_i$, this implies $x \equiv j \pmod{b_i}$ with $x = (j-x')(\Lambda+1)$. Thus, j gives the remainder, and x' determines the fractional location within the interval.

We reconstruct the current value of x via a systematic arrangement of locations and guards that force the system to traverse a piecewise-linear staircase in the continuous state space, incrementing x by one when needed. This mechanism, realised through locations $\mathsf{Loc}_{i,j}^3$ and $\mathsf{Loc}_{i,j}^4$ or alternatively Loc_i^5 and Loc_i^6.

We use c_l's to mark staircase levels for reading the current integer x, and $c_l^{i,j}$'s to shift these levels for testing remainders modulo b_i, and d_l^i's to shift further for detecting when $x \bmod b_i = a_i$ (the forbidden remainder) and triggering an increment of x purely via continuous flows.

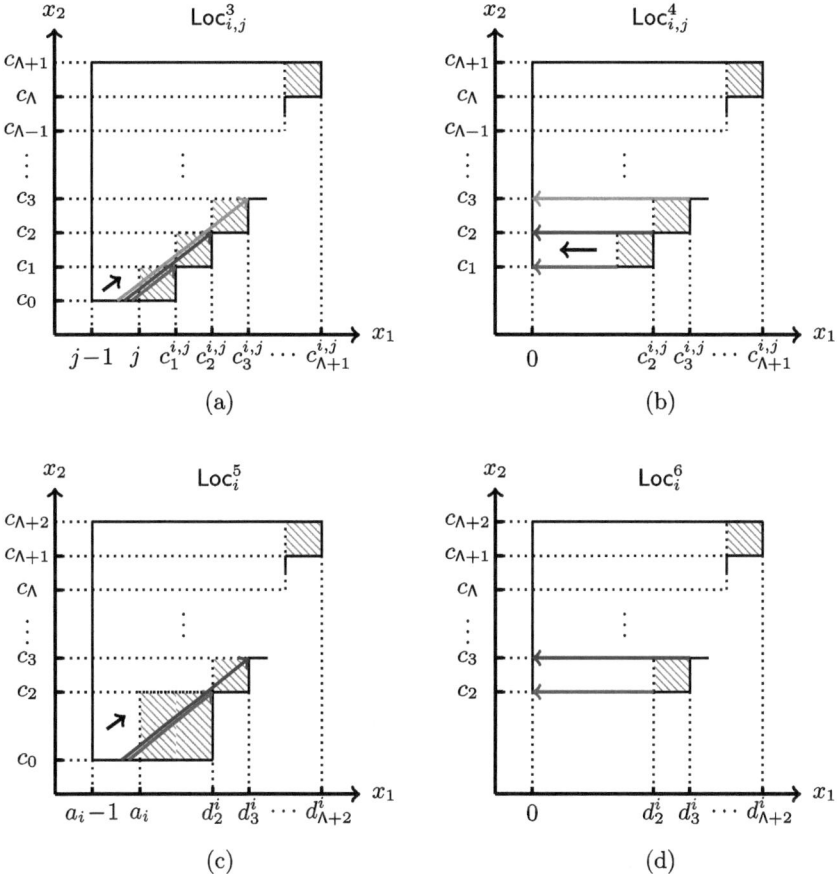

Fig. 2. (a) and (b) show how the system systematically reconstructs the current value of x from continuous dynamics; and (c) and (d) show how this mechanism extends to increment x by exactly one.

We next define $\mathsf{Loc}^3_{i,j}$ and $\mathsf{Loc}^4_{i,j}$ ($1 \leq i \leq k$; $1 \leq j \leq a_i - 1$ or $a_i + 1 \leq j \leq b_i$). These help reconstruct x and are shown in Fig. 2(a,b):

1. $\mathsf{Loc}^3_{i,j} = \bigcup_{l=0}^{\Lambda} [j-1, c^{i,j}_{l+1}] \times [c_l, c_{l+1})$ with flow $\boldsymbol{x} = (1, 1)$;
2. $\mathsf{Loc}^4_{i,j} = \bigcup_{l=1}^{\Lambda} [0, c^{i,j}_{l+1}] \times [c_l, c_{l+1})$ with flow $\boldsymbol{x} = (-1, 0)$.

For reconstructing x and incrementing it by one when needed, we use Loc^5_i and Loc^6_i ($1 \leq i \leq k$), shown in Fig. 2(c,d):

1. $\mathsf{Loc}^5_i = [a_i - 1, d^i_2] \times [c_0, c_2) \cup \bigcup_{l=2}^{\Lambda+1} [a_i - 1, d^i_{l+1}] \times [c_l, c_{l+1})$ with flow $\boldsymbol{x} = (1, 1)$;
2. $\mathsf{Loc}^6_i = \bigcup_{l=2}^{\Lambda+1} [0, d^i_{l+1}] \times [c_l, c_{l+1})$ with flow $\boldsymbol{x} = (-1, 0)$.

The guard conditions are summarised in Table 1, and all the resets are defined as the identity relationship.

Table 1. Guard conditions for $1 \leq i \leq k$, and $1 \leq j \leq a_i - 1$ or $a_i + 1 \leq j \leq b_i$.

Name	Guard	Current location	Transition to
$G_i^{1,2}$	$\{b_i\} \times [c_0, c_{\Lambda+1})$	Loc_i^1	Loc_i^2
$G_i^{2,1}$	$\{0\} \times [c_0, c_{\Lambda+1})$	Loc_i^2	Loc_i^1
$G_{i,j}^{1,3}$	$(j-1, j) \times \{c_0\}$	Loc_i^1	$\mathsf{Loc}_{i,j}^3$
$G_{i,j}^{3,4}$	$\bigcup_{l=1}^{\Lambda}[c_l^{i,j}, c_{l+1}^{i,j}] \times \{c_l\}$	$\mathsf{Loc}_{i,j}^3$	$\mathsf{Loc}_{i,j}^4$
$G_{i,j}^{4,1}$	$\{0\} \times [c_1, c_{\Lambda+1})$	$\mathsf{Loc}_{i,j}^4$	Loc_{i+1}^1
$G_i^{1,5}$	$(a_i-1, a_i) \times \{c_0\}$	Loc_i^1	Loc_i^5
$G_i^{5,6}$	$\bigcup_{l=2}^{\Lambda+1}[d_l^i, d_{l+1}^i] \times \{c_l\}$	Loc_i^5	Loc_i^6
$G_i^{6,1}$	$\{0\} \times [c_2, c_{\Lambda+1})$	Loc_i^6	Loc_1^1
$G_i^{6,0}$	$\{0\} \times [c_{\Lambda+1}, c_{\Lambda+2})$	Loc_i^6	Loc^0

Now the reachability problem for $\mathcal{H}(\mathcal{S})$ is defined as follows: Does a trajectory starting at $\boldsymbol{x}_0 = (0, 1 - \frac{1}{\Lambda+1}) \in \mathsf{Loc}_1^1$ reach $\boldsymbol{x}_f = (1, c_{\Lambda+1}) \in \mathsf{Loc}^0$?

We can verify if the current x, where $0 \leq x \leq \Lambda$, is a solution to the simultaneous incongruences problem for the given set $\mathcal{S}$ as it is provided below.

1. If at some Loc_i^1, the trajectory reaches a point $(x', 0) \in (0, b_i) \times \{0\}$ then:
   (a) If $x \not\equiv a_i \pmod{b_i}$, then guards $G_{i,j}^{1,3}$, $G_{i,j}^{3,4}$ and $G_{i,j}^{4,1}$ cause a transition to $\mathsf{Loc}_{i,j}^3$, then $\mathsf{Loc}_{i,j}^4$, then Loc_{i+1}^1 to proceed to the next pair in $\mathcal{S}$;
   (b) If $x \equiv a_i \pmod{b_i}$, then guards $G_i^{1,5}$, $G_i^{5,6}$ and $G_i^{6,1}$ trigger a transition to Loc_i^5, then Loc_i^6, then back to Loc_i^1, incrementing x by one.
2. If all checks pass without satisfying the congruences, the trajectory eventually reaches $\boldsymbol{x}_f$; otherwise, it halts earlier.

Thus, $\mathcal{H}(\mathcal{S})$ reduces the simultaneous incongruences problem to reachability in a 2-RHPCD.

4.2 Reduction to the Mortality Problem for 2-RHPCD

Now we extend the reduction of the simultaneous incongruences problem to the reachability problem introduced in Sect. 4.1 to the mortality problem by introducing locations Loc_k^7 and Loc_k^8 as depicted in Fig. 3.

Both locations Loc_k^7 and Loc_k^8 are 2-RPCD systems, each consisting of three regions:

1. Location Loc_k^7 consists of the following rectangular regions:
   (a) $r_1^7 = [0, a_k-1) \times [0, 1 - \frac{1}{\Lambda+1})$ with flow $\boldsymbol{x} = (1, 0)$;
   (b) $r_2^7 = [a_k-1, a_k - \frac{1}{\Lambda+1}] \times [0, 1 - \frac{1}{\Lambda+1})$ with flow $\boldsymbol{x} = (1, 1)$;
   (c) $r_3^7 = [0, a_k - \frac{1}{\Lambda+1}] \times [1 - \frac{1}{\Lambda+1}, 1]$ with flow $\boldsymbol{x} = (-1, 0)$.
2. Location Loc_k^8 consists of the following rectangular regions:
   (a) $r_1^8 = [a_k, b_k) \times [0, 1 - \frac{1}{\Lambda+1})$ with flow $\boldsymbol{x} = (1, 0)$;

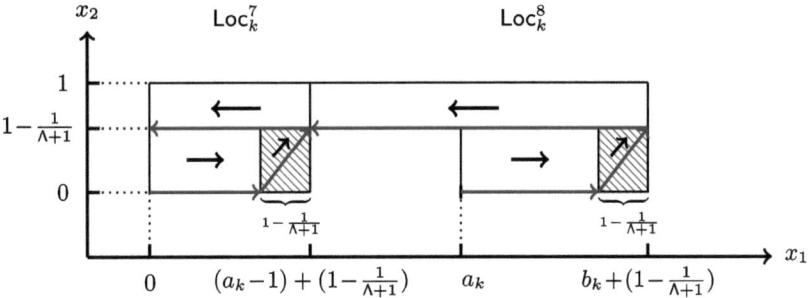

Fig. 3. The locations Loc_k^7 and Loc_k^8 are used to compute trajectories returning from the current state to the initial state $x_0 = \left(0,\ 1 - \frac{1}{\Lambda+1}\right) \in \text{Loc}_1^1$.

(b) $r_2^8 = [b_k,\ b_k + 1 - \frac{1}{\Lambda+1}] \times [0,\ 1 - \frac{1}{\Lambda+1})$ with flow $x = (1,1)$;
(c) $r_3^8 = [a_k - \frac{1}{\Lambda+1},\ b_k + 1 - \frac{1}{\Lambda+1}] \times [1 - \frac{1}{\Lambda+1},\ 1]$ with flow $x = (-1, 0)$.

The locations Loc_k^7 and Loc_k^8 are used to compute trajectories returning from the current state—located either in $[0, a_{k-1}] \times \{0\}$ or in $[a_k, b_k] \times \{0\}$ within Loc_k^1—to the initial state $x_0 = \left(0,\ 1 - \frac{1}{\Lambda+1}\right) \in \text{Loc}_1^1$. Now if the simultaneous incongruences problem for the set S has a solution, then the mortality problem for the 2-RHPCD system has none: $\mathcal{H}(S)$ admits at least one infinite trajectory and therefore is not mortal.

4.3 Computational Complexity of the Reduction

In this section, we analyse the complexity of our reduction from the simultaneous incongruences problem to the reachability (and hence mortality) problems for 2-RHPCD systems. We assume that both the given instance $S = \{(a_1, b_1), \ldots, (a_k, b_k)\}$ and the constructed 2-RHPCD are represented in binary.

We begin by recalling that 3-SAT, one of the classical NP-complete problems, admits a polynomial-time reduction to the simultaneous incongruences problem.

Theorem 2 ([8,14]). *There is a polynomial-time reduction from 3-SAT to the simultaneous incongruences problem. Hence the latter is NP-complete.*

Consequently, we may assume without loss of generality that in any instance S arising from a 3-SAT formula with n variables, both $k = |S|$ and each b_i are bounded by a polynomial in n [8,14]. It therefore suffices to consider a set S containing k pairs, where all b_i are integers polynomial in k.

Theorem 3. *Let $S = \{(a_1, b_1), \ldots, (a_k, b_k)\}$ be an instance of the simultaneous incongruences problem in which each b_i is polynomially bounded in k. Then there exists a polynomial-time reduction from S to the reachability problem for the 2-RHPCD system $\mathcal{H}(S)$.*

Proof. The binary encoding of $\mathcal{S}$ requires $\sum_{i=1}^{k}(\log a_i + \log b_i)$ bits, which is polynomial in k, since each b_i is polynomially bounded.

The least common multiple $\Lambda = \mathrm{lcm}(b_1, \ldots, b_k)$ satisfies $\Lambda \leq b_1 \cdot b_2 \cdots b_k$, so its binary representation requires $O(\log(b_1 b_2 \cdots b_k)) = O\left(\sum_{i=1}^{k} \log b_i\right)$ bits, which is again polynomial in k. Therefore, all coordinates of the initial and final states, $\boldsymbol{x}_0 = \left(0, 1 - \frac{1}{\Lambda+1}\right)$ and $\boldsymbol{x}_f = (1, c_{\Lambda+1})$, can be represented using polynomial space.

Computationally, the 2-RHPCD $\mathcal{H}(\mathcal{S})$ can be specified by lists of regions, guards, and flows. Each b_i contributes guards and regions proportional to b_i, and the total number of such elements is $O(\sum_i b_i)$, which is polynomial in k. Furthermore, representing each such element requires a number of bits polynomial in k.

Finally, by Proposition 1, computing Λ itself takes $O(k \cdot \log(\max\{b_i\}))$ time, which is polynomial in the size of the input. Hence, the entire reduction, from $\mathcal{S}$ to the reachability instance $(\mathcal{H}(\mathcal{S}), \boldsymbol{x}_0, \boldsymbol{x}_f)$, runs in polynomial time. □

Since the mortality problem differs from reachability only in the condition on infinite executions, the same complexity analysis applies to the reduction to mortality.

5 Complexity of the Reachability and Mortality Problems

In this section, we establish our main results, demonstrating the co-NP-hardness of the reachability and mortality problems for 2-RHPCD. In the following we assume a set of pairs $\mathcal{S} = \{(a_1, b_1), \ldots, (a_k, b_k)\}$, where $k \geq 1$. Furthermore, as discussed in Sect. 4.3, we can assume without loss of generality that all b_i's are polynomial in k.

5.1 The Reachability Problem for 2-RHPCD Is Co-NP-Hard

In this section, we prove that the complexity of the reachability problem for 2-RHPCD is co-NP-hard.

Theorem 4. *The reachability problem for bounded 2-RHPCD is co-NP-hard.*

Proof. We encode the simultaneous incongruences problem for the set $\mathcal{S} = \{(a_1, b_1), \ldots, (a_k, b_k)\}$ into the 2-RHPCD $\mathcal{H}(\mathcal{S})$ as described in Sect. 4.1. The complexity of the construction is polynomial in k, as detailed by Theorem 3.

By construction, the reachability problem for the 2-RHPCD $\mathcal{H}(\mathcal{S})$ has a solution if and only if there is no solution to the simultaneous incongruences problem for $\mathcal{S}$. Since the simultaneous incongruences problem is NP-complete, this implies that the reachability problem for 2-RHPCD is co-NP-hard. □

5.2 The Mortality Problem for 2-RHPCD Is Co-NP-Hard

In this section, we prove that the complexity of the mortality problem for 2-RHPCD is co-NP-hard.

Theorem 5. *The mortality problem for bounded 2-RHPCD is co-NP-hard.*

Proof. We construct the 2-RHPCD $\mathcal{H}(\mathcal{S})$ such that the system is mortal if and only if there is no solution to the simultaneous incongruences problem for $\mathcal{S} = \{(a_1, b_1), \ldots, (a_k, b_k)\}$ as provided in Sect. 4.2. Further specifics are as follows.

1. If there exists a solution to the simultaneous incongruences problem for $\mathcal{S}$, then starting from $\boldsymbol{x}_0 = (0, 1 - \frac{1}{\Lambda+1}) \in \mathsf{Loc}_1^1$, the trajectory will eventually reach some point $\boldsymbol{x} \in \mathsf{Loc}_k^1$, and it will be reset to the initial point $\boldsymbol{x}_0$. That is, there is at least one infinite trajectory. Therefore, $\mathcal{H}(\mathcal{S})$ is not mortal.
2. If the simultaneous incongruences problem for $\mathcal{S}$ has no solution, then starting from any point, the trajectory will eventually reach the final point $\boldsymbol{x}_f = (1, c_{\Lambda+1}) \in \mathsf{Loc}^0$. We assume that there is no outgoing transitions from Loc^0. That is, if there is no solution to the simultaneous incongruences problem for $\mathcal{S}$, then regardless of where the trajectory starts, it will eventually halt. Therefore, $\mathcal{H}(\mathcal{S})$ is mortal.

Since the simultaneous incongruences problem is NP-complete, this implies that the mortality problem for 2-RHPCD is co-NP-hard. □

6 Conclusion

This study establishes that both the reachability and mortality problems for two-dimensional restricted hierarchical piecewise constant derivative systems (2-RHPCD) are co-NP-hard. In particular, our results resolve the open question posed in [4] concerning the computational complexity of the mortality problem for 2-RHPCD systems.

A natural direction for future research is to investigate whether, for n-dimensional RHPCD systems, the mortality problem with $n > 2$ lie in PSPACE or is PSPACE-complete. Closing this complexity gap between the co-NP-hardness lower bound (as shown here) and the PSPACE upper bound would enhance our understanding of the computational complexity of RHPCD systems.

References

1. Asarin, E., Maler, O., Pnueli, A.: Reachability analysis of dynamical systems having piecewise-constant derivatives. Theoret. Comput. Sci. **138**(1), 35–65 (1995)
2. Asarin, E., Mysore, V., Pnueli, A., Schneider, G.: Low dimensional hybrid systems - decidable, undecidable, don't know. Inf. Comput. **211**, 138–159 (2012)

3. Bazille, H., Bournez, O., Gomaa, W., Pouly, A.: On the complexity of bounded time reachability for piecewise affine systems. In: Ouaknine, J., Potapov, I., Worrell, J. (eds.) RP 2014. LNCS, vol. 8762, pp. 20–31. Springer, Cham (2014). https://doi.org/10.1007/978-3-319-11439-2_2
4. Bell, P.C., Chen, S., Jackson, L.: On the decidability and complexity of problems for restricted hierarchical hybrid systems. Theoret. Comput. Sci. **652**, 47–63 (2016)
5. Bell, P.C., Potapov, I., Semukhin, P.: On the mortality problem: From multiplicative matrix equations to linear recurrence sequences and beyond. In: 44th International Symposium on Mathematical Foundations of Computer Science, MFCS 2019, 26–30 August 2019, Aachen, Germany. LIPIcs, vol. 138, pp. 83:1–83:15. Schloss Dagstuhl - Leibniz-Zentrum für Informatik (2019)
6. Ben-Amram, A.M.: Mortality of iterated piecewise affine functions over the integers: Decidability and complexity (extended abstract). In: Portier, N., Wilke, T. (eds.) 30th International Symposium on Theoretical Aspects of Computer Science, STACS 2013, Germany. LIPIcs, vol. 20, pp. 514–525. Schloss Dagstuhl - Leibniz-Zentrum für Informatik (2013)
7. Cormen, T.H., Leiserson, C.E., Rivest, R.L., Stein, C.: Introduction to Algorithms, 3rd edn. MIT Press, Cambridge, MA (2009)
8. Garey, M.R., Johnson, D.S.: Computers and Intractability: A Guide to the Theory of NP-Completeness. W. H. Freeman and Co, New York, NY, USA (1979)
9. Henzinger, T.A., Kopke, P.W., Puri, A., Varaiya, P.: What's decidable about hybrid automata? J. Comput. Syst. Sci. **57**(1), 94–124 (1998)
10. Lafferriere, G., Pappas, G., Yovine, S.: A new class of decidable hybrid systems. In: Hybrid Systems: Computation and Control, Second International Workshop, pp. 137–151 (1999)
11. Maler, O., Pnueli, A.: Reachability analysis of planar multi-linear systems. In: CAV, pp. 194–209 (1993)
12. de Oliveira Oliveira, M., Tveretina, O.: Mortality and edge-to-edge reachability are decidable on surfaces. In: 25th ACM International Conference on Hybrid Systems: Computation and Control (2022)
13. Sandler, A., Tveretina, O.: Deciding reachability for piecewise constant derivative systems on orientable manifolds. In: Filiot, E., Jungers, R., Potapov, I. (eds.) RP 2019. LNCS, vol. 11674, pp. 178–192. Springer, Cham (2019). https://doi.org/10.1007/978-3-030-30806-3_14
14. Stockmeyer, L.J., Meyer, A.R.: Word problems requiring exponential time: preliminary report. In: Proceedings of the 5th Annual ACM Symposium on Theory of Computing, pp. 1–9. ACM (1973)
15. Sutner, K.: On the computational complexity of finite cellular automata. J. Comput. Syst. Sci. **50**(1), 87 (1995)
16. Tveretina, O.: On the complexity of reachability and mortality for bounded piecewise affine maps. In: Kovács, L., Sokolova, A. (eds.) Reachability Problems - 18th International Conference, RP 2024. LNCS, vol. 15050, pp. 141–153. Springer (2024)

Uppaal Coshy: Automatic Synthesis of Compact Shields for Hybrid Systems

Asger Horn Brorholt[1]([✉]), Andreas Holck Høeg-Petersen[1], Peter Gjøl Jensen[1], Kim Guldstrand Larsen[1], Marius Mikučionis[1], Christian Schilling[1], and Andrzej Wasowski[2]

[1] Aalborg University, 9220 Aalborg, Denmark
{asgerhb,ahhp,pgj,kgl,marius,christianms}@cs.aau.dk
[2] IT University of Copenhagen, 2300 Copenhagen, Denmark
wasowski@itu.dk

Abstract. We present UPPAAL COSHY, a tool for automatic synthesis of a safety strategy—or *shield*—for Markov decision processes over continuous state spaces and complex hybrid dynamics. The general methodology is to partition the state space and then solve a two-player safety game [9], which entails a number of algorithmically hard problems such as reachability for hybrid systems. The general philosophy of UPPAAL COSHY is to approximate hard-to-obtain solutions using simulations. Our implementation is fully automatic and supports the expressive formalism of UPPAAL models, which encompass stochastic hybrid automata.

The precision of our partition-based approach benefits from using finer grids, which however are not efficient to store. We include an algorithm called CAAP to efficiently compute a compact representation of a shield in the form of a decision tree, which yields significant reductions.

Keywords: Shield synthesis · UPPAAL · Decision tree

1 Introduction

In prior work, we proposed an algorithm to synthesize *shields* (i.e., nondeterministic safety strategies) for Markov decision processes with hybrid dynamics [9]. The algorithm partitions the state space into finitely many cells and then solves a two-player safety game, where it uses approximation through simulation to efficiently tackle algorithmically hard problems. In this tool paper, we present our implementation UPPAAL COSHY, which is fully integrated in UPPAAL, offering an automatic tool[1] that supports the expressive UPPAAL modeling formalism, including reinforcement learning under a shield.

Our algorithm represents a shield by storing the allowed actions for each cell individually, which results in a large data structure. Since many neighboring cells allow the same actions in practice, as a second contribution, we propose a new

---

[1] Available at https://uppaal.org/features/#coshy.

algorithm called CAAP to compute a compact representation in the form of a decision tree. We demonstrate that this algorithm leads to significant reductions as part of the workflow in UPPAAL COSHY.

An extended version of this paper is available online [8].

1.1 Related Tools for Shield Synthesis and Compact Representation

Shielding. Shields are obtained by solving games, for which there exist a wide selection of tools for discrete state spaces [10,11,19]. Notably, TEMPEST [21] synthesizes shields for discrete systems and facilitates learning through integration with PRISM [20]. UPPAAL TIGA synthesizes shields for timed games [4].

In contrast, our tool applies to a richer class of models, including stochastic hybrid systems with non-periodic control and calls to external C libraries.

One benefit of our tool is the full integration with UPPAAL STRATEGO [12] to directly use the synthesized shield in reinforcement learning.

Decision Trees. Encoding strategies as decision trees is a popular approach to achieving compactness and interpretability [2,3,6,16,22]. However, these works focus on creating approximate representations from tabular data. For a fixed set of predicates, the smallest possible tree can be obtained by enumeration techniques [13,14]. In contrast, our method transforms a given decision tree into an *equivalent* decision tree. Our method is specifically designed to efficiently cope with strategies of many axis-aligned decision boundaries.

2 Shield Synthesis for Hybrid Systems

In this section, we recall a general shield synthesis algorithm for hybrid systems outlined in prior work [9]. We start by recalling the formalism for control systems.

2.1 Euclidean Markov Decision Processes

Definition 1 (Euclidean Markov decision process [18]). *A k-dimensional Euclidean Markov decision process (EMDP) is a tuple $\mathcal{M} = (S, A, T)$ where*

- *$S \subseteq \mathbb{R}^k$ is a closed and bounded subset of the k-dimensional Euclidean space,*
- *A is a finite set of actions, and*
- *$T: S \times A \to (S \to \mathbb{R}_{\geq 0})$ maps each state-action pair (s, a) to a probability density function over S, i.e., we have $\int_{s' \in S} T(s,a)(s')ds' = 1$.*

For simplicity, the state space S is continuous. However, the extension to discrete variables, e.g., locations of hybrid components, is straightforward. Since optimizing strategies is not our focus, we do not formally introduce the notion of cost and rely on the reader's intuition. (See [9] for details.)

A *run* π of an EMDP is an alternating sequence $\pi = s_0 a_0 s_1 a_1 \ldots$ of states and actions such that $T(s_i, a_i)(s_{i+1}) > 0$ for all $i \geq 0$. A (memoryless) stochastic *strategy* for an EMDP is a function $\bar{\sigma}: S \to (A \to [0, 1])$, mapping a state to a

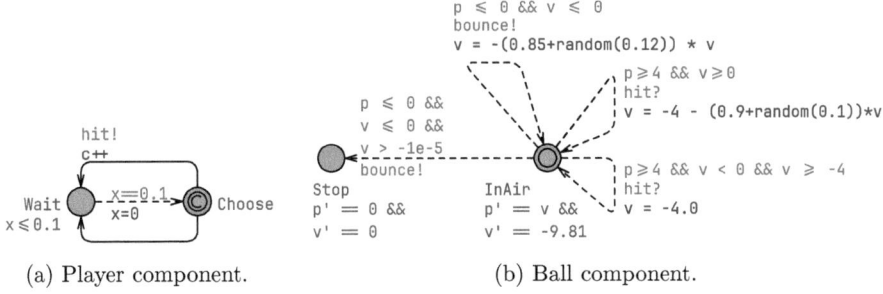

Fig. 1. The *bouncing ball* modeled in UPPAAL.

probability distribution over the actions. A run $\pi = s_0 a_0 s_1 a_1 \ldots$ is an *outcome* of $\bar{\sigma}$ if $\bar{\sigma}(s_i)(a_i) > 0$ for all $i \geq 0$. Similarly, a (memoryless) nondeterministic strategy is a function $\sigma \colon S \to 2^A$, mapping a state to a set of actions. A run $\pi = s_0 a_0 s_1 a_1 \ldots$ is an outcome of σ if $a_i \in \sigma(s_i)$ for all $i \geq 0$.

A *safety property* (or invariant) φ is a set of states $\varphi \subseteq S$. A run $\pi = s_0 a_0 s_1 a_1 \ldots$ is *safe* with respect to φ if $s_i \in \varphi$ for all $i \geq 0$. A nondeterministic strategy σ is a *shield* with respect to φ if all outcomes of σ are safe.

2.2 Running Example (Bouncing Ball)

We introduce our running example: a *bouncing ball* that can be hit by a player to keep it bouncing [9,18]. We shortly explain our two-component UPPAAL model. The player component is shown in Fig. 1(a). In the (initial) location Choose, there are two available control actions (solid lines). The player chooses every 0.1 s (enforced by the clock x). The hit action (upper edge) attempts to hit the ball, and increments the cost counter c to be used for reinforcement learning in Sect. 5.1. The other action (lower edge) does not attempt to hit the ball.

The ball component, shown in Fig. 1(b), is described by two state variables, position p and velocity v, which evolve according to the ordinary differential equations shown below the initial location InAir. The two dashed edges on the right model a successful hit action, which is only triggered if the ball is high enough (four meters or higher above the ground); they differ in whether the ball is currently jumping up or falling down. The two dashed edges on the left model a bounce on the ground. The ball bounces back up with a random dampening (upper edge) or goes to the state Stop if the velocity is very low (lower edge). In the following, we shall see how to obtain a shield that enforces the safety property that Stop is never reached, i.e., $\varphi = \{s \mid \text{Ball is not in Stop in } s\}$.

2.3 Partition-Based Shield Synthesis

Since an EMDP consists of infinitely many states, we employ a finite-state abstraction. For that, we partition the state space $S \subseteq \mathbb{R}^k$ with a regular *rectangular* grid. (In [9], we only allowed a grid of uniform size in all dimensions.) Formally, given a (user-defined) granularity vector $\gamma \in \mathbb{R}^k$ and offset vector $\omega \in \mathbb{R}^k$, we partition the state space into disjoint *cells* of equal size. Each cell C is the Cartesian product of half-open intervals $[\omega_i + p_i\gamma_i,\ \omega_i + (p_i+1)\gamma_i[$ in each dimension i, for cell index $p \in \mathbb{N}^k$. We define the *grid* as the set $\mathcal{P}_\gamma^\omega = \{C \mid C \cap S \neq \emptyset\}$ of all cells that overlap with the bounded state space. Note the number of cells will depend on γ. For each $s \in S$, $[s]_{\mathcal{P}_\gamma^\omega}$ denotes the unique cell containing s.

An EMDP $\mathcal{M}$, a granularity vector γ and offset vector ω induce a finite labeled transition system $\mathcal{T}_{\mathcal{M},\gamma,\omega} = (\mathcal{P}_\gamma^\omega, A, \rightarrow)$, where

$$C \xrightarrow{a} C' \iff \exists s \in C.\ \exists s' \in C'.\ T(s,a)(s') > 0. \quad (1)$$

Given a safety property $\varphi \subseteq S$ and a grid $\mathcal{P}_\gamma^\omega$, let $\mathcal{C}_\varphi^0 = \{C \in \mathcal{P}_\gamma^\omega \mid C \subseteq \varphi\}$ denote those cells that are safe in zero steps. We define the set of *safe cells* as the maximal set $\mathcal{C}_\varphi$ such that

$$\mathcal{C}_\varphi = \mathcal{C}_\varphi^0 \cap \{C \in \mathcal{P}_\gamma^\omega \mid \exists a \in A.\ \forall C' \in \mathcal{P}_\gamma^\omega.\ C \xrightarrow{a} C' \implies C' \in \mathcal{C}_\varphi\}. \quad (2)$$

Given the finiteness of $\mathcal{P}_\gamma^\omega$ and monotonicity of Eq. 2, $\mathcal{C}_\varphi$ may be obtained in a finite number of iterations using Tarski's fixed-point theorem [23].

A (nondeterministic) strategy for $\mathcal{T}_{\mathcal{M},\gamma,\omega}$ is a function $\nu : \mathcal{P}_\gamma^\omega \rightarrow 2^A$. The most permissive shield ν_φ (i.e., safe strategy) obtained from $\mathcal{C}_\varphi$ [5] is given by

$$\nu_\varphi(C) = \{a \in A \mid \forall C' \in \mathcal{P}_\gamma^\omega.\ C \xrightarrow{a} C' \implies C' \in \mathcal{C}_\varphi\}.$$

A shield ν for $\mathcal{T}_{\mathcal{M},\gamma,\omega}$ induces a shield σ for $\mathcal{M}$ in the standard way [9]:

Theorem 1. *Given an EMDP $\mathcal{M}$, a safety property $\varphi \subseteq S$, and a grid $\mathcal{P}_\gamma^\omega$, if ν is a shield for $\mathcal{T}_{\mathcal{M},\gamma,\omega}$, then $\sigma(s) = \nu([s]_{\mathcal{P}_\gamma^\omega})$ is a shield for $\mathcal{M}$.*

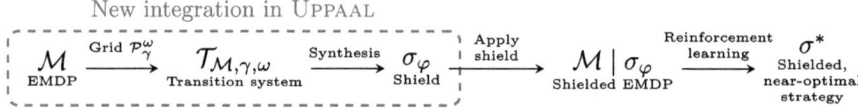

Fig. 2. Workflow for obtaining a near-optimal shielded strategy in UPPAAL.

Figure 2 shows the overall workflow of the shield synthesis and how the shield can later be used to (reinforcement-) learn a near-optimal strategy *under this shield*. The green box marks the steps that we newly integrated in UPPAAL.

For the *bouncing ball*, we will obtain the shield shown in Fig. 3(a). To effectively implement the aforementioned approach, there are additional challenges which we address in the following section.

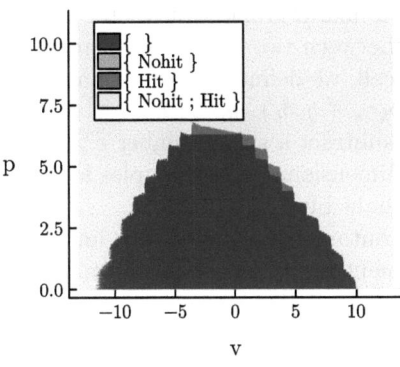 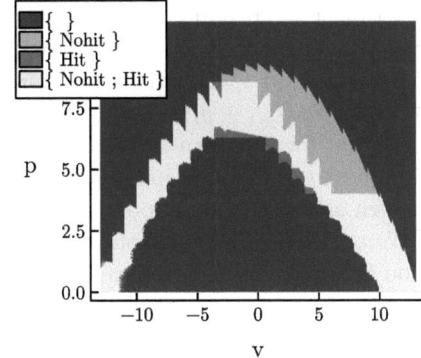

(a) It is safe to leave the bounds.   (b) It is unsafe to leave the bounds.

Fig. 3. Two shields for the *bouncing ball*. Colors represent the allowed actions in the corresponding state of velocity v and position p while in location `InAir`.

3 Effective Implementation of Shield Synthesis

In this section, we discuss our implementation of the approach to synthesize a shield as outlined in Sect. 2 in UPPAAL COSHY. In particular, a practical implementation faces the following two main challenges.

First, we receive the safety property φ in the form of a user query (see Sect. 5.1). Thus, the definition of the cells $\mathcal{C}_\varphi^0$ that are immediately safe generally requires symbolic reasoning, which is not readily available. Instead, we check a finite number of states within each cell, which we describe in Sect. 3.1.

Second, determining Eq. (1) requires to solve reachability questions for infinitely many states. While this can be done for simple classes of systems, we deal with very general systems (e.g., nonlinear hybrid dynamics), for which reachability is undecidable [15]. This motivated us to instead compute an approximate solution, which we outline in Sect. 3.2.

Thanks to the above design decisions, our implementation is fully automatic and supports the expressive formalism of general UPPAAL models (e.g., stochastic hybrid automata with calls to general C code).

We also identified further practical challenges, which we address in the later parts of this section. Definition 1 requires a bounded state space, but it is for instance difficult to determine upper bounds for the position and velocity of the *bouncing ball*; in Sect. 3.3, we explain how we treat such cases in practice. In Sect. 3.4, we discuss an optimization to omit redundant dimensions.

3.1 Determining Initial Safe Cells

We apply *systematic sampling* from a cell, i.e., samples are not drawn at random. Rather, we uniformly cover the cell with n^k samples, where $n \in \mathbb{N}, n \neq 0$ is a user-defined parameter. Recall from Sect. 2.3 that a cell C of a grid $\mathcal{P}_\gamma^\omega$ is rectangular and defined by an index vector p, an offset ω and a granularity vector γ, all of dimension k. Let $\delta_i = \frac{\gamma_i}{n-1}$ be the distance between two samples in dimension i when $n > 1$, and $\delta_i = 0$ otherwise. For any cell, we define the corresponding set of samples as $\{(\omega_1 + p_1\gamma_1 + q_1\delta_1, \ldots, \omega_k + p_k\gamma_k + q_k\delta_k) \mid q_i \in \{0, 1, \ldots, n-1\}\}$. To account for the open upper bounds, we subtract a small number $\epsilon > 0$ from the highest samples. An example of a two-dimensional set of samples for $n = 4$ is shown as the dark blue points inside the light blue cells in Fig. 4.

We note that the above only applies to continuous variables. Our implementation treats discrete variables (e.g., component locations) in the natural way.

Finally, to approximate the set $\mathcal{C}_\varphi^0$, we draw samples from each cell and check for each sample whether it violates the specification. A cell is added to $\mathcal{C}_\varphi^0$ only if all samples in that cell satisfy the specification.

For the *bouncing ball*, the ball should never be in the Stop location. Since the location is a discrete variable, and each cell only belongs to one location, checking a single sample from a cell C already determines whether $C \in \mathcal{C}_\varphi^0$. Thus, our approach is exact and efficient in the common case where the safety property is given via an error location.

3.2 Determining Reachability

We approximate cell reachability $C \xrightarrow{a} C'$, as defined in Eq. (1), similarly to [9] but adapted to work in UPPAAL. In a UPPAAL model, actions $a \in A$ correspond to controllable edges (indicating that the controller can act).

For each cell C and action $a \in A$, we iterate over all sampled states s (as described before) and select the edge corresponding to a, which gives us a new state s'; starting from s', we simulate the environment (using the built-in simulator in UPPAAL) until a state s'' is reached in which the controller has the next choice (i.e., multiple action edges are enabled) again.[2] Thus, s'' is a witness to add the corresponding cell $[s'']_{\mathcal{P}_\gamma^\omega}$ to the transition relation $C \xrightarrow{a} [s'']_{\mathcal{P}_\gamma^\omega}$. Assuming the simulator is numerically sound, the resulting transition system underapproximates $\mathcal{T}_{\mathcal{M},\gamma,\omega}$. As observed in [9], the more simulations are run, the more likely do we obtain the true solution. To check whether this underapproximation is sufficiently accurate, the existing queries for statistical model checking in UPPAAL can be used, as we shall see in Sect. 5.

In general, two simulations starting in the state s may not yield the same state s'' due to stochasticity. In [9], stochasticity was treated as additional dimensions over which to sample (systematically). This was possible by manually crafting the reachability sampling for each model. Detecting stochastic behavior in UPPAAL models automatically turned out to be difficult due to the rich formalism. Instead, we decided to simply let the simulator sample from the stochastic distribution. As a side effect, this new design allows us to support stochasticity with general distributions, particularly with unbounded support.

---

[2] Where [9] required a fixed control period, UPPAAL COSHY supports non-periodic control. This is demonstrated in [8, Appendix B].

Since this design may generally miss some corner-case behavior, we expose a user-defined parameter m to control the number of times sampling is repeated.

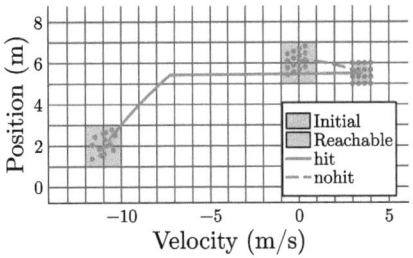
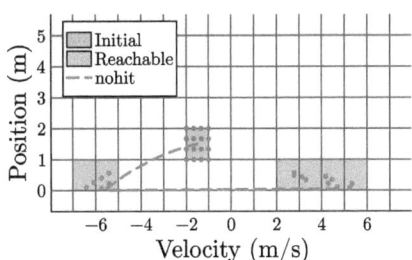
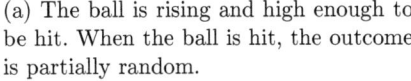

(a) The ball is rising and high enough to be hit. When the ball is hit, the outcome is partially random.

(b) The ball is too low to be hit, but it bounces off the ground. The velocity loss upon a bounce is partially random.

Fig. 4. Example of a grid for the *bouncing ball*. By sampling from the initial cell (blue) and simulating the dynamics, we discover reachable cells (green). (Color figure online)

We illustrate the reachability approximation for the *bouncing ball* in Fig. 4 for $n = 4$ (number of samples per dimension) and $m = 1$ (number of re-sampling). When the ball moves through the air, it behaves deterministically. In Fig. 4(a), when the ball is not hit, we obtain successor states that keep a regular "formation" (top right green dots). When the ball is hit, the successor states are affected by randomness (bottom left green dots). Figure 4(b) shows a similar randomized effect when the ball touches the ground.

3.3 Generalization to Unbounded State Spaces

Definition 1 requires the state space to be bounded, but bounds can be hard to determine for some systems. This includes the *bouncing ball*, for which upper bounds for position and velocity are not immediately clear. Indeed, if we consider the bounded state space where $p \in [0, 11]$ and $v \in [-13, 13]$, the system dynamics do not guarantee that the ball stays within these bounds. If we plot velocity against position, as in Fig. 3, then a falling ball near the left end of the plot may leave the bounds on the left (because it becomes too fast).

Conceptually, our implementation deals with out-of-bounds situations by modifying the transition system. All samples leading to a state outside the specified bounds go to a dummy cell C_{out}, for which all transitions lead back to itself. A user-defined option with the following choices determines the behavior:

1. Raise an error when reaching C_{out} during simulation (default behavior).
2. Include C_{out} in C_φ^0, i.e., leaving the bounds is always safe.
3. Exclude C_{out} from C_φ^0, i.e., leaving the bounds is always unsafe.
4. Automatically choose between options 2 and 3 using sampling.

With Option 4, samples are taken outside the specified bounds, similar to Sect. 3.1. For the *bouncing ball*, our tool samples states such as ($v = 26$, $p = 22$, Ball.Stop), even though these states may not be reachable in practice. If any sample state is found to be unsafe, C_{out} is considered unsafe, and safe otherwise. The result of synthesizing a shield with this option is shown in Fig. 3(b). In particular, that shield forbids to hit the ball when it is too fast, which ensures that it does not leave the bounds. Alternatively, we obtain a more permissive shield by choosing Option 2, as shown in Fig. 3(a).

3.4 Omitting Variables from Consideration

As emphasized in [1], a shield can be obtained from an abstract model that only simulates behaviors relevant to the safety specification. For example, cost variables may only be relevant during learning. While every variable in a model can be included in the partitioning, this is computationally demanding.

Therefore, we allow that variables are omitted from the grid specification. However, this raises a new challenge when sampling a state from a cell, since a concrete state requires a value for each variable. To address that, we set each omitted variable to the unique value of the initial state, which must always be specified in a UPPAAL model. Hence, the user must define the initial state such that the values of omitted variables are sensible defaults. (Note that the initial state is ignored by the shield synthesis in all other respects.)

The choice not to include a variable in the grid must be made carefully, as this can change the behavior of the transition system and potentially lead to an unsound shield. As a rule of thumb, it is appropriate to omit variables if they always have the same value when actions are taken, or if they are only relevant for keeping track of a performance value such as cost.

For the *bouncing ball*, the player (Fig. 1(a)) is always in the location Choose when taking an action. By setting Choose as the initial location, this component's location is not relevant to keep track of in the partitioning. Moreover, the variable c is used to keep track of cost and does not matter to safety. Lastly, the clock variable x is used to measure time until the next player action. It is always 0 when it is time for the player to act, and so it can also be omitted.

4 Obtaining a Compact Shield Representation

In this section, we present a new technique for obtaining a compact representation of shields that stem from an axis-aligned state-space partitioning (as described in Sect. 2.3). Here, we choose to represent the shield as a decision tree. We note that we aim for a functionality-preserving representation, i.e., we transform a grid-based shield to an equivalent decision-tree-based shield.

Recall that each cell prescribes a set of allowed actions. Let two cells be *similar* if the shield assigns the same set of actions to them. Our goal is to form (hyper)rectangular clusters of similar cells, which we call *regions*; in other words, we aim to find a coarser partitioning. In a nutshell, our approach works as follows. Initially, we start from the finest partitioning where each cell is a separate region. Then, we iteratively merge neighboring regions of similar cells, thereby obtaining a coarser partitioning, such that the resulting region is rectangular again. We call our algorithm CAAP (**C**oarsify **A**xis-**A**ligned **P**artitionings).

4.1 Representation of Partitionings and Regions

We start by noting that an axis-aligned partitioning of a state space $S \subseteq \mathbb{R}^k$ can be represented by a binary decision tree $\mathcal{T}$ where each leaf node is a set of actions and each inner node splits the state space with a predicate of the form $\rho(s) = s_i < c$, where s is a state vector, s_i is a state dimension, and $c \in \mathbb{R}$. Given a state s, the tree evaluation, written $\mathcal{T}(s)$, is defined as usual: Start at the root node. At an inner node, evaluate the predicate $\rho(s)$. If $\rho(s) = \top$, descend to the left child; otherwise, descend to the right child. At a leaf node, return the corresponding set of actions. We denote the partitioning induced by a decision tree $\mathcal{T}$ as $\mathcal{P}_\mathcal{T}$. Our goal in this section is: given a decision tree $\mathcal{T}$ inducing a partitioning $\mathcal{P}_\mathcal{T}$, find an equivalent but smaller decision tree.

Given a tree $\mathcal{T}$, we store all bounds c of the predicates $s_i < c$ in a matrix M of k rows where the i-th row contains the bounds associated with state dimension s_i in ascending order. For example, consider the bounds in Fig. 5(a) and M on the right.

	1	2	3	4
s_1	0	2	3	4
s_2	0	2	3	4

We extract a bounds vector from M via an index vector $p \in \mathbb{N}^k$ such that the i-th entry of p contains the column index for the i-th row. In other words, the resulting vector consists of the values M_{i,p_i}. For instance, $p = (1,3)$ yields the vector $s^p = (0,3)$ (row s_1 column 1 and row s_2 column 3). We can view this vector as a state in the state space given as $s^p = (M_{1,p_1}, \ldots, M_{k,p_k})$.

We define a region R in terms of two index vectors $(p^{\min}, p^{\max})$ representing the minimal and maximal corner in each dimension. Then, increasing $p_i^{\max}$ corresponds to expanding R in dimension i.

4.2 Expansion of Rectangular Regions

For an expansion to be legal, it must satisfy the following three *expansion rules*:

Definition 2 (Expansion rules). *Let R' be a candidate region for a new partitioning $\mathcal{P}'$ derived from $\mathcal{P}_\mathcal{T}$. Then R' is legal if it satisfies these three rules:*

1. *All cells in region R' have the same action set,*
2. *Region R' does not intersect with other regions in $\mathcal{P}'$,*
3. *Region R' does not cut any other region R from the original partitioning $\mathcal{P}_\mathcal{T}$ in two, i.e., the difference $R \setminus R'$ is either empty or rectangular.*

The first two cases are directly related to the definition of the problem, i.e., the produced partitioning should respect $\mathcal{T}$ and only have non-overlapping regions (see Fig. 5(b) and Fig. 5(c)). The third case is required in order to ensure that

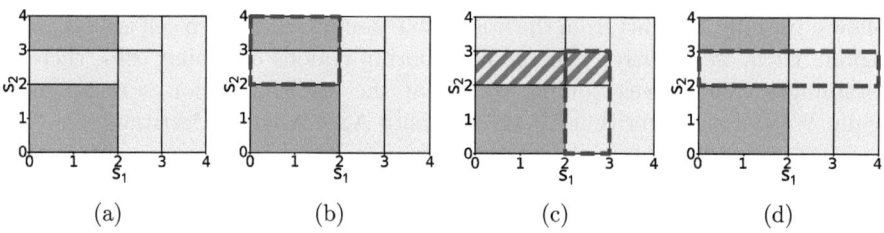

Fig. 5. Expansion example. Yellow and purple denote distinct actions. Striped regions have been fixed in previous iterations. The dashed border is the new candidate region R'. (a) An input partitioning. (b) A violation of Rule 1, since the expanded region contains different actions. (c) A violation of Rule 2, since the expanded region overlaps with a striped area. (d) A violation of Rule 3, since the expansion cuts the rightmost region into two new regions. (Color figure online)

in each iteration, the algorithm does not increase the overall number of regions when adding a region from the original partitioning to the new partitioning. To appreciate this, consider the visualization in Fig. 5(d). The candidate expansion cuts the rightmost region (given by $(3,0)$ and $(4,4)$) in two such that the remainder would have to be represented by *two* regions—one given by $((3,0),(4,2))$ and one given by $((3,3),(4,4))$. Clearly, all three expansion rules of Definition 2 can be checked in time linear in the number of nodes of $\mathcal{P}_\mathcal{T}$.

To determine the expansion of regions, we propose the following greedy approach: let $(p^{\min}, p^{\max})$ define a region. We then want to find a vector $\Delta_p \in \mathbb{Z}^k$ such that $(p^{\min}, p^{\min} + \Delta_p)$ defines a region that obeys the three expansion rules and is (locally) maximal, in the sense that increasing it in any dimension would violate at least one of the expansion rules. Note that a vector $\Delta_p = p^{\max} - p^{\min}$ satisfies the expansion rules trivially but is possibly not maximal. Thus, a solution is guaranteed to exist. However, note that there is not necessarily a unique maximal solution, and that the set of solutions is not convex, i.e., there may exist solutions Δ_p^1 and Δ_p^2 such that $\Delta_p^1 \leq \Delta_p^2$ but no other Δ_p' with $\Delta_p^1 \leq \Delta_p' \leq \Delta_p^2$ satisfies the expansion rules. Formally:

Definition 3 (Expansion vector Δ_p). *Given $p^{\min} \in \mathbb{Z}^k$, a decision tree $\mathcal{T}$ over a k-dimensional state space, and a set $\mathcal{P}$ of fixed regions, $\Delta_p \in \mathbb{Z}^k$ is a vector such that for $p^{\max} = p^{\min} + \Delta_p$ the region $R = (p^{\min}, p^{\max})$ does not violate any of the expansion rules in Definition 2 and for any vector $\Delta_p' = (\Delta_{p_1}, \ldots, \Delta_{p_i} + 1, \ldots, \Delta_{p_k})$ at least one of the rules is violated.*

Our greedy approach to finding Δ_p starts with $\Delta_p = p^{\max} - p^{\min}$ for some region $R = (p^{\min}, p^{\max})$. It then iteratively selects a dimension d by a uniformly random choice and attempts to increment the d-th entry of Δ_p. For that, we define the candidate region $R' = (p^{\min}, p^{\min} + \Delta_p)$ and check the rules 1 and 2. If any of them is violated, we mark the corresponding dimension d as exhausted, roll back the increment, and continue with a new dimension not marked as exhausted yet, until none is left.

Table 1. Queries run on the *bouncing ball* model. New query type highlighted. All statistical results are given with a 99% confidence interval.

#	Query	Result
1	`strategy efficient = minE(c) [<=120]` `{} -> {v, p} : <> time>=120`	✓
2	`simulate [<=120]{ p, v } under efficient`	✓
3	`E[<=120;100] (max: c) under efficient`	≈ 0
4	`Pr[<=120;10000] (<> Ball.Stop) under efficient`	[0.9995, 1]
5	`strategy shield = acontrol: A[] !Ball.Stop` `{ v[-13, 13]:1300, p[0, 11]:550, Ball.location }`	✓
6	`saveStrategy("/shield.json", shield)`	✓
7	`strategy compact_shield = loadStrategy("/compact.json")`	✓
8	`simulate [<=120]{ p, v } under compact_shield`	✓
9	`strategy shielded_efficient = minE(c) [<=120]` `{} -> {v, p} : <> time>=120 under compact_shield`	✓
10	`simulate [<=120]{ p, v } under shielded_efficient`	✓
11	`E[<=120;100] (max: c) under shielded_efficient`	34.6 ± 0.6
12	`Pr[<=120;10000] (<> Ball.Stop) under shielded_efficient`	[0, 0.00053]

As mentioned above, the set of solutions is not convex. Correspondingly, if Rule 3 is violated, the algorithm initiates an attempt at *repairing* the candidate expansion by continuing the expansion to the largest bound in the expansion dimension of any of the broken regions. This way, we check whether the violation can be overcome by simply expanding more aggressively. When all dimensions have been exhausted, Δ_p adheres to Definition 3.

We note that the algorithm is not guaranteed to find a local optimum. One reason is that the repair only expands in one dimension. This choice is deliberate to keep the algorithm efficient and avoid a combinatorial explosion. A more detailed description including pseudocode can be found in [8, Appendix C].

5 Case Studies and Evaluation

In this section, we evaluate our implementation of UPPAAL COSHY and CAAP. In Sect. 5.1, we demonstrate a typical application. In Sect. 5.2, we benchmark the implementations on several models.

5.1 A Complete Run of the Bouncing Ball

Table 1 shows a typical usage of UPPAAL with a sequence of queries on the *bouncing ball* example to produce a safe and efficient strategy (cf. Fig. 2). Documentation of the new query syntax is available online and in [8, Appendix A].[3]

In Query 1, we train a strategy called `efficient`, which is only concerned with cost and does not consider safety. Such a strategy is trivial: simply never pick the `hit` action. This is seen in Query 2, which simulates a single run of 120 s. It outputs position p and velocity v, which are visualized in Fig. 6(a). Query 3 statistically evaluates the strategy in 100 runs to estimate the expected value of c. The result "≈ 0" indicates that only this value was observed. Query 4 estimates the probability of a run being unsafe to be in the interval [0.9995, 1] with 99% confidence; in this case, as expected, all 10 000 runs were unsafe.

Query 5 synthesizes a shield `shield`. The shield matches the one shown in Fig. 3(a). In queries 6 and 7, the shield is converted to a compact representation by saving it to a file, calling the CAAP implementation, and loading the result back into UPPAAL. The shield is simulated in Query 8, for which any of the allowed actions is selected randomly (this happens implicitly); while safe, this shielded but randomized strategy is not efficient and hits the ball more often than needed, as visualized in Fig. 6(b).

In Query 9, we learn a strategy `shielded_efficient` under the shield using UPPAAL STRATEGO [12]. This strategy keeps the ball in the air without excessive hitting, as shown by the output of Query 10 in Fig. 6(c). The result of Query 11 shows the expected cost, and Query 12 shows that the safety property holds with high confidence: None of the 10 000 runs were unsafe.

5.2 Further Examples

State-space transformations can be used to synthesize a shield more efficiently [7]. Since UPPAAL supports function calls, transformations can be applied by modifying the model. Details can be found in [8, Appendix D].

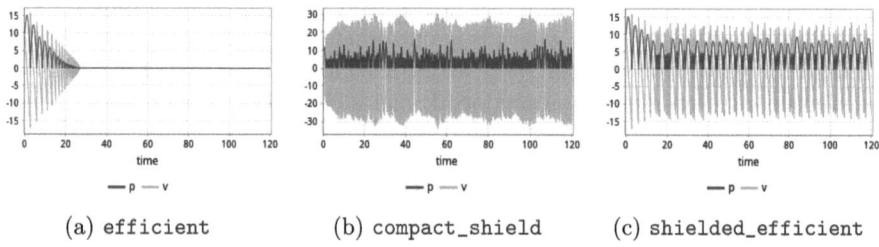

(a) `efficient` (b) `compact_shield` (c) `shielded_efficient`

Fig. 6. *Bouncing ball* simulations (position, velocity) under different strategies.

---

[3] https://docs.uppaal.org/language-reference/query-syntax/controller_synthesis/#approximate-control-queries.

Table 2. Computation time and sizes for synthesizing and reducing shields for three models. The original size is the number of cells, whereas the reduced size is the number of regions. All shields were statistically evaluated to be at least 99.47% safe with a confidence interval of 99% (no unsafe runs observed).

Model	n	m	Synthesis time	Size	Reduction time	Reduced size
Bouncing ball	3	1	218 s	1 430 000	53 s	2972
Boost converter	3	20	1 430 s	136 800	21 s	571
Random walk	4	20	82 s	40 000	1.5 s	60
Water tank	3	1	0.1 s	168	0.1 s	24

Next, we show quantitative results of the shield synthesis and subsequent shield reduction, for which we also use three additional models. Firstly, the *boost converter* [9] models a real circuit for stepping up the voltage of a direct current (DC) input. The controller must keep the voltage close to a reference value, without exceeding safe bounds for the voltage and current. The state space is continuous, with significant random variation in the outcome of actions.

In the *random walk* model [9,17], the player must travel a certain distance before time runs out by choosing between a fast but expensive and a slow but cheap action. The state space is continuous and the outcomes of actions follow uniform distributions.

In the *water tank* model inspired from [1], a tank must be kept from overflowing or running dry. Water flows from the tank at a rate that varies periodically. At each time step, the player can control the inflow by switching a pump on or off. The state space is discrete.

We show results for computing and reducing shields in Table 2. The *water tank* is fully deterministic, and the *bouncing ball* only has low-variance stochastic behavior. The *boost converter* and *random walk* have a high variance in action outcomes, which is why we use $m = 20$ simulation runs per sampled state. We evaluated the shields statistically and found no unsafe runs in 10 000 trials. The reduction yields significantly smaller representations at acceptable run time.

6 Conclusion

We have described our implementation of the shield synthesis algorithm from [9] in the tool UPPAAL COSHY. Our tool can work with rich inputs modeled in UPPAAL. We have also presented the CAAP algorithm to reduce the shield representation significantly, which is crucial for deployment on an embedded device.

We see several directions for future integration into UPPAAL. As discussed, our implementation does not apply *systematic* sampling for random dynamics; however, we think that many sources of randomness in UPPAAL models can be handled systematically. Currently, the reduction algorithm CAAP is implemented as a standalone tool, but it would be useful to also integrate it directly with UPPAAL. During development, we found it helpful to visualize shields, as in Fig. 3, which could be offered in the user interface. In the same line, an explanation why a state is marked unsafe in a shield would help in debugging a model.

Acknowledgments. This research was partly supported by the Independent Research Fund Denmark under reference number 10.46540/3120-00041B and the Villum Investigator Grant S4OS under reference number 37819.

References

1. Alshiekh, M., Bloem, R., Ehlers, R., Könighofer, B., Niekum, S., Topcu, U.: Safe reinforcement learning via shielding. In: McIlraith, S.A., Weinberger, K.Q. (eds.) AAAI, pp. 2669–2678. AAAI Press (2018). https://doi.org/10.1609/AAAI.V32I1.11797
2. Ashok, P., Jackermeier, M., Jagtap, P., Kretínský, J., Weininger, M., Zamani, M.: dtControl: decision tree learning algorithms for controller representation. In: Ames, A.D., Seshia, S.A., Deshmukh, J. (eds.) HSCC, pp. 17:1–17:7. ACM (2020). https://doi.org/10.1145/3365365.3382220
3. Ashok, P., Jackermeier, M., Křetínský, J., Weinhuber, C., Weininger, M., Yadav, M.: dtControl 2.0: explainable strategy representation via decision tree learning steered by experts. In: TACAS 2021. LNCS, vol. 12652, pp. 326–345. Springer, Cham (2021). https://doi.org/10.1007/978-3-030-72013-1_17
4. Behrmann, G., Cougnard, A., David, A., Fleury, E., Larsen, K.G., Lime, D.: UPPAAL-Tiga: time for playing games! In: Damm, W., Hermanns, H. (eds.) CAV 2007. LNCS, vol. 4590, pp. 121–125. Springer, Heidelberg (2007). https://doi.org/10.1007/978-3-540-73368-3_14
5. Bernet, J., Janin, D., Walukiewicz, I.: Permissive strategies: from parity games to safety games. RAIRO Theor. Inf. Appl. **36**(3), 261–275 (2002). https://doi.org/10.1051/ita:2002013
6. Breiman, L., Friedman, J.H., Olshen, R.A., Stone, C.J.: Classification and Regression Trees. Wadsworth (1984)
7. Brorholt, A.H., Høeg-Petersen, A.H., Larsen, K.G., Schilling, C.: Efficient shield synthesis via state-space transformation. In: Steffen, B. (ed.) AISoLA. LNCS, vol. 15217, pp. 206–224. Springer, Heidelberg (2024). https://doi.org/10.1007/978-3-031-75434-0_14
8. Brorholt, A.H., et al.: Uppaal coshy: automatic synthesis of compact shields for hybrid systems (2025). https://arxiv.org/abs/2508.16345
9. Brorholt, A.H., Jensen, P.G., Larsen, K.G., Lorber, F., Schilling, C.: Shielded reinforcement learning for hybrid systems. In: Steffen, B. (ed.) AISoLA. LNCS, vol. 14380, pp. 33–54. Springer, Heidelberg (2023). https://doi.org/10.1007/978-3-031-46002-9_3
10. Chatterjee, K., Henzinger, T.A., Jobstmann, B., Radhakrishna, A.: GIST: a solver for probabilistic games. In: Touili, T., Cook, B., Jackson, P. (eds.) CAV 2010. LNCS, vol. 6174, pp. 665–669. Springer, Heidelberg (2010). https://doi.org/10.1007/978-3-642-14295-6_57
11. Chatterjee, K., Henzinger, T.A., Jobstmann, B., Singh, R.: QUASY: quantitative synthesis tool. In: Abdulla, P.A., Leino, K.R.M. (eds.) TACAS 2011. LNCS, vol. 6605, pp. 267–271. Springer, Heidelberg (2011). https://doi.org/10.1007/978-3-642-19835-9_24

12. David, A., Jensen, P.G., Larsen, K.G., Mikučionis, M., Taankvist, J.H.: UPPAAL STRATEGO. In: Baier, C., Tinelli, C. (eds.) TACAS 2015. LNCS, vol. 9035, pp. 206–211. Springer, Heidelberg (2015). https://doi.org/10.1007/978-3-662-46681-0_16
13. Demirovic, E., et al.: Murtree: optimal decision trees via dynamic programming and search. J. Mach. Learn. Res. **23**, 26:1–26:47 (2022). https://jmlr.org/papers/v23/20-520.html
14. Demirović, E., Schilling, C., Lukina, A.: In search of trees: decision-tree policy synthesis for black-box systems via search. In: AAAI, pp. 27250–27257. AAAI Press (2025). https://doi.org/10.1609/aaai.v39i26.34934
15. Doyen, L., Frehse, G., Pappas, G.J., Platzer, A.: Verification of hybrid systems. In: Handbook of Model Checking, pp. 1047–1110. Springer, Cham (2018). https://doi.org/10.1007/978-3-319-10575-8_30
16. Du, M., Liu, N., Hu, X.: Techniques for interpretable machine learning. Commun. ACM **63**(1), 68–77 (2020). https://doi.org/10.1145/3359786
17. Jaeger, M., Bacci, G., Bacci, G., Larsen, K.G., Jensen, P.G.: Approximating euclidean by imprecise markov decision processes. In: Margaria, T., Steffen, B. (eds.) ISoLA 2020. LNCS, vol. 12476, pp. 275–289. Springer, Cham (2020). https://doi.org/10.1007/978-3-030-61362-4_15
18. Jaeger, M., Jensen, P.G., Guldstrand Larsen, K., Legay, A., Sedwards, S., Taankvist, J.H.: Teaching stratego to play ball: optimal synthesis for continuous space MDPs. In: Chen, Y.-F., Cheng, C.-H., Esparza, J. (eds.) ATVA 2019. LNCS, vol. 11781, pp. 81–97. Springer, Cham (2019). https://doi.org/10.1007/978-3-030-31784-3_5
19. Kwiatkowska, M., Norman, G., Parker, D., Santos, G.: PRISM-games 3.0: stochastic game verification with concurrency, equilibria and time. In: Lahiri, S.K., Wang, C. (eds.) CAV 2020. LNCS, vol. 12225, pp. 475–487. Springer, Cham (2020). https://doi.org/10.1007/978-3-030-53291-8_25
20. Kwiatkowska, M., Norman, G., Parker, D.: PRISM 4.0: verification of probabilistic real-time systems. In: Gopalakrishnan, G., Qadeer, S. (eds.) CAV 2011. LNCS, vol. 6806, pp. 585–591. Springer, Heidelberg (2011). https://doi.org/10.1007/978-3-642-22110-1_47
21. Pranger, S., Könighofer, B., Posch, L., Bloem, R.: TEMPEST - synthesis tool for reactive systems and shields in probabilistic environments. In: Hou, Z., Ganesh, V. (eds.) ATVA 2021. LNCS, vol. 12971, pp. 222–228. Springer, Cham (2021). https://doi.org/10.1007/978-3-030-88885-5_15
22. Quinlan, J.R.: Learning decision tree classifiers. ACM Comput. Surv. **28**(1), 71–72 (1996). https://doi.org/10.1145/234313.234346
23. Tarski, A.: A lattice-theoretical fixpoint theorem and its applications. Pacific J. Math. **5**(2), 285–309 (1955). https://www.projecteuclid.org/journalArticle/Download?urlId=pjm%2F1103044538

Weighing Obese Timed Languages

Eugene Asarin[1](✉)[iD], Aldric Degorre[1][iD], Cătălin Dima[2][iD],
and Bernardo Jacobo Inclán[1][iD]

[1] Université Paris Cité, CNRS, IRIF, Paris, France
{asarin,adegorre,jacoboinclan}@irif.fr
[2] Université Paris-Est Créteil, LACL, Créteil, France
dima@u-pec.fr

Abstract. The bandwidth of a timed language characterizes the quantity of information per time unit (with a finite observation precision ε). Obese timed automata have an unbounded frequency of events and produce information at the maximal possible rate. In this article, we compute the bandwidth of any such automaton in the form $\approx \alpha/\varepsilon$. Our approach reduces the problem to computing the best reward-to-time ratio in a weighted timed graph constructed from the given timed automaton, with weights corresponding to the entropy of auxiliary finite automata.

1 Introduction

One of the first articles on automata [10] showed that the number of words of length n in a regular language asymptotically grows as 2^{nH}, and computed the rate H; also the rate of polynomial growth (for the case when $H = 0$) was fully characterized in [22]. Later, the growth (exponential, polynomial, and even intermediary) has been explored for context-free languages [9,15], finitely generated groups [12], and in different terms (entropy rate of subshifts) in symbolic dynamics [20]. Growth analysis of automata can be seen as a quantitative version of reachability: counting the number of paths instead of verifying their existence.

An important motivation for studies of growth rate consists in its interpretation as information content (entropy, capacity). In the 40 s, [21] related word counts to the capacity of a discrete noiseless channel; in the 60 s, [18] conceptualized the counting ("combinatorial") approach to the quantity of information, as compared to the probabilistic and the algorithmic ones. According to this approach, an element of a finite set S conveys $\log_2 \#S$ bits of information, and in most cases, asymptotics of this amount w.r.t. some size parameter is studied. Nowadays, counting-based analysis of formal languages provides a background to the theory of codes [7], and to protocols [14, Chapter 6] implemented in every hard disk drive and DVD [13]. In a nutshell, one can encode a source language with a growth rate H into a channel with growth rate H' only if $H \leq H'$. Also, words of such a language can be "zipped" with a compression rate at most H.

This work was funded by ANR project MAVeriQ ANR-CE25-0012.

© The Author(s), under exclusive license to Springer Nature Switzerland AG 2026
P. Ganty and A. Mansutti (Eds.): RP 2025, LNCS 16230, pp. 112–125, 2026.
https://doi.org/10.1007/978-3-032-09524-4_8

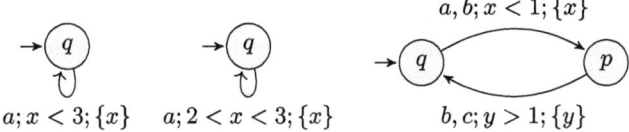

Fig. 1. Three timed automata and their asymptotic bandwidths (when ε is small), respectively $\frac{1}{\varepsilon}$ (obese); $\frac{1}{2}\log(\frac{1}{\varepsilon})$ (normal); and 2 (meager).

Timed languages [1], widely used to represent behaviors of real-time and cyber-physical systems, extend usual formal languages with timing information of each event. In the long run, we aim to port the growth rate/information content analysis to timed languages and automata with theoretical and practical ambitions (e.g. coding and compression). For information quantity **per event**, we have defined and computed the growth rate in [2,3]. However, we believe that practically it would be more relevant to consider growth with respect **to time**, not to the number of events. Our main challenge can thus be stated as follows:

*for a timed automaton, define and compute the **amount of information in bits per time unit** conveyed by its accepted words (observed with a precision $\varepsilon > 0$).* (1)

Previously, we made several steps toward the solution of (1). In [16], we formalize the notion of amount of information for timed languages, coin the term *bandwidth* for it, and prove that it is relevant to constrained channel coding with bounded delay. The timed language being continuously infinite, even when the time is bounded, its elements cannot be counted. Instead, we follow the approach from [19] and count the elements of an ε-net, approximating the timed words with a finite precision $\varepsilon > 0$, and use this count to define the bandwidth.

In [5], we classify languages of timed automata into three classes characterized by structural properties of the automata, that differ by the way that information is conveyed (see the examples on Fig. 1). In **meager** automata, the bandwidth is $O(1)$. Information mostly comes from discrete choices, since all timed transitions are very constrained. In [6], we characterized the exact value of the bandwidth of meager automata as the spectral radius of a finite-state abstraction. In **normal** ones, the bandwidth is $\alpha \log 1/\varepsilon$, and the information is mostly conveyed through free choices of some duration every couple of time units. In typical cases, the coefficient α is the number of "degrees of freedom" (continuous duration choices) per time unit in some optimal cycle. However, the question of computing α for the most general case remains open. **Obese** automata have a bandwidth $\Omega(1/\varepsilon)$. In some parts of such automata, events can happen with high frequency (without a lower bound on delays), yielding a huge bandwidth. Since very different bandwidth production mechanisms come into play in the three cases, bandwidth computation techniques also differ a lot.

In this paper, we concentrate on obese automata and compute the main asymptotic term of their bandwidth in the form α/ε. We proceed as follows: after a preprocessing, we identify within the automaton all "spots" that admit very fast information production, and compute the information rate $\alpha_\mathcal{D}$ of each such spot $\mathcal{D}$ (as the logarithm of the spectral radius of a matrix). Next, we transform the obese automaton into an "abstract" weighted timed graph, with states corresponding to spots, weights corresponding to their rates $\alpha_\mathcal{D}$, and transitions to switching modes and resetting some clocks. Finally, the overall rate α corresponds to the maximal reward-per-time ratio in the abstract timed graph, computable due to [8]. We prove that α/ε is indeed an upper and a lower bound for the bandwidth, and that α is the logarithm of an algebraic number.

An example of a planetary robot modeled with an obese timed automaton is provided in Fig. 2. The robot records all events, observed with time granularity ε, in a log file. The bandwidth α/ε corresponds to the disk space (in bits per hour of the robot's life) required to store the log file.

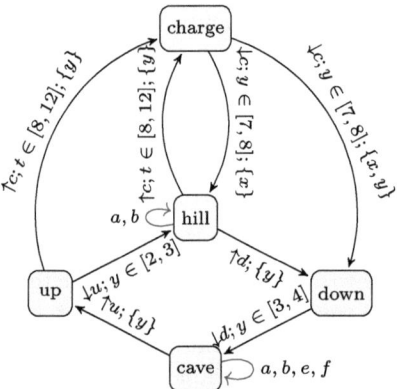

All transitions have guards: $x \leq 20, t \leq 25$;
All states have self-loops: $m, t = 25, \{t\}$.

	clocks
x	battery charge $= 20 - x$
t	daytime clock
y	time elapsed in current action
	events
a, b, e, f	exploration events
$\uparrow u, \downarrow u$	start and end going uphill
$\uparrow d, \downarrow d$	start and end going downhill
$\uparrow c, \downarrow c$	start and end charging
m	midnight

Fig. 2. A Martian robot modeled by an obese TA. The robot explores two spots: a cave and a hill, can travel between them, and must charge the batteries during daytime (starting from 8 to 12, for 7 to 8 h) when uphill. The charge suffices for 20 h, day duration is 25 h. Events a, b, e, f happen at a high speed and lead to obesity.

The paper is structured as follows. In Sect. 2, we recall some useful notions on timed automata and introduce an operation of squeezing on regular languages. In Sect. 3, we recall our definition of bandwidth and state the problem of bandwidth computation for timed regular languages. In main Sect. 4, we compute the bandwidth of obese timed languages. We conclude and briefly discuss the complexity aspects and perspectives in Sect. 5. Due to space limitations most of the proofs are provided in the full version of this article on ArXiV.

2 Background on Timed and Finite Automata

$\mathcal{P}(X)$ denotes the powerset of X and $\mathcal{P}_+(X)$ removes the empty set from $\mathcal{P}(X)$. We employ words, languages, and automata on both alphabets $\mathcal{P}(\Sigma)$ and Σ. Each word $w = a_1 \ldots a_n \in \Sigma^*$ can be also seen as $\{a_1\} \ldots \{a_n\} \in \mathcal{P}(\Sigma)^*$. Abusing set notation, we omit commas between braces, e.g., shortening $\{a,b\}\{b,c\}$ to $\{ab\}\{bc\}$. We denote the set of all letters in w by $\ell(w) \in \mathcal{P}(\Sigma)$. Finally, $\mathbb{L} \subseteq \mathbb{R}$ stands for the set of logarithms of algebraic numbers.

Definition 1. *Given Σ, a finite alphabet of discrete events, a* timed word *over Σ is an element from $(\Sigma \times \mathbb{R}_+)^*$ of the form $w = (a_1, t_1) \ldots (a_n, t_n)$, with $0 \leq t_1 \leq t_2 \cdots \leq t_n$. We denote $\mathrm{dur}(w) = t_n$. A* timed language *over Σ is a set of timed words over the same alphabet. Given a timed language L and $T \in [0, \infty)$, the corresponding* time-restricted language *is $L_T = \{w \in L \mid \mathrm{dur}(w) \leq T\}$.*

For a set of variables X, let G_X be the set of finite conjunctions of constraints of the form $x \sim b$ and $x \sim y + b$ with $x, y \in X$, $\sim \in \{<, \leq, >, \geq\}$ and $b \in \mathbb{N}$. We rely here on the following form of timed automata:

Definition 2 ([1], variant from [5]). *A TA is a tuple $(Q, X, \Sigma, \Delta, S, I, F)$, with Q the finite set of* locations, *X the finite set of* clocks, *Σ a finite alphabet, $S, I, F: Q \to G_X$ resp. the* starting, initial, *and* final clock constraints, *and the* transition relation *(set of* edges*) $\Delta \subseteq Q \times Q \times \Sigma \times G_X \times 2^X$.*

A TA is *bounded* whenever there is a constant M such that all the starting conditions $S(q)$ and guards $\mathfrak{g}_\delta$ require all clock values to be smaller than M.

The semantics of a TA is a *timed transition system* whose states are tuples $(q, \mathbf{x})$ composed of a location $q \in Q$ and a clock valuation (vector) $\mathbf{x} \in [0, \infty)^X$. We denote $\mathbf{x}[\mathfrak{r}]$ the operation of resetting the clocks in $\mathfrak{r} \subseteq X$. Each edge $\delta = (q, q', a, \mathfrak{g}, \mathfrak{r}) \in \Delta$ generates timed transitions $(q, \mathbf{x}) \xrightarrow{\delta, d} (q', \mathbf{x}')$ where $\mathbf{x} \models S(q)$, $\mathbf{x} + d \models \mathfrak{g}$, $\mathbf{x}' = (\mathbf{x} + d)[\mathfrak{r}]$ and $\mathbf{x}' \models S(q')$. Here, $\mathbf{x} + d = (x_1 + d, \ldots, x_n + d)$.

Paths are sequences of edges that agree on intermediary locations and *runs* are sequences of timed transitions that agree on intermediary states. An *accepting run* is a run $\rho = (q_0, \mathbf{x}_0) \xrightarrow{(\delta_1, d_1)} (q_1, \mathbf{x}_1) \cdots \xrightarrow{(\delta_n, d_n)} (q_n, \mathbf{x}_n)$, in which $\mathbf{x}_0 \models I(q_0)$ and $\mathbf{x}_n \models F(q_n)$. Given a run ρ as above, we define $\mathrm{path}(\rho) \triangleq \delta_1 \ldots \delta_n$. Furthermore, if $\delta_i = (q_i, q_{i+1}, a_i, \mathfrak{g}_i, \mathfrak{r}_i)$, then the timed word associated with ρ is defined as $\mathrm{word}(\rho) \triangleq (a_1, d_1)(a_2, d_1 + d_2) \ldots (a_n, \sum_{i \leq n} d_i)$. The *duration* of ρ is $\mathrm{dur}(\rho) = \sum d_i$. The language $L(\mathcal{A})$ of a TA $\mathcal{A}$ is the set of timed words associated with some accepting run.

A *timed graph* is a tuple (Q, X, Δ, S) with Q, X, S as in Def. 2 and $\Delta \subseteq Q \times Q \times G_X \times 2^X$. It is *time-divergent* whenever the number of edges in runs of duration $\leq T$ is bounded for every T.

Clock vectors $\mathbf{x}$ and $\mathbf{y}$ are *region-equivalent* [1] if for any clock c, $\lfloor x_c \rfloor = \lfloor y_c \rfloor$ and $\{x_c\} = 0$ iff $\{y_c\} = 0$; and for any two clocks c_1, c_2, $\{x_{c_1}\} \leq \{x_{c_2}\}$ iff $\{y_{c_1}\} \leq \{y_{c_2}\}$ (where $\{x\}$ denotes the fractional part of x). Equivalence classes w.r.t. region-equivalence (called *regions*) are simplices of dimension $d \leq \#X$.

Optimal Reward-to-Time Ratio in Finite and Timed Graphs. Consider a finite directed graph where each edge e is associated with a *reward* $w(e)$ and *time* $t(e)$. For a path π, its reward $w(\pi)$ is the sum of rewards of its edges, similarly for time. A classical optimization problem is maximizing the reward-to-time ratio $w(\pi)/t(\pi)$ over long or infinite paths. The maximal asymptotic ratio can be attained by iterating some optimal simple cycle. We refer the reader to [11] for a survey of exact and approximation algorithms for finding the optimal cycle and/or computing the best ratio.

The authors of [8] have extended the theory of reward-to-time ratio to timed graphs. Let us phrase their result in a form suitable for our investigations. A *weighted timed graph* (WTG) is just a timed graph together with a reward function $w : Q \to \mathbb{R}_+$. The reward of a transition $p \xrightarrow{\delta,t} q$ (or rather of a stay of t time units in p) is $w(p)t$; the reward $w(\rho)$ of a run ρ is the sum of rewards of all its transitions, its ratio is $w(\rho)/\operatorname{dur}(\rho)$.

Lemma 1 ([8], variant). *Given a bounded time-divergent WTG, one can compute the optimal ratio $\alpha \geq 0$ and a constant $C > 0$ such that for all runs ρ, $w(\rho) \leq \alpha \operatorname{dur}(\rho) + C$; and for any $T > 0$ there exists a run ρ with $\operatorname{dur}(\rho) \leq T$ and $w(\rho) > \alpha T - C$.*

Careful analysis of the algorithm in [8] shows that, when all $w(e), t(e) \in \mathbb{L}$, the resulting α belongs to $\mathbb{L}$, and can be computed both using some representation of the elements of $\mathbb{L}$ or as a real number with arbitrary precision.

Growth Rate of Regular Languages and Squeezing. For a language L on alphabet Σ its (logarithmic) *growth rate* is defined as follows [10]:

$$\mathcal{G}L = \limsup_{n \to \infty} \left(\log \#L_n \right)/n \qquad (\text{where } L_n = L \cap \Sigma^n).$$

For a finite automaton $\mathcal{A} = (Q, \Sigma, \Delta, I, F)$, with $Q = \{q_1, \ldots, q_n\}$, we define its $n \times n$ *adjacency matrix* $M_\mathcal{A} = (m_{ij})$ with $m_{ij} = \#\{a \in \Sigma : (q_i, a, q_j) \in \Delta\}$.

Theorem 1 ([10]). *Let $\mathcal{A}$ be a trim[1] deterministic finite automaton. Then $\mathcal{G}L(\mathcal{A})$ equals the logarithm of the spectral radius[2] of $M_\mathcal{A}$.*

A word W over an alphabet $\mathcal{P}(\Sigma)$ can be *squeezed*, yielding a word over $\mathcal{P}(\Sigma)$:
- first factorize it arbitrarily: $W = W_1 W_2 \ldots W_k$ where $W_i \in \mathcal{P}(\Sigma)^*$;
- then transform each factor W_i into the union of its letters (as subsets of Σ).

Given a language $L \subseteq \mathcal{P}(\Sigma)^*$, we define its squeezing $\mathcal{S}L \subseteq \mathcal{P}(\Sigma)^*$, a language consisting of all squeezings of all the words in L. We also use a notation for its n-letter fragment defining $\mathcal{S}_n L = \mathcal{S}L \cap \mathcal{P}(\Sigma)^n$. Hence, $\{abc\}\{b\}\emptyset\{abc\}\{a\} \in \mathcal{S}(\{ab\}\{ac\}\{b\}\{c\}\{abc\}\emptyset\{a\}\{a\})$. We abuse notation and apply squeezing also to languages over Σ, hence $\{ab\}\{abc\}\emptyset\{a\}\{ab\} \in \mathcal{S}(abacccbacbaba)$.

If L is regular, then $\mathcal{S}L$ is regular too. The following construction, illustrated in Fig. 3, yields an automaton recognizing $\mathcal{S}L$ from one recognizing L:

Construction 1. *Given $\mathcal{A} = (Q, \mathcal{P}(\Sigma), \Delta, I, F,)$, $\mathcal{S}L(\mathcal{A})$ is accepted by $\mathcal{S}\mathcal{A} = (Q, \mathcal{P}(\Sigma), \Delta', I, F)$, where $(p, B, q) \in \Delta'$ whenever $\mathcal{A}$ has a path from p to q with the union of its labels equal to B. In particular, $\mathcal{S}\mathcal{A}$ has self-loops labeled by $\emptyset$ on each state.*

---
[1] with all the states reachable from an initial and co-reachable to a final one.
[2] the max of the eigenvalue moduli.

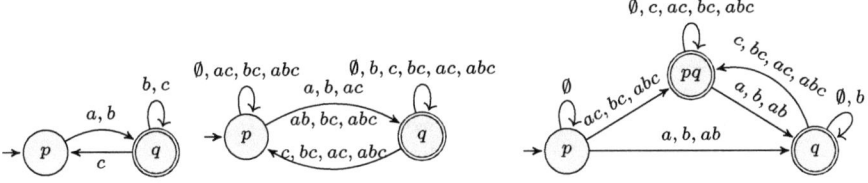

Fig. 3. (Left) an automaton for $(a+b)(b+c+ca+cb)^*$; (middle) an automaton for its squeezing (sets are written without braces); (right) its determinization.

Proposition 1. *Given a finite automaton $\mathcal{A}$, the growth rate of its squeezed language $\mathcal{GSL}(\mathcal{A})$ belongs to $\mathbb{L}$ and is computable from $\mathcal{A}$.*

So, for the automaton $\mathcal{A}$ on the left of Fig. 3 (used later for bandwidth computation), the growth rate of $\mathcal{SL}(\mathcal{A})$ is the logarithm of the spectral radius of the adjacency matrix of the automaton on the right:

$$\mathcal{GSL}(\mathcal{A}) = \log \operatorname{rad} \begin{pmatrix} 1 & 3 & 3 \\ 0 & 2 & 4 \\ 0 & 3 & 5 \end{pmatrix} = \log((7+\sqrt{57})/2) \approx 2.863.$$

3 Bandwidth of Timed Languages and the Main Problem

Pseudo-Distance on Timed Words. The notion of bandwidth of timed languages relies on a pseudo-distance on timed words defined in [4]. Here we extend it to $\mathcal{P}(\Sigma)$ as follows (where $\min \emptyset = \infty$):

Definition 3. *Given $w = (A_1, t_1)\ldots(A_n, t_n)$ and $v = (B_1, s_1)\ldots(B_m, s_m)$ two timed words over $\mathcal{P}(\Sigma)$, the pseudo-distance $d(w,v)$ is defined as follows:*

$$\overrightarrow{d}(w,v) \triangleq \max_{\substack{i\in\{1..n\}\\a\in A_i}} \min_{\substack{j\in\{1..m\}\\b\in B_j}} \{|t_i - s_j| : a = b\}; \quad d(w,v) \triangleq \max(\overrightarrow{d}(w,v), \overrightarrow{d}(v,w)).$$

Computing d is illustrated in Fig. 4. Intuitively, an observer who tries to distinguish timed words u and v will pair each timing of an event in u to the closest timing in v "emitting" the same event, and deduce that u and v are close when all these pairs of timings are close enough. We symmetrize this distance, similarly to [4]. Also, d may fail to distinguish timed words, that is $d(w_1, w_2) = 0$ but $w_1 \neq w_2$, which is the reason why d is only a pseudo-distance.

ε-capacity and ε-entropy. These two closely related notions from [19] characterize the quantity of information needed to describe (with precision ε) any element of a given space. We first recall a couple of definitions (adapted to pseudo-distances).

Definition 4. *Let (X, d) be a (pseudo-)metric space, a subspace of some space Y. A subset $M \subseteq X$ is called ε-separated whenever each $x, y \in M$ with $x \neq y$ satisfy $d(x,y) > \varepsilon$. A set $N \subseteq Y$ is an ε-net for X (in Y) whenever for all $x \in X$ there exists $y \in N$ with $d(x,y) \leq \varepsilon$.*

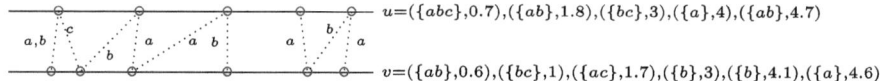

Fig. 4. Pseudo-distance between two timed words over 2^Σ, $\vec{d}(u,v) = 0.2$. Dotted edges represent the closest position for matching letter.

For a totally bounded space [17] X and any $\varepsilon > 0$, a finite ε-net must exist within X itself, and cardinalities of ε-separated sets are bounded, which justifies:

Definition 5 ([19]). *Given a totally bounded (pseudo-)metric space $X \subseteq Y$ let $\mathcal{M}_\varepsilon(X)$ be the maximal size of an ε-separated subset of X. Then the ε-capacity of X is defined as $\mathcal{C}_\varepsilon(X) = \log \mathcal{M}_\varepsilon(X)$. Let now $\mathcal{N}_\varepsilon^Y(X)$ be the minimal size of an ε-net for X (in Y). Then the ε-entropy of X (in Y) is $\mathcal{H}_\varepsilon^Y(X) = \log \mathcal{N}_\varepsilon^Y(X)$.*

As shown in [19], ε-entropy and ε-capacity are related by inequalities

$$\mathcal{H}_\varepsilon^Y(X) \leq \mathcal{C}_\varepsilon(X) \leq \mathcal{H}_{\varepsilon/2}^Y(X). \qquad (2)$$

Problem Statement. Let us introduce the central notion for this article:

Definition 6 ([16]). *The ε-entropic bandwidth and ε-capacitive bandwidth of a timed language L are defined respectively as $\mathcal{BH}_\varepsilon(L) = \limsup_{T\to\infty} \mathcal{H}_\varepsilon^U(L_T)/T$ and $\mathcal{BC}_\varepsilon(L) = \limsup_{T\to\infty} \mathcal{C}_\varepsilon(L_T)/T$.*

Here U stands for the universal timed language. Remark that (L_T, d) is totally bounded ([4]), hence $\mathcal{H}_\varepsilon^U(L_T)$ and $\mathcal{C}_\varepsilon(L_T)$ are well defined. A couple of simple properties should be mentioned: it follows from (2) that $\mathcal{BH}_\varepsilon(L) \leq \mathcal{BC}_\varepsilon(L) \leq \mathcal{BH}_{\varepsilon/2}(L)$; from [4, Thm. 2] follows the upper bound valid for any timed language L (over a fixed alphabet Σ): $\mathcal{BC}_\varepsilon(L) = O(1/\varepsilon)$.

We have shown in [5] that there are three classes of timed regular languages: meager with bandwidth $O(1)$, normal with bandwidth $\Theta(\log 1/\varepsilon)$, and obese with the maximal possible bandwidth $\Theta(1/\varepsilon)$. In the rest of the paper, we deal with a particular case of (1): the computation of the bandwidths of obese TA.
Main Problem. *Given an obese TA $\mathcal{A}$, compute a real number α such that $\mathcal{BC}_\varepsilon(L(\mathcal{A})) = (1 + o(1))\alpha/\varepsilon$ for small ε.*

We are equally interested in the asymptotic behavior of $\mathcal{BH}$.

4 Bandwidth Computation for Obese Automata

4.1 Trivially Timed Languages

A TA is called *trivially timed* if all its guards and starting conditions are **true**. Such automata yield the simplest examples of obese languages.

For a TA $\mathcal{A} = (Q, X, \Sigma, \Delta, S)$, we define its *support* as a finite automaton $\operatorname{supp}\mathcal{A} = (Q, \Sigma, \Delta')$, where for each edge $(q, q', a, \mathfrak{g}, \mathfrak{r}) \in \Delta$ we put (q, q', a) in Δ'. For any TA $\mathcal{A}$, we call the *trivially timed version* of it, the automaton $\widetilde{\mathcal{A}}$ obtained by replacing all guards and starting conditions by **true**.

The following notation will simplify our statements on asymptotic behaviors up to a multiplicative or additive constant, where $\varepsilon \to 0$ and uniformity w.r.t. all other parameters is required.

Definition 7 (Asymptotic notation). *Given two positive-valued functions $f(\varepsilon, x)$ and $g(\varepsilon, x)$ (with any kind of arguments x), we write $f \lesssim g$ whenever*

$$\forall \varkappa > 0 \; \exists c, \varepsilon_0 > 0, \forall \varepsilon < \varepsilon_0, \forall x : f(\varepsilon, x) < c g^{1+\varkappa}(\varepsilon, x).$$

For real-valued $f(\varepsilon, x)$ and $g(\varepsilon, x)$, we write $f \lessapprox g$ whenever

$$\forall \varkappa > 0 \; \exists c, \varepsilon_0 > 0, \forall \varepsilon < \varepsilon_0, \forall x : f(\varepsilon, x) < c + (1 + \varkappa) g(\varepsilon, x).$$

We write $f \approx g$ if $f \lesssim g$ and $g \lesssim f$, and $f \approxeq g$ iff $f \lessapprox g$ and $g \lessapprox f$.

We can now characterize the entropy and capacity of trivially timed languages, which will play a key role for the general case.

Proposition 2. *Let $\mathcal{A}$ be trivially timed with $\operatorname{supp} \mathcal{A}$ strongly connected. Denote $\alpha = \mathcal{GS}(\operatorname{supp} \mathcal{A})$. Then entropy and capacity of $L_t(\mathcal{A})$ satisfy $\mathcal{H}^U_{\varepsilon/2}(L_t) \approxeq \mathcal{C}_\varepsilon(L_t) \approxeq \alpha t / \varepsilon$ for $t > 0$. Also, α is independent of the choice of initial or final locations.*

As a consequence, L is obese, the bandwidth satisfies $\mathcal{BH}_{\varepsilon/2}(L) \approxeq \mathcal{BC}_\varepsilon(L) \approxeq \alpha / \varepsilon$.

4.2 Bandwidth-Preserving Preprocessing of Timed Automata

Given a TA (possibly obese), we will first proceed with a couple of bandwidth-preserving transformations, as well as the classification of its transitions into red ones and black ones. Informally, high bandwidth is produced within red SCCs ("spots"), and black transitions are used to connect them and to loop.

Adding Heartbeat and Urgency Clock.

Construction 2. *Given a TA, we add two features:*

Heartbeat: a new clock h and a new letter b; next, we add to each transition the constraint $h \leq 1$, and to each location p a self-loop $p \xrightarrow{b, h-1, \{h\}} p$.

Urgency clock: a new clock u, reset at every transition, never tested in guards.

A heartbeat is emitted every time unit and contains high-frequency runs within 1 time unit intervals. At the next stage of the algorithm, the urgency clock will help discriminating urgent transitions. The two features do not change the bandwidth; the latter even preserves the language.

Region-Splitting and Bounding. This step is a language-preserving transformation of a TA into a form similar to the region automaton [1], but typed as a TA. We give a nondeterministic version of the definition from [3,5,6].

Definition 8. *A region-split TA (or* RsTA*) is a TA* $(Q, X, \Sigma, \Delta, S, I, F)$, *which is bounded and such that, for any location* $q \in Q$: *(1)* $S(q)$ *defines a non-empty region, called the* starting region *of* q; *(2) all states in* $\{q\} \times S(q)$ *are reachable from an initial state and co-reachable to a final one; (3) for any edge* $(q, q', a, \mathfrak{g}, \mathfrak{r}) \in \Delta$, $(\{S(q) + d \mid d \in \mathbb{R}_+\} \cap \mathfrak{g})[\mathfrak{r}] = S(q')$, *where we utilize the* $(\cdot)[\mathfrak{r}]$ *operator lifted to sets of clock valuations.*

Any TA having an upper bound on some clock at every transition (a heartbeat h in our case) can be brought into a bounded region-split form, with the same clocks but an exponentially larger set of locations.

When a TA with urgency clock u is put into a region-split form, all 0-time transitions (leading to a region with $u = 0$) are separated from those taking a positive time and leading to a region with $u > 0$, which makes the next transformation possible.

Eliminating Zeros. The next transformation removes all urgent transitions.

Definition 9. *In a TA (with urgency clock* u), *a transition* $\delta = (p, a, \mathfrak{g}, \mathfrak{r}, q)$ *is* urgent *whenever* $\mathfrak{g} \implies (u = 0)$. *An* RsTA *is called* 0-free *when none of its transitions is urgent.*

Given an RsTA $\mathcal{A}$ (with urgency clock) over an alphabet Σ, it is possible to construct a 0-free RsTA $\nu \mathcal{A}$ over $\mathcal{P}_+(\Sigma)$ with the same bandwidth.

Construction 3 (0-elimination, automata). *For each path* $q_0 \xrightarrow{\delta_0} q_1 \cdots \xrightarrow{\delta_k} q_{k+1}$ *with non-urgent* δ_0 *and urgent* $\delta_1..\delta_k$ *such that* $\delta_i = (q_i, q_{i+1}, a_i, \mathfrak{g}_i, \mathfrak{r}_i)$, *we add a compound transition* $\delta' = \left(q_0, q_{k+1}, \{a_0, \ldots, a_n\}, \mathfrak{g}_0, \bigcup_{i=0}^{k} \mathfrak{r}_i\right)$. *Then, we remove the urgent transitions. We also make initial all the states reachable from an initial one by urgent transitions only.*

The Red and the Black. After preprocessing, we now have a 0-free RsTA with the heartbeat and urgency clock, over alphabet $\mathcal{P}_+(\Sigma)$, and we will call it *standard-form timed automaton* (SFA). Our next aim is to identify its parts producing unbounded frequency information (red) and others (black).

Definition 10. *A cycle in an SFA is* fast *whenever it can be traversed twice in less than 1 time unit. A transition is* red *whenever it belongs to a fast cycle. It is* black *otherwise.*

Red transitions are illustrated by Fig. 5, left.

Only fast cycles may produce symbols with unbounded frequency, and it is immediate from [5, Thm. 4] that any obese SFA has a fast cycle. Given an automaton with black and red transitions, we consider the graph of red transitions and refer to strongly connected components therein as *red SCCs*.

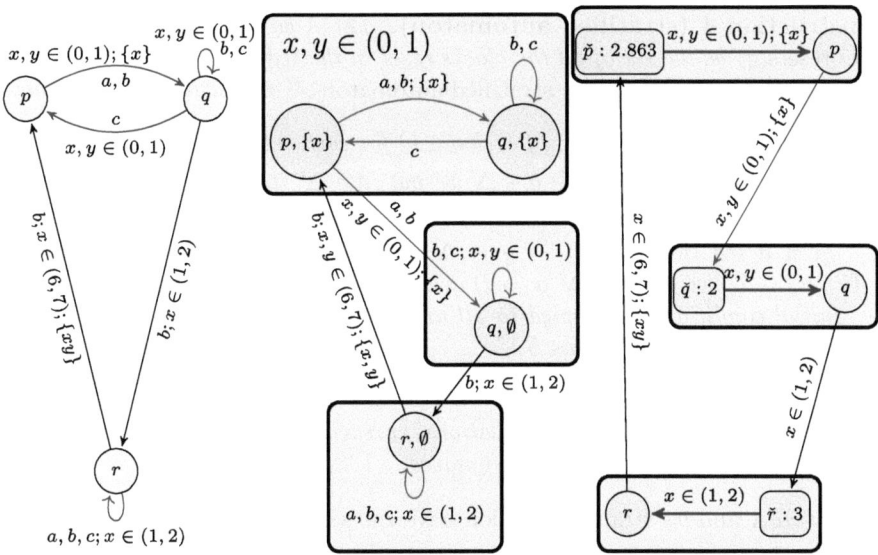

Fig. 5. Example of bandwidth computation. Left: a timed automaton. Middle: its stratified version, pink rectangles correspond to spots, sink states omitted. Right: its abstraction, blue arrows are abstract transitions. (Color figure online)

Splitting the red component. In the next step, we build a *stratified* automaton, in which we materialize different *strategies* for producing bandwidth when staying inside a given red SCC. The example on Fig. 5, left, illustrates strategies. One strategy for the red SCC $\{p, q\}$ consists of wandering with high frequency over all its red transitions (as we will see, this yields an ε-capacity of $\approx 2.863 t/\varepsilon$ in time $t \leq 1$). At the end of time t, the clock x (recently reset) has a value close to 0, and it takes 6 more time units to travel back to p with the clock y reset. Another strategy consists of staying in q while emitting with a high frequency events b and c. This yields a smaller capacity (only $2t/\varepsilon$), but the advantage is that x is not reset, and at the end equals t. Thus, the travel back to p takes $6 - t$ time units, less than in the first strategy. The bandwidth computation should thus consider and compare all possible strategies in all red SCCs (and switching between them).

In the following construction of the stratified automaton, we represent such strategies as subsets Z of the clocks (to be reset, as explained below). For each location q and set of clocks $Z \subseteq X$, we make a location (q, Z) referred to as Z-avatar of q. Every Z-avatar is committed to resetting all the clocks in Z and no others before taking a black transition. Switching from a Z-avatar to a $Z \setminus \{x\}$-avatar (and switching strategies) is possible when taking a transition that resets x. Finally, we leave the red SCC by taking a black transition from some $\emptyset$-avatar.

Construction 4 (stratified automaton). *Let $\mathcal{A}$ be an SFA. For a location q, let $\mathrm{resets}(q)$ be the set of all the clocks reset in the red SCC containing q (or $\emptyset$ if q is not in a red SCC). The stratified automaton $\mathcal{A}'$ is constructed as follows:*

- *Q' contains avatars (q, Z) of each $q \in Q$ for each set of clocks $Z \subseteq \mathrm{resets}(q)$;*
- *for a black transition $p \xrightarrow{\delta} q \in \Delta$ we put into Δ' its avatars $(p, \emptyset) \xrightarrow{\delta} (q, Z)$ for all Z;*
- *given a red transition $p \xrightarrow{\delta} q \in \Delta$, for all sets of clocks Z, U satisfying $U \subseteq \mathfrak{r}_\delta \subseteq Z$, we put into Δ' a (red) avatar of δ, going from (p, Z) to $(q, Z \setminus U)$;*
- *initial conditions are copied to all avatars: $I'(q, Z) = I(q)$ for all Z; similarly for starting conditions S;*
- *final conditions are copied in a more restricted way: $F'(q, \emptyset) = F(q)$.*

Finally, the automaton is trim: locations unreachable from an initial one and not co-reachable to a final one are removed.

Any SFA and its stratified version have the same languages, see the ArXiV version.

Speedy Timed Automata. We remark now that while the black subgraph of both our automata (SFA and its stratified version) can have any form, the red one is quite specific. To describe its components, we introduce a new class of TA, *speedy* ones. Such TA exhibit "obese-like" behavior during a time interval.

Definition 11. *Given $Z \subsetneq X$, we say that a strongly connected RsTA $\mathcal{D}$ with at least one transition is Z-speedy (or just speedy) whenever clocks not in Z are never reset; and all the guards have the form (for some constants d_y)*

$$\mathfrak{g}_\mathcal{D} = \bigwedge_{z \in Z}(0 < z < 1) \wedge \bigwedge_{y \notin Z}(d_y < y < d_y + 1).$$

The set of all clocks reset by transitions in $\mathcal{D}$ is denoted $\mathfrak{r}_\mathcal{D} \subseteq Z$.

Let us characterize the size of languages of speedy automata.

Lemma 2. *Let $\mathcal{D}$ be a speedy RsTA, with $\alpha = \mathcal{GS}(\mathrm{supp}\,\mathcal{D})$. Let L be the timed language of $\mathcal{D}$ with some nonempty sets of initial and final locations. Then for $t \in (0, 1]$ the language L_t has entropy and capacity $\mathcal{H}_{\varepsilon/2}(L_t) \approx \mathcal{C}_\varepsilon(L_t) \approx \frac{\alpha t}{\varepsilon}$.*

The proof is based on a comparison of $\mathcal{D}$ with its trivially timed version $\widetilde{\mathcal{D}}$. The following lemma justifies speedy automata:

Lemma 3. *In an SFA, any red SCC is Z-speedy (with Z the set of clocks being reset within it). In a stratified automaton, in every red SCC, all locations are avatars (q, Z) with a same Z, and such SCC is Z-speedy.*

Definition 12. *Red SCCs in a stratified automaton are called spots, the set thereof is denoted $\mathbb{D}$. The growth rate of a spot $\mathcal{D}$ is defined as $\alpha_\mathcal{D} = \mathcal{GS}(\mathrm{supp}\,\mathcal{D})$.*

4.3 Abstraction of a Stratified Automaton

We transform now the stratified automaton $\mathcal{A}$ into an "abstract" WTG $\widehat{\mathcal{A}}$ by processing (abstracting) every spot $\mathcal{D}$. All red transitions within $\mathcal{D}$ are removed and runs within it are replaced by first waiting in a state of weight α_D and next taking a special abstract transition. All other states weigh 0. More precisely:

Construction 5. *The stratified automaton* $\mathcal{A} = (Q, X, \mathcal{P}_+(\Sigma), \Delta, S, I, F)$, *is transformed into the* abstract *WTG* $\widehat{\mathcal{A}} = (\widehat{Q}, X, \widehat{\Delta}, \widehat{S}, w)$ *with*

- $\widehat{Q} = Q \cup \widecheck{Q}$ *with* $\widecheck{Q} = \{\widecheck{q} : q \in \cup \mathbb{D}\}$, *i.e., we add a copy* $\widecheck{q}$ *("abstract location") for each q within a spot (to represent entering and evolving in this spot)*
- $\widehat{\Delta}$ *is constructed as follows:*
  - all the transitions in Δ except red ones within the same spot are preserved (labels are removed);
  - for a red transition $p \xrightarrow{a,\mathfrak{g},\mathfrak{r}} q$ with $p \notin \mathcal{D}$ and $q \in \mathcal{D}$ for some spot $\mathcal{D}$, and for a black transition $p \xrightarrow{a,\mathfrak{g},\mathfrak{r}} q$ with $q \in \mathcal{D}$, a copy $p \xrightarrow{\mathfrak{g},\mathfrak{r}} \widecheck{q}$ is created;
  - for each spot $\mathcal{D}$ and locations $p, q \in \mathcal{D}$, a new abstract transition $\widecheck{p} \xrightarrow{\mathfrak{g}_\mathcal{D}, \mathfrak{r}_\mathcal{D}} q$ is introduced, with $\mathfrak{g}_\mathcal{D}$ and $\mathfrak{r}_\mathcal{D}$ as in Def. 11.
- For $q \in Q$, *the starting conditions are* $\widehat{S}(q) = \widehat{S}(\widecheck{q}) = S(q)$.
- For all locations p in a spot $\mathcal{D}$ we put $w(\widecheck{p}) = \alpha_D$, *which is the growth rate of the spot. All other weights in* $\widehat{\mathcal{A}}$ *are 0.*

An important property of the abstract WTG $\widehat{\mathcal{A}}$ is its time-divergence, and thanks to it, Lem. 1 can be applied to compute the optimal reward-to time ratio. We will show that this ratio equals the bandwidth coefficient that we aim to compute.

4.4 Main Result and an Example of Computation

Let $\alpha \in \mathbb{L}$ be the maximal reward-to-time ratio for $\widehat{\mathcal{A}}$ (computable from $\mathcal{A}$).

Theorem 2. *Either* $\alpha = 0$ *and $\mathcal{A}$ is not obese, or* $\mathcal{BC}_\varepsilon(\mathcal{A}) \approx \mathcal{BH}_{\varepsilon/2}(\mathcal{A}) \approx \alpha/\varepsilon$.

Proof (ideas). To prove that $\mathcal{BH}_{\varepsilon/2}(\mathcal{A}) \approx \alpha/\varepsilon$ we consider a small $\varepsilon/2$-net Net for all runs in the WTG $\widehat{\mathcal{A}}$ of duration $\leq T$ and build an $\varepsilon/2$-net from it for $L_T(\mathcal{A})$ by replacing, in each run, each abstract transition by an $\varepsilon/2$-net for the corresponding speedy language (with growth rate at most α/ε).

Towards the opposite inequality, we take in the graph $\widehat{\mathcal{A}}$ a run $(\delta_1, d_1) \ldots (\delta_n, d_n)$ of duration T with almost optimal reward-to-time ratio. We replace each abstract transition δ_j by the ε-separated set in the corresponding speedy language of duration d_j (its size is $\gtrsim 2^{\alpha_j d_j/\varepsilon}$ due to Lem. 2). □

An **example of bandwidth computation** given in Fig. 5 is slightly simplified (with respect to the algorithm) for the sake of clarity. Pink rectangles in the stratified automaton correspond to spots (to be abstracted at the next step). Growth rate of the upper spot is $\log((7+\sqrt{57})/2) \approx 2.863$ (as shown in

Sect. 2), it is 2 and 3 for the two others. In the abstract WTG, all the red arrows within each spot are replaced by an abstract blue transition from the input to the output states. Weights of input states correspond to the growth rates of their respective spots. The optimal cycle in the WTG has duration 7 and spends one time unit in both $\check{p}$ and $\check{r}$, with the ratio $\approx (3 + 2.863)/7 \approx 0.838$. This is better than the slightly faster cycle of duration 6 spending one time unit in both $\check{q}$ and $\check{r}$, with the ratio $(2+3)/6 \approx 0.833$. Thus the bandwidth is $\approx 0.838/\varepsilon$.

5 Conclusions

We have shown that the bandwidth of an obese TA can be computed as the logarithm of an algebraic number, similarly to finite automata. The algorithm for computing it requires a number of preprocessing steps: splitting into regions, dealing with 0-transitions (which required introducing a generalization of the pseudo-distance in [4]), separating "high-frequency" components (called here *spots*) from "normal-frequency" components, computing the bandwidth of spots, transforming the timed automaton into a WTG where weights are the bandwidths of the various spots, and finally applying results from [8] for computing the maximal reward-per-time ratio of the resulting WTG.

As for complexity, we remark that by reduction from reachability, similar to that of [5, Thm. 6], deciding whether $\alpha = 1$ or 2 (even knowing that one of those holds), is PSPACE-hard. We leave a more precise complexity analysis as a future work. We also plan to address the problem of computing the bandwidth of the last remaining class of timed automata, normal ones, by characterizing the coefficient α such that $\mathcal{BC}_\varepsilon(L(\mathcal{A})) \approxeq \alpha \cdot \log(1/\varepsilon)$ as the best reward-to-time ratio of some finite abstraction of the timed automaton. We intend to apply these results to the approximate coding and compression of timed data.

References

1. Alur, R., Dill, D.L.: A theory of timed automata. Theoret. Comput. Sci. **126**, 183–235 (1994). https://doi.org/10.1016/0304-3975(94)90010-8
2. Asarin, E., Basset, N., Béal, M.-P., Degorre, A., Perrin, D.: Toward a Timed Theory of Channel Coding. In: Jurdziński, M., Ničković, D. (eds.) FORMATS 2012. LNCS, vol. 7595, pp. 27–42. Springer, Heidelberg (2012). https://doi.org/10.1007/978-3-642-33365-1_4
3. Asarin, E., Basset, N., Degorre, A.: Entropy of regular timed languages. Inf. Comput. **241**, 142–176 (2015). https://doi.org/10.1016/j.ic.2015.03.003
4. Asarin, E., Basset, N., Degorre, A.: Distance on timed words and applications. In: Jansen, D.N., Prabhakar, P. (eds.) FORMATS 2018. LNCS, vol. 11022, pp. 199–214. Springer, Cham (2018). https://doi.org/10.1007/978-3-030-00151-3_12
5. Asarin, E., Degorre, A., Dima, C., Jacobo Inclán, B.: Bandwidth of timed automata: 3 classes. In: Proc. FSTTCS. LIPIcs, vol. 284, pp. 10:1–10:17 (2023). https://doi.org/10.4230/LIPICS.FSTTCS.2023.10, full version https://arxiv.org/abs/2310.01941

6. Asarin, E., Degorre, A., Dima, C., Jacobo Inclán, B.: Computing the bandwidth of meager timed automata. In: Proceedings of CIAA. Lecture Notes in Computer Science, vol. 15015, pp. 19–34. Springer (2024). https://doi.org/10.1007/978-3-031-71112-1_2, full version https://arxiv.org/abs/2406.12694
7. Berstel, J., Perrin, D., Reutenauer, C.: Codes and Automata. Cambridge University Press (2009). https://doi.org/10.1017/CBO9781139195768
8. Bouyer, P., Brinksma, E., Larsen, K.G.: Optimal infinite scheduling for multi-priced timed automata. Formal Methods Syst. Des. **32**(1), 3–23 (2008). https://doi.org/10.1007/S10703-007-0043-4
9. Bridson, M.R., Gilman, R.H.: Context-free languages of sub-exponential growth. J. Comput. Syst. Sci. **64**(2), 308–310 (2002). https://doi.org/10.1006/jcss.2001.1804
10. Chomsky, N., Miller, G.A.: Finite state languages. Inf. Control **1**(2), 91–112 (1958). https://doi.org/10.1016/S0019-9958(58)90082-2
11. Dasdan, A., Irani, S.S., Gupta, R.K.: Efficient algorithms for optimum cycle mean and optimum cost to time ratio problems. In: Proceedings of the DAC, pp. 37–42. ACM (1999). https://doi.org/10.1145/309847.309862
12. Grigorchuk, R.: Degrees of growth of finitely generated groups, and the theory of invariant means. Mathematics of the USSR — Izvestiya **25**(2), 259 (1985). https://doi.org/10.1070/IM1985v025n02ABEH001281
13. Immink, K.: EFMPlus: the coding format of the multimedia compact disc. IEEE Trans. Consum. Electron. **41**(3), 491–497 (1995). https://doi.org/10.1109/30.468040
14. Immink, K.: Codes for Mass Data Storage Systems. Shannon Foundation Publ. (2004)
15. Incitti, R.: The growth function of context-free languages. Theoret. Comput. Sci. **255**(1–2), 601–605 (2001). https://doi.org/10.1016/S0304-3975(00)00152-3
16. Jacobo Inclán, B., Degorre, A., Asarin, E.: Bounded delay timed channel coding. In: Proceedings of the FORMATS 2022. LNCS, vol. 13465, pp. 65–79. Springer (2022). https://doi.org/10.1007/978-3-031-15839-1_4
17. Kelley, J.L.: General Topology. Dover Publications (2017)
18. Kolmogorov, A.N.: Three approaches to the quantitative definition of information. Int. J. Comput. Math. **2**(1-4), 157–168 (1968). https://doi.org/10.1080/00207166808803030
19. Kolmogorov, A.N., Tikhomirov, V.M.: ε-entropy and ε-capacity of sets in function spaces. Uspekhi Matematicheskikh Nauk **14**(2), 3–86 (1959). https://doi.org/10.1090/trans2/017
20. Lind, D., Marcus, B.: An Introduction to Symbolic Dynamics and Coding. Cambridge University Press (1995). https://doi.org/10.1017/CBO9780511626302
21. Shannon, C.E.: A mathematical theory of communication. Bell Syst. Tech. J. **27**(3), 379–423 (1948). https://doi.org/10.1002/j.1538-7305.1948.tb01338.x
22. Szilard, A., Yu, S., Zhang, K., Shallit, J.: Characterizing regular languages with polynomial densities. In: Havel, I.M., Koubek, V. (eds.) MFCS 1992. LNCS, vol. 629, pp. 494–503. Springer, Heidelberg (1992). https://doi.org/10.1007/3-540-55808-X_48

Box-Reachability in Vector Addition Systems

Shaull Almagor[1](✉)[iD], Itay Hasson[1], Michał Pilipczuk[2][iD],
and Michael Zaslavski[3]

[1] Department of Computer Science, Technion, Israel
itay.h@campus.technion.ac.il, michal.pilipczuk@mimuw.edu.pl
[2] University of Warsaw, Warsaw, Poland
[3] Haifa, Israel

Abstract. We consider a variant of reachability in Vector Addition Systems (VAS) dubbed *box reachability*, whereby a vector $v \in \mathbb{N}^d$ is box-reachable from $\mathbf{0}$ in a VAS $\mathcal{V}$ if $\mathcal{V}$ admits a path from $\mathbf{0}$ to v that not only stays in the positive orthant (as in the standard VAS semantics), but also stays below v, i.e., within the "box" whose opposite corners are $\mathbf{0}$ and v.

Our main result is that for two-dimensional VAS, the set of box-reachable vertices almost coincides with the standard reachability set: the two sets coincide for all vectors whose coordinates are both above some threshold W. We also study properties of box-reachability, exploring the differences and similarities with standard reachability.

Technically, our main result is proved using powerful machinery from convex geometry.

1 Introduction

Vector Addition Systems (VAS) are a well-established formalism for modelling and reasoning about concurrent systems, hardware and software analysis, biological and chemical processes, and many more. In particular, VAS and Petri-Nets are equivalent (in a sense), with the latter being a long-studied model [17].

Formally, a VAS is given by a finite set of vectors $\mathcal{V} \subseteq \mathbb{Z}^d$. A *trace* in $\mathcal{V}$ is then a sequence of vectors obtained by summing vectors from $\mathcal{V}$, provided that

---

M. Zaslavski—Independent author.
The full version can be found on https://arxiv.org/abs/2508.12853.
S. Almagor is supported by the ISRAEL SCIENCE FOUNDATION (grant No. 989/22). The work of Mi. Pilipczuk on this manuscript is a part of project BOBR that has received funding from the European Research Council (ERC) under the European Union's Horizon 2020 research and innovation programme (grant agreement no. 948057).

© The Author(s), under exclusive license to Springer Nature Switzerland AG 2026
P. Ganty and A. Mansutti (Eds.): RP 2025, LNCS 16230, pp. 126–139, 2026.
https://doi.org/10.1007/978-3-032-09524-4_9

all coordinates remain non-negative. The latter requirement is dubbed the *VAS Semantics*. A vector v is *reachable* if there is a trace that starts[1] at $\mathbf{0}$ and ends in v.

The fundamental problem for VAS is *reachability*: given a vector v, decide whether it is reachable in $\mathcal{V}$. This problem has a rich history, and its exact complexity has been only recently settled as ACKERMANN-Complete [6,13], with better bounds known for dimensions 1 and 2. A more involved problem is characterizing the set of reachable vectors. In general, there is no simple characterization (which is understandable, given the high complexity of reachability). However, in dimension up to five this set is *semilinear* [10].

From a modelling perspective, the coordinates of the vectors in a VAS typically correspond to some resources (e.g., memory locks, battery level, queue size, molecule concentration, etc.). Then, reachability amounts to the question of whether a certain configuration of the resources can be reached. In the standard reachability question, no attention is paid to *how* the target configuration is reached. In particular, it may be the case that in order to reach a certain target configuration, the trace must go through configurations with a higher number of certain resources than the target. Such traces may be less useful to the designer, or might not fit with their intent. For example, consider a chemical solution of three chemical compounds A, B, C, whose respective quantities are modelled as (x_A, x_B, x_C). Suppose we start at $(1,2,2)$ and wish to reach $(10,10,10)$. It may be the case that due to reactions between the materials, one first needs to flood the solution with compound A (going above 10), and only then the reactions can lead down to $(10,10,10)$. In case A is expensive, dangerous or volatile, this trace might not be useful for actual reachability. Similar settings can occur with any types of resources, where one does not want to go over the amount of the resources in the target before reaching it.

To capture this notion, we introduce *box-reachability*: a vector $x = (x_1, \ldots, x_d) \in \mathbb{N}^d$ is *box-reachable* in a VAS $\mathcal{V}$ if there is a trace from $\mathbf{0}$ to x such that throughout the entire trace, each coordinate i remains at most x_i.

Example 1. Consider a 2-VAS $\mathcal{V} = \{(-1,2), (2,-1), (10,10)\}$. Let us examine the configurations $t_1 = (11, 11)$ and $t_2 = (21, 21)$.

Observe that $(-1,2) + (2,-1) = (1,1)$, and the path $(-1,2), (2,-1)$ can be taken from every vector $(x,y) \geq (1,0)$. Thus, both t_1 and t_2 are reachable in $\mathcal{V}$, via the paths $\pi_1 : \mathbf{0} \xrightarrow{(10,10)(-1,2)(2,-1)}_{\mathbb{N}} (11,11)$ and $\pi_2 : \mathbf{0} \xrightarrow{(10,10)((-1,2)(2,-1))^{11}}_{\mathbb{N}} (21,21)$, respectively. However, both π_1 and π_2 exceed their respective boxes. Indeed, the penultimate vector reached by π_1 is $(9, 12)$ (exceeding $(11,11)$), and by π_2 is $(19, 22)$ (exceeding $(21,21)$).

It is not hard to see that $t_1 \notin \mathsf{BoxReach}(\mathcal{V})$. Indeed, the first vector that must be taken in any path in $\mathcal{V}$ is $(10, 10)$, and from there any added vector in $\mathcal{V}$ exceeds the $(11, 11)$ box.

---

[1] Often a starting vector is also given, but the problem has the same "flavour" also when starting from $\mathbf{0}$.

However, $t_2 \in$ BoxReach($\mathcal{V}$), via the run $\pi_3 : \mathbf{0} \xrightarrow{(10,10)(-1,2)(2,-1)(10,10)}_\mathbb{N}$ $(21,21)$. Moreover, a similar path can be taken to box-reach any vector (k,k) with $k \geq 20$, via the path $\mathbf{0} \xrightarrow{(10,10)((-1,2)(2,-1))^{k-20}(10,10)}_\mathbb{N} (k,k)$.

This example illustrates two central characteristics of the box-reachable set. First, we notice that the classical reachability set does not coincide with box-reachability. Indeed, as we show above, $(11,11)$ is reachable but not box-reachable. An even more obvious example of this is that $(30,0)$ is reachable in $\mathcal{V}$ (via the path $(10,10)(2,-1)^{10}$), but is not box-reachable, since all paths start in $(10,10)$ and in particular go above 0 in the y-coordinate.

The second characteristic is that vectors that are "deep" enough in the positive quadrant, e.g., $(21,21)$, allow us enough wiggle room along the path to apply all the "unsafe" vectors that exceed their box (e.g., $(-1,2),(2,-1)$) without leaving the overall box. This suggests that perhaps the box-reachability set and the classical reachability set do coincide for vectors with large-enough entries. Indeed, this is the main contribution of this paper, for dimension 2. □

Contribution. We focus on dimension 2 (see Example 2 and the full version for context on this). In Theorem 1 we characterize the box-reachable set as follows: for every 2-VAS there exists a (polynomial size) $W \in \mathbb{N}$ such that the reachability set and the box-reachability sets coincide on $[W, \infty)^2$.

Technically, the proof of Theorem 1 uses insights on the geometry of the reachable set in the context of the cone it spans. We rely on nontrivial machinery from convex geometry (specifically, Steinitz's lemma [8] and the Deep-in-the-cone Lemma [5]; both are introduced in Sect. 4.1).

Focusing on 2-VAS may seem limited. To justify this, we show that our result no longer holds neither for 3-VAS nor for 1-VASS (i.e., Vector Addition Systems with States).

In addition, we study the shape of the box-reachable set, showing that it is semilinear for 2-VAS as well as 1-VASS, and drawing a connection between box reachability in d-VAS and standard reachability in $2d$-VAS.

In a broader context, our work can be viewed as a form of *boundedness* constraint on the VAS, in the sense that box-reaching runs are bounded by their target. Unlike boundedness by a constant, this still allows counters to become arbitrarily large.

In terms of understanding the geometry of VAS, our contribution stands in an interesting contrast to standard methods of reasoning about VAS: the most common approach in reasoning about low-dimensional VAS (and VASS) is that of Linear Path Schemes, which intuitively allow to iterate a small number of cycles many times. However, LPS are "highly non-box safe", in the sense that LPS typically reach their target by taking one cycle many times, making some counter very large, and then correcting this by iterating another cycle, and so on. Thus, LPS guide the path far outside the box and then back. In our work, we need to exactly avoid this type of behaviors.

Related Work. Reachability in VAS and VASS have received much attention in recent years. We refer the reader to [17] for a survey, and to [6,13] for more recent works and references therein. More closely related to this work are studies concerning boundedness and geometrical understanding of the reachability set. Boundedness has been studied in early works on VAS [11,15], as it is essential to the study of coverability. Later studies [1,7,12,16] refine the notion of boundedness or look at variants of the model, as well as obtain improved bounds for coverability.

On the geometry front, several works have explicitly tried to provide a better understanding of the reachability sets of VASS, e.g., [2,4] where the geometry of the continuous variant of VASS is considered, and [9] which explicitly aims to characterize geometric properties of the reachability set.

Paper Organization. In Sect. 3 we present the notion of box reachability, and show why we focus on 2-VAS. In Sect. 4 we prove that reachability and box-reachability eventually coincide. This is split into several parts: Sect. 4.1 presents the geometric tools we use, and Sect. 4.2 prove the theorem by addressing two different cases, with different techniques. In Sect. 5 we show the semilinearity of box-reachability for 2-VAS and for 1-VASS. We conclude with a discussion in Sect. 6.

Due to space constraints, most proofs appear in the full version.

2 Preliminaries

Let $\mathbb{N} = \{0, 1, 2, \dots\}$ denote the naturals and $\mathbb{Z}$ the integers. For $i \leq j \in \mathbb{Z}$ we denote by $[i, j] = \{i, i+1, \dots, j\}$ the interval between i and j. We extend the notation to $[i, \infty] = \{m \in \mathbb{Z} \mid m \geq i\}$. We denote vectors in $\mathbb{Z}^d$ in bold (e.g., $\boldsymbol{u}$), and refer to their components by subscripts: $\boldsymbol{u} = (\boldsymbol{u}_1, \dots, \boldsymbol{u}_d)$. We denote $\|\boldsymbol{u}\| = \max_{i \in [1, d]} \{|\boldsymbol{u}_i|\}$ the infinity-norm of $\boldsymbol{u}$.

For two vectors $\boldsymbol{u}, \boldsymbol{v} \in \mathbb{Z}^d$ we write $\boldsymbol{u} \leq \boldsymbol{v}$ if $\boldsymbol{u}_i \leq \boldsymbol{v}_i$ for all $i \in [1, d]$. For a set S we denote by S^* (resp. S^+) the set of finite sequences (resp. non-empty finite sequences) of elements of S. For a vector $\boldsymbol{v} \in \mathbb{R}^2$, we typically refer to its coordinates as $\boldsymbol{v} = (\boldsymbol{v}_x, \boldsymbol{v}_y)$.

A *vector addition system of dimension d* (d-VAS, for short) is a finite set of vectors $\mathcal{V} \subseteq \mathbb{Z}^d$. We denote by $\|\mathcal{V}\| = d \cdot \sum_{\boldsymbol{v} \in \mathcal{V}} \|\boldsymbol{v}\|$. A *path* of $\mathcal{V}$ is a finite sequence of vectors $\pi = \boldsymbol{v_1}, \boldsymbol{v_2}, \dots, \boldsymbol{v_n} \in \mathcal{V}^*$ (we sometimes omit the commas for brevity). The *length* of π is $|\pi| = n$. For $i, j \leq n$ we denote the infix $\pi[i, j] = \boldsymbol{v}_i, \dots, \boldsymbol{v}_j$ (if $j < i$ the infix is empty). We also denote $\pi[j, \dots] = \pi[j, |\pi|]$ and $\pi[j] = \pi[j, j]$.

We introduce some notation for various properties of π: the *effect* of π is $\text{eff}(\pi) = \sum_{i=1}^{n} \boldsymbol{v}_i$. Fix a coordinate $1 \leq k \leq d$, we denote by $\text{eff}_k(\pi) = \text{eff}(\pi)_k$ the effect in coordinate k. The *drop* and *peak* of π in coordinate k are $\text{drop}_k(\pi) = |\min_{j \in [0, n]} \text{eff}_k(\pi[1, j])|$ and $\text{peak}_k(\pi) = \max_{j \in [0, n]} \text{eff}_k(\pi[1, j])$. Observe that $\text{drop}_k(\pi) \geq 0$ and $\text{peak}_k(\pi) \geq 0$, since for $j = 0$ we have that $\pi[1, 0]$ is empty, and thus has effect $\boldsymbol{0}$.

A starting vector s and a path π as above induce the *trace* $s_0, s_1, \ldots, s_n$ where $s_0 = s$ and for every $0 < i \le n$ we have $s_i = s_{i-1} + v_i$. We then write $s \xrightarrow{\pi} s_n$. If $s_i \in \mathbb{N}^d$ for all $i \in [0, n]$, we write $s \xrightarrow{\pi}_\mathbb{N} s_n$, and if we wish to emphasize that some coordinates may be negative, we write $s \xrightarrow{\pi}_\mathbb{Z} s_n$.

A vector t is *reachable* in $\mathcal{V}$ if there exists a path π such that $\mathbf{0} \xrightarrow{\pi}_\mathbb{N} t$. We say that t is $\mathbb{Z}$-*reachable* if $\mathbf{0} \xrightarrow{\pi} t$, but not necessarily via a non-negative path. The *reachability set of* $\mathcal{V}$ is then $\mathsf{Reach}(\mathcal{V}) = \{t \in \mathbb{N}^d \mid \exists \pi \in \mathcal{V}^* \text{ s.t. } \mathbf{0} \xrightarrow{\pi}_\mathbb{N} t\}$.

The fundamental problem regarding VAS is the *reachability problem*: given a VAS $\mathcal{V}$ and vector t, decide whether $t \in \mathsf{Reach}(\mathcal{V})$. The complexity bounds on this problem were recently tightened to ACKERMANN-Complete [6,13]. For general dimensions d, the reachability set can be complicated. For dimension up to 5, however, this set is always effectively *semilinear* [10]. We remark that this is also known for 2-VASS [3,14].

3 Box Reachability

We start by introducing our main object of study, namely box-reachable vectors. Consider a d-VAS $\mathcal{V}$ and a vector $t \in \mathbb{N}^d$. We say that t is *box-reachable in* $\mathcal{V}$, denoted $\mathbf{0} \xrightarrow{\pi}_\square t$ if there is a path $\pi \in \mathcal{V}^*$ such that $\mathbf{0} \xrightarrow{\pi}_\mathbb{N} t$ and in addition, for every $1 \le i \le |\pi|$ it holds that $\mathsf{eff}(\pi[1, i]) \le t$. That is, the trace induced by π from $\mathbf{0}$ remains within the "box" whose opposite corners are $\mathbf{0}$ and t. The *box-reachability set* of $\mathcal{V}$ is then $\mathsf{BoxReach}(\mathcal{V}) = \{v \in \mathbb{N}^d \mid v \text{ is box-reachable in } \mathcal{V}\}$.

In contrast with the reachability problem for VAS, the corresponding *box-reachability problem*, namely deciding whether a vector t is box-reachable in $\mathcal{V}$, is far less involved. Indeed, there are at most $\|t\|^d$ vectors that can be traversed in order to box-reach t, and since a shortest path to t does not visit the same vector twice, we can easily limit the search space. However, as discussed in Sect. 1, given a VAS $\mathcal{V}$, we would like to understand the general shape of the box-reachable set of configurations.

We can now state our main result, which shows that for vectors with large-enough entries (i.e., deep enough in the first quadrant), reachability coincides with box-reachability (see Sect. 4).

Theorem 1. *reachandboxcoincide   For every 2-VAS $\mathcal{V}$, there exists an effectively-computable $W \in \mathbb{N}$ such that $\mathsf{BoxReach}(\mathcal{V}) \cap [W, \infty]^2 = \mathsf{Reach}(\mathcal{V}) \cap [W, \infty]^2$.*

Our focus on 2-VAS may seem restrictive. However, as we now demonstrate (see also Sect. 6), Theorem 1 cannot be extended to richer models, namely to 3-VAS or to 1-VASS (Vector Addition Systems with States).

Example 2. Consider the 3-VAS $\mathcal{V} = \{(0, 1, 1), (1, 2, -1), (1, -1, 2)\}$ and targets of the form $(2n, n+1, n+1)$. Note that these targets are arbitrarily "deep" in the positive octant, and are reachable via the path $(0, 1, 1), ((1, 2, -1), (1, -1, 2))^n$.

However, these targets are not box-reachable: in order to start any path, the vector $(0, 1, 1)$ must be taken first. Then, it is easy to see that the number of

$(1, 2, −1)$ and $(1, −1, 2)$ used must be equal, and equal to n. Therefore, after taking $(0, 1, 1)$, the rest of any reaching path consists of vectors with negative entries, and in particular is not box-reaching.

For 1-VASS, it suffices to have two states q_0, q_1 such that q_1 is only reachable via a negative number (but is reachable). Then any path ending in q_1 cannot be box-reaching.

4 Reachability and Box Reachability Eventually Coincide

In this section we prove Theorem 1. The proof relies on two results regarding cones and lattices: the first is the *Steinitz Lemma* [8], which intuitively allows us to reorder paths so that they do not diverge wildly, and instead stay within some bounded "corridor". The second is the *Deep-in-the-Cone Lemma* [5], which connects $\mathbb{N}$-reachability with $\mathbb{Z}$-reachability. We start with some definitions and known results.

4.1 Reachability, Cones and Lattices

Consider a set of vectors $D = \{v_1, v_2, \ldots, v_n\} \subseteq \mathbb{Z}^d$ (we do not think of it as a VAS at this point). The *cone* spanned by D is the set of vectors in $\mathbb{R}^d$ expressible as non-negative combinations of the vectors in D. Similarly, we define the *integer cone* where we restrict attention to non-negative integer combinations, and the *lattice*, where the coefficients are all integers:

$$\mathsf{cone}(D) := \{\lambda_1 v_1 + \ldots + \lambda_n v_n \mid \forall i.\ \lambda_i \in \mathbb{R}_{\geq 0}\} \subseteq \mathbb{R}^d$$
$$\mathsf{intCone}(D) := \{\lambda_1 v_1 + \ldots + \lambda_n v_n \mid \forall i.\ \lambda_i \in \mathbb{N}\} \subseteq \mathbb{Z}^d$$
$$\mathsf{lattice}(D) := \{\lambda_1 v_1 + \ldots + \lambda_n v_n \mid \forall i.\ \lambda_i \in \mathbb{Z}\} \subseteq \mathbb{Z}^d$$

The set D can be thought of as a representation of $\mathsf{cone}(D)$, typically called the *V-representation* (where V stands for "Vertex"). A classical result by Weyl [18] shows that we can compute from D a set $\mathcal{F} \subseteq \mathbb{Z}^d$ such that[2] $\mathsf{cone}(D) = \{v \in \mathbb{R}^d \mid \forall f \in \mathcal{F} : \langle f, v \rangle \geq 0\}$. The set $\mathcal{F}$ is called the *H-representation* of $\mathsf{cone}(D)$ (where H stands for "Halfspace"). Intuitively, the set $\mathcal{F}$ can be understood as comprising vectors that are perpendicular to the boundaries of the cone. Each vector $f \in \mathcal{F}$ defines a (positive) halfspace in $\mathbb{R}^d$, and the inequality $\langle f, v \rangle \geq 0$ states that the vector v lies on the "correct" side of this hyperplane.

In general, computing the H-representation from D may take exponential time. However, for cones in $\mathbb{R}^2$ an H-representation has at most two vectors $\mathcal{F} = \{f_1, f_2\}$, and it can be computed in polynomial time. Moreover, $\|f_1\|$ and $\|f_2\|$ are polynomial in $\|D\| = \max\{\|v\| \mid v \in D\}$. Indeed, this follows by finding two "extremal" vectors that span the cone (or by first detecting that the cone is all of $\mathbb{R}^2$, a single halfspace, or a single ray – easier cases commented on in the

---
[2] $\langle \cdot, \cdot \rangle$ is the standard inner product.

following), which is in turn done by sorting $v_1,\ldots,v_n$ by their angle with the x-axis, and then the facets are defined by normals to these two generating vectors. In particular, the entries of the normals are integers, and their description size is identical to that of the vectors, e.g., the normal we take to (x,y) is either $(-y,x)$ or $(y,-x)$, depending on the required direction of the halfspace.

For an H-representation $\mathcal{F}$ of cone(D), each $\boldsymbol{f} \in \mathcal{F}$ induces a measure of how "far" a vector $\boldsymbol{v}$ is from the respective facet of cone(D), namely the inner product $\langle \boldsymbol{f}, \boldsymbol{v} \rangle$. Specifically, given some $M \in \mathbb{N}$, we say that a vector $\boldsymbol{v} \in$ cone(D) is M-deep in the cone if $\langle \boldsymbol{f}, \boldsymbol{v} \rangle \geq M$ for every $\boldsymbol{f} \in \mathcal{F}$. Our first tool is the following lemma from [5] (which actually follows from existing literature, see references in [5]), which connects reachability in the integer cone with reachability in the lattice.

Lemma 1 (Deep-in-the-Cone, Lemma 16 in [5]). *Given a set of vectors $D = \{v_1, v_2, \ldots, v_n\} \subseteq \mathbb{Z}^d$ and the H-representation $\mathcal{F}$ of cone(D), there exists a constant $M \in \mathbb{N}$, depending on D and $\mathcal{F}$, such that if $v \in$ cone$(D) \cap \mathbb{Z}^d$ is M-deep in the cone, then $v \in$ intCone(D) if and only if $v \in$ lattice(D).*

Moreover, M is polynomial in $\|D\|, \|\mathcal{F}\|$ and exponential in d.

Since in our setting the dimension is $d = 2$, and as mentioned above, $\|\mathcal{F}\| = \|\mathcal{D}\|$, we have that M in Lemma 1 is polynomial in $\|D\|$.

Our second tool is the classical *Steinitz Lemma*.

Lemma 2 (Steinitz, as stated in [8]). *Let $v_1,\ldots,v_k$ be a non-empty sequence of vectors in $\mathbb{R}^d$. Let $v = \sum_{j=1}^{k} v_j$ and $I = \max_{1 \leq j \leq k} \|v_j\|_\infty$. There exists a permutation σ of $\{1,\ldots,k\}$ such that for every $n \in \{d,\ldots,k\}$, we have:*

$$\left\| \sum_{j=1}^{n} v_{\sigma(j)} - \frac{n-d}{k} v \right\|_\infty \leq d \cdot I$$

In the context of 2-VAS, this lemma states that if a vector v is reachable via some path π, then the vectors of π can be rearranged so that the resulting path, dubbed a *Steinitz path* does not stray too far from the straight line that connects the origin to v. Specifically, the guaranteed bound (dubbed the *Steinitz constant*) is $d \cdot I = 2 \|\mathcal{V}\|$. The form of Steinitz paths guarantees in particular that their drop and peak are not too large, which is particularly useful for us. More precisely, we have the following (see the full version for the proof).

Lemma 3. *Consider a Steinitz path π with eff$(\pi) = (x,y) \geq (0,0)$, then* drop$_x(\pi)$, drop$_y(\pi) \leq 2 \|\mathcal{V}\|$ *and* peak$_x(\pi) \leq x + 2 \|\mathcal{V}\|$ *and* peak$_y(\pi) \leq y + 2 \|\mathcal{V}\|$.

4.2 Proof of Theorem 1

We are now ready for the main result, which we prove in the remainder of this section.

For $W \in \mathbb{N}$, denote the "L-shape" $\mathbb{L}_W = ([0, W-1] \times \mathbb{N}) \cup (\mathbb{N} \times [0, W-1])$ and its complement $\overline{\mathbb{L}}_W = [W, \infty]^2$. Also denote the real positive quadrant by $\mathbb{R}^2_{\geq 0}$.

Note that $\mathsf{BoxReach}(\mathcal{V}) \subseteq \mathsf{Reach}(\mathcal{V})$ trivially holds, and this inclusion remains valid when restricted to $\overline{\mathbb{L}}_W$. Therefore, it suffices to demonstrate the reverse inclusion: $\mathsf{Reach}(\mathcal{V}) \cap \overline{\mathbb{L}}_W \subseteq \mathsf{BoxReach}(\mathcal{V}) \cap \overline{\mathbb{L}}_W$.

Fix a 2-VAS $\mathcal{V} = \{\boldsymbol{v_1}, \ldots, \boldsymbol{v_k}\}$, and denote by M the Deep-in-the-Cone constant guaranteed by Lemma 1. We start by filtering out some degenerate cases, which would allow us to make some simplifying assumptions on $\mathcal{V}$.

We can assume without loss of generality that not all vectors in $\mathcal{V}$ are of the form $(\leq 0, \leq 0)$ (i.e., (x,y) such that $x \leq 0, y \leq 0$). Indeed, if this were the case, then $\mathsf{Reach}(\mathcal{V}) = \{\mathbf{0}\}$ and the theorem trivially holds. Similarly, it cannot hold that every vector $(x,y) \in \mathcal{V}$ has a negative coordinate, as again we would have $\mathsf{Reach}(\mathcal{V}) = \{\mathbf{0}\}$. Our next assumption is that the vectors in $\mathcal{V}$ are not all linearly dependent. Indeed, if this is the case then $\mathsf{Reach}(\mathcal{V})$ is one-dimensional, and the setting is simpler: in this case all vectors in $\mathcal{V}$ are scalar multiplications of some vector $(x,y) \in \mathbb{N}^2$. Then, reachability and box-reachability can be reasoned about in one of the components, so the setting is that of 1-VAS. We handle this in the full version.

Having set these assumptions, our first step is to find a vector reachable in $\mathcal{V}$ that has strictly positive coordinates, i.e., of the form $(>0, >0)$, and that is box reachable. Note that if $\mathcal{V}$ has a single generating vector of the form $(>0, >0)$, then it satisfies this requirement. If no generator of $\mathcal{V}$ is of this form, then there must be vectors in $\mathcal{V}$ of the form $(>0, 0)$ or $(0, >0)$ (otherwise we have $\mathsf{Reach}(\mathcal{V}) = \{\mathbf{0}\}$ as shown above). Without loss of generality, assume there is a vector $\boldsymbol{u_1} = (x, 0)$ where $x > 0$. Since not all vectors in $\mathcal{V}$ are linearly dependent, there exists a vector $\boldsymbol{u_2} = (x', y')$ with $y' > 0$. Indeed, otherwise all such vectors have $y' < 0$, so they cannot be taken after $(x, 0)$, and can therefore be discarded from $\mathcal{V}$ without changing the reachability set.

Since we assume there is no $(>0, >0)$ vector in $\mathcal{V}$, it follows that $x' \leq 0$. We now define $\boldsymbol{s} = (-2x'+1)\boldsymbol{u_1} + \boldsymbol{u_2} = ((-2x'+1)x+x', y') = (x+x'(1-2x), y')$, and notice that $x + x'(1-2x) > 0$, since $1-2x < 0$ and $x' \leq 0$. Thus, $\boldsymbol{s}$ is of the form $(>0, >0)$. Moreover, we claim that $\boldsymbol{s}$ is box-reachable. Indeed, consider the path $\zeta = (x,0)^{-x'}(x',y')(x,0)^{-x'+1}$ then $\mathsf{eff}(\zeta) = \boldsymbol{s}$. Also, $\mathsf{drop}_x(\zeta) = \mathsf{drop}_y(\zeta) = 0$ since the only negative coordinate is x', and that is taken only after $(x,0)^{-x'}$. Finally, $\mathsf{peak}_y(\zeta) = y' = \mathsf{eff}_y(\zeta)$ and $\mathsf{peak}_x(\zeta) = -x'x + x' + (-x'x) = \mathsf{eff}_x(\zeta)$.

We now define $\boldsymbol{s}_\nearrow = 2\|\mathcal{V}\|\boldsymbol{s}$. Notice that $\|\boldsymbol{s}_\nearrow\| = 2\|\mathcal{V}\| \|((-2x'+1)x + x', y')\| \leq 2\|\mathcal{V}\|(2\|\mathcal{V}\|^2 + 2\|\mathcal{V}\|) \leq 8\|\mathcal{V}\|^3$. That is, the representation of $\boldsymbol{s}_\nearrow$ is polynomial in $\|\mathcal{V}\|$. Moreover, we have $\boldsymbol{s} \geq (1,1)$ as it is positive and integer. Therefore, $\boldsymbol{s}_\nearrow \geq (2\|\mathcal{V}\|, 2\|\mathcal{V}\|)$. Also, since $\boldsymbol{s}$ is box-reachable, so is $\boldsymbol{s}_\nearrow$. We summarize the properties of $\boldsymbol{s}_\nearrow$ for later reference.

Observation 1. $\boldsymbol{s}_\nearrow$ is box-reachable in $\mathcal{V}$, satisfies $\boldsymbol{s}_\nearrow \geq (2\|\mathcal{V}\|, 2\|\mathcal{V}\|)$ and $\|\boldsymbol{s}_\nearrow\| \leq 8\|\mathcal{V}\|^3$.

In the following we split our analysis to two cases, according to whether $\mathsf{cone}(\mathcal{V})$ contains the positive quadrant $\mathbb{R}^2_{\geq 0}$ or not. The cases are depicted in Fig. 1. Specifically, since we assume that $\mathsf{cone}(\mathcal{V})$ is not one-dimensional or all of $\mathbb{R}^2$ (Remark 1), we have that $\mathsf{cone}(\mathcal{V}) = \mathsf{cone}(\boldsymbol{\chi_1}, \boldsymbol{\chi_2})$, where $\boldsymbol{\chi_1}, \boldsymbol{\chi_2}$ are the two extremal vectors whose cone spans all the vectors in cone($\mathcal{V}$), as discussed in Sect. 4.1 (indeed, a nontrivial cone in $\mathbb{R}^2$ is spanned by at most two vectors).

We can then formulate the different cases as follows:

- If χ_1 is of the form ($\leq 0, \geq 0$) and χ_2 is of the form ($\geq 0, \leq 0$), then $\mathbb{R}^2_{\geq 0} \subseteq$ cone($\mathcal{V}$) (depicted in Fig. 1a).
- If χ_1, χ_2 are both of the form ($\geq 0, \geq 0$), then cone($\mathcal{V}$) $\subseteq \mathbb{R}^2_{\geq 0}$ (depicted in Fig. 1b).
- If χ_1 is of the form ($> 0, > 0$) and χ_2 is of the form ($\mathbb{Z}, \leq 0$) (or vice-versa), then $\mathbb{R}^2_{\geq 0} \cap$ cone($\mathcal{V}$) $\neq \emptyset$ (depicted in Figs. 1c and 1d).

Importantly, there are no other cases to consider. Indeed, keeping in mind that the angle between χ_1 and χ_2 is at most 180° (since the cone comprises positive combinations), a simple examination of the other configurations of χ_1 and χ_2 shows that they lead to either cone($\mathcal{V}$)$\cap \mathbb{R}^2_{\geq 0} = \{0\}$ or to a one dimensional cone, which we already handled.

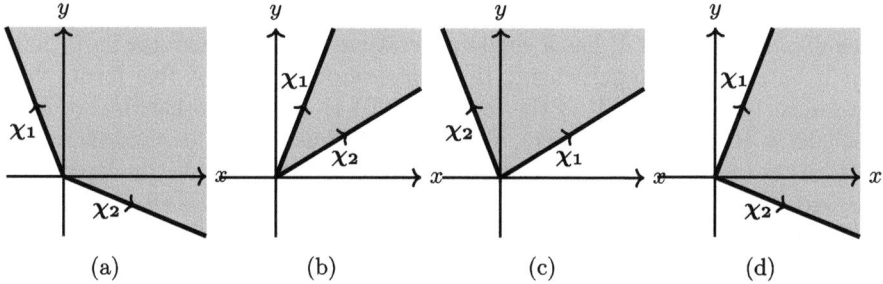

Fig. 1. Possible overlaps of the cone and the first quadrant. In Fig. 1a the cone contains the positive quadrant, while in Figs. 1b to 1d it does not. Note that Cases (c) and (d) may intersect the *negative* quadrant, even though this is not explicitly depicted.

First Case: The Cone Contains the Positive Quadrant. We present a sketch of the proof in this case (see Fig. 2 for a depiction). The full details appear in the full version.

Recall the box-reachable, strictly positive vector $s_\nearrow$ in $\mathcal{V}$ (as per Observation 1). We choose W large enough so that any target $v \in \overline{\mathbb{L}}_W$ is so far from the axes that $v - 2s_\nearrow$ is M-deep-in-the-cone (for the appropriate M obtained from Lemma 1). Note that this relies on the fact that the cone contains the first quadrant, so that being deep in the first quadrant also implies being deep-in-the-cone. Intuitively, the vector $v - 2s_\nearrow$ captures the fact that we consider a path to v starting and ending with $s_\nearrow$.

Assume that v is $\mathbb{N}$-reachable in $\mathcal{V}$. In particular, $v \in$ intCone($\mathcal{V}$). By adding $-2s_\nearrow$ to v (note the negative coefficient -2) we have that $v - 2s_\nearrow \in$ lattice($\mathcal{V}$). We can now invoke Lemma 1 and get that $v - 2s_\nearrow \in$ intCone($\mathcal{V}$). Then, we obtain a Steinitz path ρ with effect $\boldsymbol{v} - 2\boldsymbol{s}_{\nearrow}$ as per Lemma 2. By Lemma 3 and our careful choice of $\boldsymbol{s}_{\nearrow}$, we can show that the path $\boldsymbol{s}_{\nearrow} \cdot \rho \cdot \boldsymbol{s}_{\nearrow}$ box reaches $\boldsymbol{v}$. Intuitively, this is because $\boldsymbol{s}_{\nearrow}$ gets us far enough from the origin to overcome the Steinitz corridor of ρ, and since the effect of $\boldsymbol{s}_{\nearrow} \cdot \rho$ is $\boldsymbol{v} - \boldsymbol{s}_{\nearrow}$, then this corridor also does not exceed $\boldsymbol{v}$, allowing us to complete the path with another $\boldsymbol{s}_{\nearrow}$.

Remark 1. If $\mathsf{cone}(\mathcal{V}) = \mathbb{R}^2$, the H-representation has $\mathcal{F} = \emptyset$. Thus, all vectors are deep-in-the-cone, and by Lemma 1 we have $\mathsf{lattice}(\mathcal{V}) = \mathsf{intCone}(\mathcal{V})$. Therefore, this case is also covered by the analysis above. Similarly, if $\mathsf{cone}(\mathcal{V})$ is a single half-space (e.g., spanned by $(1,0), (-1,0)$, and $(0,1)$) then $\mathcal{F}$ is a singleton, and the argument proceeds similarly.

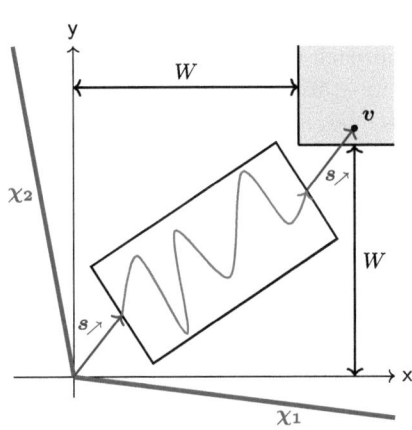

Fig. 2. The cone contains $\mathbb{R}^2_{\geq 0}$. The faces of the cone are depicted in red. The target $\boldsymbol{v}$ is inside $\overline{\mathbb{L}}_W$ (highlighted orange). The blue vectors are $\boldsymbol{s}_{\nearrow}$, and are connected by the green Steinitz path, surrounded by its bounding "corridor". (Color figure online)

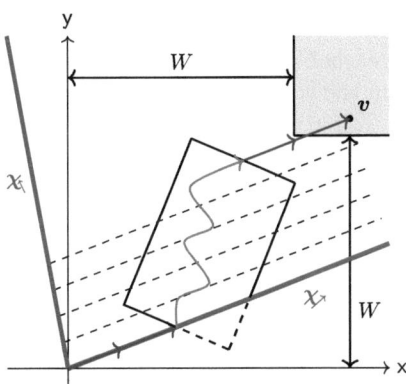

Fig. 3. The cone intersects $\mathbb{R}^2_{\geq 0}$. The faces of the cone are depicted in red. The target $\boldsymbol{v}$ is inside $\overline{\mathbb{L}}_W$ (highlighted orange). The blue vectors are $\boldsymbol{s}_{\nearrow}$, and are connected by the green Steinitz path, surrounded by its bounding "corridor". The dashed lines represent the "levels" of dot product with $\boldsymbol{f}_{\nearrow}$. (Color figure online)

Second Case: The Cone Does Not Contain the Positive Quadrant. We now consider the setting where $\mathsf{cone}(\mathcal{V})$ does not contain $\mathbb{R}^2_{\geq 0}$. Since $\boldsymbol{s}_{\nearrow} \in \mathsf{cone}(\mathcal{V})$, then $\mathsf{cone}(\mathcal{V}) \cap \mathbb{R}^2_{\geq 0} \supsetneq \{\boldsymbol{0}\}$, and therefore we fall into one of the cases depicted in Figs. 1b to 1d. Note, however, that case Fig. 1b is trivial, as every vector in $\mathcal{V}$ has non-negative coordinates, so $\mathsf{Reach}(\mathcal{V}) = \mathsf{BoxReach}(\mathcal{V})$. We are therefore left with the cases where $\mathsf{cone}(\mathcal{V}) = \mathsf{cone}(\chi_1, \chi_2)$ neither contains or is contained in the first quadrant. That is, we have that χ_1 and χ_2 are either of the form $(>0, >0)$ and $(\mathbb{Z}, \leq 0)$ (Fig. 1d), or of the form $(>0, >0)$ and $(\leq 0, \mathbb{Z})$ (Fig. 1c).

We observe that these cases are symmetric in the sense that swapping the roles of the x and y axes interchanges between them. Therefore, it suffices to analyze subcase 1c, i.e., where $\mathsf{cone}(\mathcal{V})$ contains the y-axis. We denote the extremal vectors as $\chi_\nearrow = (x_1, y_1)$ with $x_1, y_1 > 0$ and $\chi_\nwarrow = (x_2, y_2)$ with $x_2 \leq 0, y_2 \in \mathbb{Z}$. The respective H-representation is then $\mathcal{F} = \{f_\nearrow, f_\nwarrow\}$ with $f_\nearrow = (-y_1, x_1)$ and $f_\nwarrow = (y_2, -x_2)$. We again sketch the proof, with the details in the full version.

Set W large enough, and consider a vector $v \in \mathsf{Reach}(\mathcal{V}) \cap \overline{\mathbb{L}}_W$. We need to prove that $v \in \mathsf{BoxReach}(\mathcal{V})$. As before, we focus on $v - 2s_\nearrow$.

If we are lucky, then $v - 2s_\nearrow$ is M-deep-in-the-cone (spanned by $\mathcal{V}$). In this case we apply the same reasoning as in the first case: we first use the lattice-reachability of $v - 2s_\nearrow$ to conclude that $v - 2s_\nearrow \in \mathsf{intCone}(\mathcal{V})$. Then, we build the Steinitz path ρ' from $s_\nearrow$ to $v - s_\nearrow$, and finally show that the path $s_\nearrow \cdot \rho' \cdot s_\nearrow$ box-reaches v.

Unfortunately, unlike the first case, $v - 2s_\nearrow$ is not necessarily M-deep-in-the-cone. Indeed, the facet of the cone defined by $\chi_\nearrow$ intersects $\overline{\mathbb{L}}_W$, so there are reachable vectors that are on the facet itself, and are therefore not at all deep in the cone. Thus, we take a different approach (see Fig. 3 for a depiction).

If $v - 2s_\nearrow$ is not deep in the cone, we start by showing that it is close to the facet $\chi_\nearrow$ (which is not completely trivial, since the depiction in Fig. 1c is misleading, in that $\chi_\nwarrow$ could be in the negative quadrant). We then notice that all vectors in $\mathcal{V}$ are either parallel to $\chi_\nearrow$, or increase the inner product with $f_\nwarrow$ by at least 1 (due to the integer coordinates). Thus, if $v - 2s_\nearrow$ is far enough in the quadrant but close to $\chi_\nearrow$, reaching it requires a lot of vectors parallel to $\chi_\nearrow$ (otherwise the path strays too far from $\chi_\nearrow$, and cannot come back). We can use these vectors as a "box-safe padding", and use a Steinitz path in the middle, to obtain an overall box-reaching path. More precisely, we first advance along $\chi_\nearrow$ until we are far enough above the x-axis, and far enough below v_y. We then use a Steinitz path to simulate most of the original path to v apart from some more vectors parallel to $\chi_\nearrow$, while staying well below v_y. We then complete the path by staying parallel to $\chi_\nearrow$ until reaching v. □

5 On the Semilinearity of Box Reachability

5.1 From Box Reachability in d-VAS to Reachability in $2d$-VAS

Having established the eventual-equivalence of $\mathsf{BoxReach}(\mathcal{V})$ and $\mathsf{Reach}(\mathcal{V})$ for 2-VAS in $\overline{\mathbb{L}}_W$, it is natural to ask what happens in $\mathbb{L}_W$, where the two sets do not coincide. In the following, we show that $\mathsf{BoxReach}(\mathcal{V})$ is semilinear and in particular is semilinear in the restriction to $\mathbb{L}_W$. In fact, we show a more general result, linking box-reachability to standard reachability, at the cost of doubling the dimension (see the full version).

Theorem 2. *Consider a d-VAS $\mathcal{V}$, and define the $2d$-VAS $\mathcal{V}'$ as*

$$\mathcal{V}' = \{(x_1, \ldots, x_d, -x_1, \ldots, -x_d) \mid (x_1, \ldots, x_d) \in \mathcal{V}\} \cup \{e_{d+i} \mid 1 \leq i \leq d\}$$

where e_{d+i} is the unit vector with 1 at coordinate $d + i$. Then

$$\mathsf{BoxReach}(\mathcal{V}) = \{(v_1, \ldots, v_d) \in \mathbb{N}^d \mid (v_1, \ldots, v_d, 0, \ldots, 0) \in \mathsf{Reach}(\mathcal{V}')\}.$$

By [10], the reachability set of 4-VAS is semilinear, and in particular the box-reachability set is also semilinear as a projection. By Theorem 2 we therefore have:

Corollary 1. *For a* 2-VAS $\mathcal{V}$, *the set* BoxReach($\mathcal{V}$) *is semilinear.*

5.2 Box Reachability in 1-VASS Is Semilinear

As demonstrated in Example 2, Theorem 1 cannot be extended to either 1-VASS or 3-VAS. Still, it is desirable to reason about box-reachability in VASS. In this section, we show that the box-reachable set in 1-VASS is semilinear.

The precise definitions and proofs are in the full version. We sketch the main idea. The fundamental tool we rely on is that for 1-VASS, reachability can be characterized by *Linear Path Schemes with One Cycle* (1-LPS). These are path schemes of the form $\alpha\beta^*\gamma$, where α, γ are paths, and β is a cycle. We observe that crucially, while paths of this form might not be box-reaching, they cannot exceed their box by too much. Indeed, they can go outside their box only as much as γ or β do. We refer to this excess as the *overshoot* of the LPS.

We therefore define the notion of a *closing suffix* – a (box reaching) path that can be concatenated after $\alpha\beta^*\gamma$ to make it box reaching. In particular, a closing suffix goes beyond the overshoot of the LPS (see Fig. 4).

We prove that we can characterize box reachability using 1-LPS with closing suffixes as follows: consider if some state-counter pair (x, q) is reachable from $(0, q_0)$ and x is large enough, then (x, q) is box reachable from $(0, q_0)$ if and only if there is some 1-LPS $\alpha\beta^*\gamma$ and a closing suffix θ such that (x, q) is box-reachable by $\alpha\beta^*\gamma\theta$, and such that α, β, γ and θ are all relatively short.

Technically, this characterization is obtained by assuming there is a very long closing suffix, and defining a sequence of indices along it, each of which is itself the beginning of a closing suffix. We then show that if the closing suffix is too long, it can be shortened.

We remark that this technique fails when LPS with more than one cycle are needed, e.g., for 2-VASS.

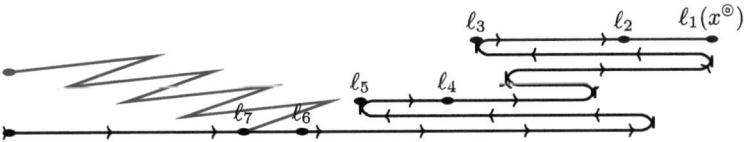

Fig. 4. The black path is a box-reaching path to $x^\odot$ (the y axis is only for readability). Each ℓ_i is the beginning of a possible closing suffix (and in particular these suffixes are box-reaching paths). The blue zig-zag is an LPS replacing the prefix up to ℓ_m. Note that the LPS is not box-reaching itself, but its "overshoot" is covered by the long closing suffix. (Color figure online)

We remark that the above also gives an alternative proof of Corollary 1 – indeed, in $\mathbb{L}_W$ we can cast the 2-VAS to a 1-VASS by using the state space to capture the bounded coordinates.

6 Discussion and Future Work

In this work we define the notion of box-reachability. Intuitively, the set of box-reachable vectors exhibit a "bounded" behavior that is more relaxed than a fixed, or even an existentially-quantified bound on some of the coordinates [7]. Conceptually, this allows us to capture models where one wishes to reach a target without using more resources than the end goal.

We prove that despite the restriction this places on reachability, for 2-VAS the set of box-reachable configurations coincides with standard reachability for all configurations beyond some threshold. As demonstrated in Example 2, extending Theorem 1 is impossible for 3-VAS and 1-VASS.

We also show that box-reachability is semilinear for 2-VAS and 1-VASS. We leave open the question of whether this can be extended to d-VAS for $3 \leq d \leq 5$ and 2-VASS, for which standard reachability is semilinear.

The introduction of box-reachability gives rise to several other natural definitions: one could consider *approximate* box reachability, where the box is allowed to be greater than the target by some fixed (additive or multiplicative) constant. Naive definitions of such extensions turn out to be not very interesting, in the sense that using the Steinitz path, even without our tools in Sect. 4, suffices to show that they coincide with reachability (since the Steinitz path already exceeds the target only by a little bit). Nonetheless, it is possible that specific modelling needs may require some novel tweaks of box-reachability notions, and we hope this research serves as a basis for reasoning about such extensions.

Acknowledgments. We thank an anonymous referee for suggesting Theorem 2, which greatly simplified our previous proof of Corollary 1.

References

1. Almagor, S., Cohen, N., Pérez, G.A., Shirmohammadi, M., Worrell, J.: Coverability in 1-VASS with disequality tests. In: 31st International Conference on Concurrency Theory (CONCUR 2020). Schloss-Dagstuhl-Leibniz Zentrum für Informatik (2020)
2. Almagor, S., Ghosh, A., Leys, T., Pérez, G.A.: The geometry of reachability in continuous vector addition systems with states. In: 48th International Symposium on Mathematical Foundations of Computer Science (MFCS 2023). Schloss Dagstuhl-Leibniz-Zentrum für Informatik (2023)
3. Blondin, M.: The reachability problem for two-dimensional vector addition systems with states. J. ACM (JACM) **68**(5), 1–43 (2021)
4. Blondin, M., Haase, C.: Logics for continuous reachability in petri nets and vector addition systems with states. In: 32nd Annual ACM/IEEE Symposium on Logic in Computer Science (LICS 2017), pp. 1–12. IEEE (2017)

5. Cslovjecsek, J., Koutecký, M., Lassota, A., Pilipczuk, M., Polak, A.: Parameterized algorithms for block-structured integer programs with large entries. TheoretiCS **4** (2025)
6. Czerwiński, W., Orlikowski, Ł.: Reachability in vector addition systems is Ackermann-complete. In: IEEE 62nd Annual Symposium on Foundations of Computer Science (FOCS 2021), pp. 1229–1240. IEEE (2022)
7. Demri, S.: On selective unboundedness of VASS. J. Comput. Syst. Sci. **79**(5), 689–713 (2013)
8. Grinberg, V.S., Sevast'yanov, S.V.: Value of the Steinitz constant. Funct. Anal. Appl. **14**(2), 125–126 (1980)
9. Guttenberg, R., Raskin, M., Esparza, J.: Geometry of reachability sets of vector addition systems. In: 34th International Conference on Concurrency Theory (CONCUR 2023). Schloss Dagstuhl-Leibniz-Zentrum für Informatik (2023)
10. Hopcroft, J., Pansiot, J.J.: On the reachability problem for 5-dimensional vector addition systems. Theor. Comput. Sci. **8**(2), 135–159 (1979)
11. Karp, R.M., Miller, R.E.: Parallel program schemata. J. Comput. Syst. Sci. **3**(2), 147–195 (1969)
12. Künnemann, M., Mazowiecki, F., Schütze, L., Sinclair-Banks, H., Węgrzycki, K.: Coverability in VASS revisited: improving Rackoff's bound to obtain conditional optimality. In: 50th International Colloquium on Automata, Languages, and Programming (ICALP 2023). Schloss Dagstuhl-Leibniz-Zentrum für Informatik (2023)
13. Leroux, J.: The reachability problem for Petri nets is not primitive recursive. In: IEEE 62nd Annual Symposium on Foundations of Computer Science (FOCS 2021), pp. 1241–1252. IEEE (2022)
14. Leroux, J., Sutre, G.: On flatness for 2-dimensional vector addition systems with states. In: Gardner, P., Yoshida, N. (eds.) CONCUR 2004. LNCS, vol. 3170, pp. 402–416. Springer, Heidelberg (2004). https://doi.org/10.1007/978-3-540-28644-8_26
15. Rackoff, C.: The covering and boundedness problems for vector addition systems. Theor. Comput. Sci. **6**(2), 223–231 (1978)
16. Rosier, L.E., Yen, H.C.: A multiparameter analysis of the boundedness problem for vector addition systems. J. Comput. Syst. Sci. **32**(1), 105–135 (1986)
17. Schmitz, S.: The complexity of reachability in vector addition systems. ACM SigLog News **3**(1), 4–21 (2016)
18. Weyl, H.: Elementare Theorie der konvexen Polyeder. Commentarii Mathematici Helvetici **7**(1), 290–306 (1934)

Knowing-How Reasoning with Budgets Recasted: Universal Reachability Problem on VASS

Stéphane Demri[1], Laurent Doyen[1], and Raul Fervari[1,2(✉)]

[1] Université Paris-Saclay, ENS Paris-Saclay, CNRS, Laboratoire Méthodes Formelles, 91190 Gif-sur-Yvette, France
[2] FAMAF, Universidad Nacional de Córdoba and CONICET, Córdoba, Argentina
rfervari@unc.edu.ar

Abstract. We investigate the decidability/complexity status of the model-checking problem for an ability-based logic expressing knowing how assertions and enriched with budget constraints. To do so, we introduce a new control-state reachability problem for complete vector addition systems with states where the transitions are labelled by letters from a finite alphabet. It is required that all runs labelled by a given word lead synchronously to the target control states. First, we show that the model-checking problem involving knowing how can be reduced to the new problem on VASS. Second, we establish Ack-completeness of both problems by properly adapting or using developments about well-structured transition systems, belief functions and length function theorems.

1 Introduction

Ubiquity of VASS for Applications. Vector addition systems with states (VASS) [23,25] are ubiquitous models, closely related to Petri nets [40], that can be viewed as Minsky machines without zero-tests [37]. A wide range of decision problems have been considered for such counter machines, see e.g. [26,35,36,39], with new problems popping up regularly, for instance to recast problems from formal verification or database theory. For example, the satisfiability problem for FO2 over data words was shown to be equivalent to the reachability problem for VASS [5,14]. Decidability of the reachability problem was shown in the seminal papers [26,36]. Further developments on this problem are made in [27,29,30,40], including its Ack-completeness in [13,31,32]. The complexity class Ack [42] has been already successful to characterise the complexity of many problems involving counters, see e.g. [22,31,38]. More precisely, Ack is the class of Ackermannian problems closed under primitive-recursive reductions introduced in [42, Sect. 2.3], corresponding to the fast-growing complexity class $\mathcal{F}_\omega$ in the extended Grzegorczyk hierarchy (see details in [42]). In contrast, the covering and boundedness problems for VASS are "only" ExpSpace-complete [10,19,35,39], as well as variants [4,17]. To illustrate briefly the wide range of applications of VASS, let us mention that the covering problem can express the thread-state reachability problem for replicated finite-state programs [24], as well as decision problems for the parameterised verification of ad-hoc networks [16].

Knowing-How Logics with Budgets. This paper stems from investigations about knowing-how logics that have emerged recently as a novel formalism to express assertions related to the ability of an agent to achieve a certain goal [44–46]. This provides formal foundations to the epistemic concept of "knowing how", dedicated to strategic reasoning and automated planning. One of the appealing features of these logics [44–46] is the simplicity in the language, allowing to abstract the properties from the concrete strategies an agent may use. It has been shown that such logics enjoy interesting computational properties, e.g. the satisfiability problem for the logic from [44] is shown to be in NP^{NP} [3], while its model-checking problem is in PSpace [18]. Interestingly, one of the logics considered therein is the extension of knowing how with numerical constraints, expressing not only that an agent is able to achieve a goal, but also that in doing so they are able to remain within a certain budget. For instance, one can express that a robot can complete an exploration mission without running out of fuel or in a certain amount of time. The need to express budget-like constraints about plans in ability-based logics is advocated in [33, Sect. 3.1] and [34]. Assuming that actions have costs, the execution of a sequence of actions requires that the agent stays always within the budget. Adding resource reasoning is a well-known paradigm in ATL-like logics [1,2,8], in energy games [6,11], and in multi-agent systems [9]. In [18], investigations can be found about the complexity of adding resource reasoning in the ability-based logic $\mathcal{L}_{kh}$ introduced in [44]; the new logic is called $\mathcal{L}_{kh}(\star)$. Though the decidability status of the model-checking problem for $\mathcal{L}_{kh}(\star)$ is therein left open, the subproblem in which the cost of actions is independent of the state, is shown to be ExpSpace-complete [18, Theorem 4].

Our Contributions. In this work, we aim at determining whether model-checking for $\mathcal{L}_{kh}(\star)$ (Sect. 2.1) is decidable and, if so, to characterize its computational complexity. To do so, we introduce the universal control-state reachability problem for complete VASS with transitions labelled by letters from a finite alphabet. This is a variant of the (existential) control-state reachability problem asking for the existence of a finite sequence of letters such that all finite runs from an initial configuration following this sequence lead to a target control state (Sect. 2.2). Our first contribution is to show that the model-checking problem is reducible to the new problem for complete VASS (Sect. 3). Then, we show that model-checking for $\mathcal{L}_{kh}(\star)$ is Ack-complete (Theorem 1), in particular it requires non-primitive recursive time. For Ack-membership, we take advantage of the notion of belief functions [38], from the underlying well-structured transition system (WSTS) [21, Sect. 3] and the analysis about the length of bad sequences [38] and relying on [20] (Sect. 4). On the other hand, the hardness proof relies on the Ack-hardness of the control-state reachability problem for VASS with resets [43] (Sect. 5).

2 Preliminaries

In this section, we present the model-checking problem for an ability-based logic with budgets, which is the starting point of this work. Its second part is dedicated to a new decision problem for complete VASS, called the *universal control-state reachability problem*, in which we ask if there exists a word σ such that all finite runs labelled by σ lead to target control states, while maintaining non-negative counters.

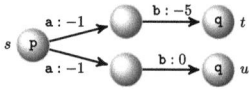

Fig. 1. A model for $\mathcal{L}_{kh}(\star)$ ($r = 1$).

2.1 Knowing-How Logics with Budgets: the Model-Checking Problem

We present below the knowing-how logic with budgets, first introduced in [18] and called herein $\mathcal{L}_{kh}(\star)$. We introduce standard notions like models, plans and strong executability (see e.g. [44–46]) and budget-related notions such as weight functions, computations and budget compatibility.

Linear Plans. Let Act be a countable set of action symbols, and Act* be the set of finite sequences over Act. Elements of Act* are called *plans*, with ε being the *empty plan*. Given $\sigma \in$ Act*, let $|\sigma|$ be the length of σ (with $|\varepsilon| \stackrel{\text{def}}{=} 0$). For $0 \leq k \leq |\sigma|$, the plan σ_k is σ's initial segment up to (and including) the kth position ($\sigma_0 \stackrel{\text{def}}{=} \varepsilon$). For $0 < k \leq |\sigma|$, the action $\sigma[k]$ is the one in σ's kth position.

Let $(R_a)_{a \in \text{Act}}$ be a family of binary relations $R_a \subseteq S \times S$, for some set S. Define $R_\varepsilon \stackrel{\text{def}}{=} \{(s,s) \mid s \in S\}$ and, for all $\sigma \neq \varepsilon \in$ Act* and $a \in$ Act, let $R_{\sigma a} \stackrel{\text{def}}{=} \{(s,t) \in S \times S \mid \exists t' \in S \text{ s.t. } (s,t') \in R_\sigma \text{ and } (t',t) \in R_a\}$. Consider a plan $\sigma \in$ Act*: for all $s \in S$, we define $R_\sigma(s) \stackrel{\text{def}}{=} \{t \in S \mid (s,t) \in R_\sigma\}$. For all $X \subseteq S$, $R_\sigma(X) \stackrel{\text{def}}{=} \bigcup_{s \in X} R_\sigma(s)$.

Models, Weight Functions and Computations. Models of $\mathcal{L}_{kh}(\star)$ are of the form $\mathcal{S} = (S, (R_a)_{a \in \text{Act}}, wf, L)$ where: (1) S is a non-empty set of states; (2) $(R_a)_{a \in \text{Act}}$ is a collection of binary relations on S ($s \xrightarrow{a} t$ denotes that $(s,t) \in R_a$); (3) $L : S \to 2^{\text{Prop}}$ is a labelling, for some set Prop of atomic propositions; and (4) $wf : S \times \text{Act} \times S \to \mathbb{Z}^r$ is a so-called *weight function* of dimension r. The value $wf(s, a, t)$ is read as the cost of executing the action a from the state s to t (or equivalently to fire the transition $s \xrightarrow{a} t$).

A *computation* is a sequence $\lambda = s_0 \xrightarrow{a_1} s_1 \cdots \xrightarrow{a_K} s_K$, and its (accumulated) *weight* is $wf(\lambda) \stackrel{\text{def}}{=} \sum_{k=1}^{K} wf(s_{k-1}, a_k, s_k)$ (empty computations have weight **0**).

Figure 1 shows an example of a model with budgets: state labels (propositions p and q) are shown inside states, and some state names are shown outside states.

Constraints on Plans. Intuitively, the works [44–46] establish that an agent knows how to achieve condition ψ given condition φ, when she has an appropriate plan that allows her to go from any state in which φ holds, only to states in which ψ holds. In order to characterise what 'appropriate' means, one can impose restrictions on what qualifies as an adequate plan.

Let $(R_a)_{a \in Act}$ be a collection of binary relations. A plan $\sigma \in Act^*$ is *strongly executable* (SE) at $s \in S$ iff for all $k \in [0, |\sigma| - 1]$ and $t \in R_{\sigma_k}(s)$, we have $R_{\sigma[k+1]}(t) \neq \varnothing$, that is every computation on a prefix of σ can be prolonged to a computation of σ. We define the set $SE(\sigma) \stackrel{def}{=} \{s \in S \mid \sigma \text{ is SE at } s\}$. Strong executability is a notion inherited from conformant planning [12].

A plan $\sigma = a_1 \cdots a_K$ is *b-compatible* at s ($b \in \mathbb{N}^r$) iff for every computation $\lambda = s_0 \xrightarrow{a_1} s_1 \cdots \xrightarrow{a_K} s_K$ with $s_0 = s$, we have for all $L \in [1, K]$, $b + wf(\lambda_{\leq L}) \geq 0$ (with $\lambda_{\leq L} \stackrel{def}{=} s_0 \xrightarrow{a_1} s_1 \cdots \xrightarrow{a_L} s_L$). The plan σ is *b-compatible* at a set $X \subseteq S$ iff it is *b*-compatible at all $s \in X$. The tuple b is understood as the *initial budget*.

The Logic $\mathcal{L}_{kh}(\star)$. Let us present below a natural way to extend the logic from [44]. We write $\mathcal{L}_{kh}(r)$ to indicate our ability-based logic with $r \geq 0$ resource types. If $r = 0$, then $\mathcal{L}_{kh}(r)$ corresponds to the logic from [44], written $\mathcal{L}_{kh}$ here. The logic $\mathcal{L}_{kh}(\star)$ denotes the version in which the number of resource types is arbitrary. For $r \geq 0$, the formulae of $\mathcal{L}_{kh}(r)$ have the following syntax:

$$\varphi ::= p \mid \neg \varphi \mid \varphi \vee \varphi \mid \mathsf{Kh}^b(\varphi, \varphi),$$

where $p \in \mathsf{Prop}$, and $b \in \mathbb{N}^r$. Formulae of the form $\mathsf{Kh}^b(\varphi, \psi)$ are read as *"when φ holds, the agent knows how to make ψ true with budget b"*.

Semantics. The satisfaction relation $\Vdash$ is defined as follows:

$$\begin{array}{ll} \mathcal{S}, s \Vdash p & \stackrel{def}{\Leftrightarrow} p \in L(s), \\ \mathcal{S}, s \Vdash \neg \varphi & \stackrel{def}{\Leftrightarrow} \mathcal{S}, s \nVdash \varphi, \\ \mathcal{S}, s \Vdash \varphi \vee \psi & \stackrel{def}{\Leftrightarrow} \mathcal{S}, s \Vdash \varphi \text{ or } \mathcal{S}, s \Vdash \psi, \\ \mathcal{S}, s \Vdash \mathsf{Kh}^b(\varphi, \psi) & \stackrel{def}{\Leftrightarrow} \text{there is a plan } \sigma \in Act^* \text{ such that} \\ & (1) \; [\![\varphi]\!]^\mathcal{S} \subseteq SE(\sigma), \quad (2) \; R_\sigma([\![\varphi]\!]^\mathcal{S}) \subseteq [\![\psi]\!]^\mathcal{S}, \text{ and} \\ & (3) \; \sigma \text{ is } b\text{-compatible at } [\![\varphi]\!]^\mathcal{S}, \end{array}$$

where $[\![\varphi]\!]^\mathcal{S} \stackrel{def}{=} \{s \mid \mathcal{S}, s \Vdash \varphi\}$. Let $\mathcal{S}$ be the model of Fig. 1, it holds that $\mathcal{S}, s \Vdash \mathsf{Kh}^6(\mathsf{p}, \mathsf{q})$, since the plan ab is 6-compatible. Also, we have that $\mathcal{S}, s \nVdash \mathsf{Kh}^5(\mathsf{p}, \mathsf{q})$, as ab is not 5-compatible: while the computation leading to u has weight -1, the one leading to t has weight -6, exhausting the available budget of 5.

Let us define the *model-checking problem for $\mathcal{L}_{kh}(\star)$*, written $MC(\mathcal{L}_{kh}(\star))$ and introduced in [18] (finite models are such that S, Act and Prop are finite). Its budget-free instance (i.e., no need of checking condition (3)) is called $MC(\mathcal{L}_{kh})$.

Input: a finite model $\mathcal{S}$, a state s and a formula φ.
Question: $\mathcal{S}, s \Vdash \varphi$?

In [18], it is shown that the conditions (1)–(2) can be encoded by a finite-state automaton of exponential size, leading to the PSpace-membership of MC($\mathcal{L}_{kh}$).

Proposition 1 ([18]). *Let $\mathcal{S} = (S, (R_a)_{a \in \mathsf{Act}}, \mathit{wf}, L)$ be a finite model and $X_1, X_2 \subseteq S$. The set of plans σ such that $X_1 \subseteq \mathrm{SE}(\sigma)$ and $R_\sigma(X_1) \subseteq X_2$ can be accepted by a deterministic finite-state automaton $\mathcal{A} = (Q, \mathsf{Act}, \delta, I, F)$ with $|Q| \leq 2^{|S|(|S|+2)}$. Moveover, each location in Q can be encoded in polynomial space, and deciding whether $q \xrightarrow{\mathsf{a}} q' \in \delta$ (resp. $q \in I$, $q \in F$) can be done in polynomial space.*

The decision problem MC($\mathcal{L}_{kh}(\star)$) witnesses a genuine complexity blow-up, partly due to condition (3) combined with (2). Indeed, adding condition (3) amounts to restrict further the class of plans to be considered to satisfy a knowing-how formula. Recall that the decidability status of MC($\mathcal{L}_{kh}(\star)$) is left open in [18] though the subproblem in which the cost of an action is independent of the states, $\mathit{wf}(s_1, \mathsf{a}, s_1') = \mathit{wf}(s_2, \mathsf{a}, s_2')$ for all $s_1, s_1', s_2, s_2' \in S$ and all $\mathsf{a} \in \mathsf{Act}$, is shown to be ExpSpace-complete [18, Theorem 4].

The rest of the paper is devoted to filling this gap and showing that the problem MC($\mathcal{L}_{kh}(\star)$) is decidable and Ack-complete. As a consequence of our developments, solving MC($\mathcal{L}_{kh}(\star)$) requires non-primitive-recursive time, even if restricted to a single resource type. This result is formally stated below.

Theorem 1. *MC($\mathcal{L}_{kh}(\star)$) is Ack-complete, even in dimension $r = 1$.*

2.2 A New Reachability Problem on VASS

To characterise the complexity of MC($\mathcal{L}_{kh}(\star)$), we introduce a new decision problem UCReach on complete VASS, which is structurally less rich than MC($\mathcal{L}_{kh}(\star)$) but for which it is quite handy to get complexity results. We show that UCReach is Ack-complete (Sects. 4 and 5) and equivalent to MC($\mathcal{L}_{kh}(\star)$) modulo exponential-time reductions (the exponential blow up occurs only in one direction), see Sect. 3.

We briefly recall that a *vector addition system with states (VASS)* [23] is a tuple $\mathcal{V} = (Q, \Sigma, r, R)$, where Q is a finite set of *locations*, $r \in \mathbb{N}$ is its *dimension*, and R is a finite set of *transitions* in $Q \times \Sigma \times \mathbb{Z}^r \times Q$ for some finite non-empty alphabet Σ. Herein, the transitions are labelled by letters from the alphabet Σ (also sometimes written Act). A VASS $\mathcal{V}$ is *complete* if for all $q \in Q$ and $\mathsf{a} \in \Sigma$, there is some $(\boldsymbol{u}, q')$ such that $(q, \mathsf{a}, \boldsymbol{u}, q') \in R$. A *configuration* (resp. *pseudo-configuration*) in a VASS $\mathcal{V}$ is a pair $(q, \boldsymbol{x}) \in Q \times \mathbb{N}^r$ (resp. in $Q \times \mathbb{Z}^r$). Given pseudo-configurations $(q, \boldsymbol{x})$, $(q', \boldsymbol{x}')$ and a transition $T = q \xrightarrow{\mathsf{a}, \boldsymbol{u}} q'$, we write $(q, \boldsymbol{x}) \xrightarrow{T} (q', \boldsymbol{x}')$ whenever $\boldsymbol{x}' = \boldsymbol{u} + \boldsymbol{x}$. A *pseudo-run* is a finite sequence $\rho = (q_0, \boldsymbol{x}_0) \xrightarrow{T_1} (q_1, \boldsymbol{x}_1) \xrightarrow{T_2} (q_2, \boldsymbol{x}_2) \cdots \xrightarrow{T_L} (q_L, \boldsymbol{x}_L)$ of pseudo-configurations, where $(q_0, \boldsymbol{x}_0)$ is the *initial* pseudo-configuration. We say that the pseudo-run ρ is labeled by $\mathsf{a}_1 \cdots \mathsf{a}_L$ if T_i is of the form $(\cdot, \mathsf{a}_i, \cdot, \cdot)$ for all $1 \leq i \leq L$. A *run* is a pseudo-run in which only configurations in $Q \times \mathbb{N}^r$ occur (negative values disallowed). We present the *universal control-state reachability problem* UCReach:

Input: A complete VASS $\mathcal{V} = (Q, \Sigma, r, R)$, $p_0 \in Q$, $\boldsymbol{y_0} \in \mathbb{N}^r$, $Q_F \subseteq Q$.
Question: Does there exist a word $\sigma \in \Sigma^*$ such that for all finite pseudo-runs $\rho = (q_0, \boldsymbol{x_0}) \xrightarrow{T_1} (q_1, \boldsymbol{x_1}) \xrightarrow{T_2} (q_2, \boldsymbol{x_2}) \cdots \xrightarrow{T_L} (q_L, \boldsymbol{x_L})$ from $(p_0, \boldsymbol{y_0})$ with label σ, we have $q_L \in Q_F$ and $\{\boldsymbol{x_0}, \ldots, \boldsymbol{x_L}\} \subseteq \mathbb{N}^r$ (hence ρ is a run)?

Observe that the input VASS in the definition of UCReach is complete because we can restrict ourselves to such VASS to reduce the model-checking problem for the logic $\mathcal{L}_{kh}(\star)$. A positive instance of UCReach can be formulated by an existential quantification over plans, followed by a universal quantification over the class of pseudo-runs whose label is precisely the witness plan. Compared to the standard control-state reachability problem for VASS, there is a universal quantification over all the pseudo-runs labelled by σ (instead of an existential one). Note also that we could equally assume that $|Q_F| = 1$ as UCReach can be easily reduced to this subproblem but the current formulation is more convenient to us. Though UCReach is quite natural in view of its standard existential variant, as far as we know, the problem UCReach was never defined in the literature. Most probably, so far, there was no need for it.

Furthermore, note that many decision problems on VASS for which we do not need to reach a precise configuration are known to be ExpSpace-complete, see e.g. the covering and boundedness problems [10,19,35,39]. The problem UCReach seems to be of the same flavour, as we need to reach the specified control states but the final counter values are not important. By contrast, the reachability problem (between configurations) is Ack-complete [13,31,32]. However, the problem UCReach has similarities with the fixed initial credit problem for specific blind games [38] (see also [15]). The crucial difference rests on the fact that UCReach requires a synchronised reachability condition whereas the problem in [38] is related to a safety property (non-termination). Moreover, the fixed initial credit problem in [38] handles a single counter. Below, we show that there is a logspace reduction from UCReach to its restriction with a single counter (Lemma 2).

3 Relationships Between MC($\mathcal{L}_{kh}(\star)$) and UCReach

First, we show that UCReach is closely related to MC($\mathcal{L}_{kh}(\star)$).

Lemma 1. *There is an exponential-time Turing reduction from* MC($\mathcal{L}_{kh}(\star)$) *to* UCReach.

Proof (Sketch). By using a standard labelling algorithm, the existence of a Turing reduction only requires to show that for all $\mathcal{S}=(S, (R_a)_{a \in \mathsf{Act}}, wf, L)$ and $s \in S$, $\mathcal{S}, s \Vdash \mathsf{Kh}^b(\mathsf{p}, \mathsf{q})$ can be reduced to an instance of UCReach. Recall that $\mathcal{S}, s \Vdash \mathsf{Kh}^b(\mathsf{p}, \mathsf{q})$ iff there is $\sigma \in \mathsf{Act}^*$ such that (1) $[\![\mathsf{p}]\!]^\mathcal{S} \subseteq \mathsf{SE}(\sigma)$, (2) $\mathsf{R}_\sigma([\![\mathsf{p}]\!]^\mathcal{S}) \subseteq [\![\mathsf{q}]\!]^\mathcal{S}$ and (3) for every $t \in [\![\mathsf{p}]\!]^\mathcal{S}$, σ is b-compatible at t. Below, w.l.o.g. we can assume that $[\![\mathsf{p}]\!]^\mathcal{S} \neq \emptyset$ because this case is obvious to handle. By Prop. 1, we can define a deterministic finite-state automaton $\mathcal{A} = (Q, \mathsf{Act}, \delta, I, F)$ that accepts exactly the plans satisfying (1)–(2) and $\mathcal{A}$ has the following properties:

- $|Q| \le 2^{|S|(|S|+2)}$,
- each location in Q can be encoded in polynomial space and deciding whether $q \xrightarrow{\mathsf{a}} q' \in \delta$ (respectively $q \in I$, $q \in F$) can be done in polynomial space.

Given $X \subseteq \mathsf{S}$, let us define the VASS $\mathcal{V}^\star[X] = (Q^\star, \mathsf{Act}, r, R^\star)$ as follows.

- $Q^\star \stackrel{\text{def}}{=} \mathsf{S} \times Q \uplus \{\mathtt{init}, \bot\}$,
- $(s,q) \xrightarrow{\mathsf{a},\boldsymbol{u}} (s',q') \in R^\star$ iff $s \xrightarrow{\mathsf{a}} s'$ in $\mathcal{S}$ and $q \xrightarrow{\mathsf{a}} q' \in \delta$ (synchronisation on the action a), and $\boldsymbol{u} = wf(s, \mathsf{a}, s')$. Obviously, the VASS $\mathcal{V}^\star[X]$ can be viewed as a synchronised product between $\mathcal{S}$ and $\mathcal{A}$.
- $(s,q) \xrightarrow{\mathsf{a},\mathbf{0}} \bot \in R^\star$ if there is no s' such that $s \xrightarrow{\mathsf{a}} s'$ in $\mathcal{S}$.
- $\bot \xrightarrow{\mathsf{a},\mathbf{0}} \bot \in R^\star$ for all $\mathsf{a} \in \mathsf{Act}$ and $\mathtt{init} \xrightarrow{\mathsf{a},\mathbf{0}} \bot \in R^\star$ for all $\mathsf{a} \in \mathsf{Act} \setminus \{\mathsf{a}^\dagger\}$ where $\mathsf{a}^\dagger$ is some distinguished action.
- $\mathtt{init} \xrightarrow{\mathsf{a}^\dagger, \boldsymbol{b}} (t, q_0) \in R^\star$ for all $t \in X$ and $q_0 \in I$.

The edges in $\mathcal{V}^\star[X]$ involving the sink location $\bot$ are designed so that $\mathcal{V}^\star[X]$ is complete but without affecting the correctness of the reduction. Indeed, one can show that there is a plan σ satisfying (1)–(3) iff $\mathcal{V}^\star[\![\mathsf{p}]\!]^{\mathcal{S}}]$, $\mathtt{init}$, $\mathbf{0}$, $\mathsf{S} \times F$ is a positive instance of the problem UCReach.

The size of $\mathcal{V}^\star[\![\mathsf{p}]\!]^{\mathcal{S}}]$ is at most exponential in the size of $\mathcal{S}$. So, the above equivalence provides a many-one reduction for checking $\mathcal{S}, s \Vdash \mathsf{Kh}^b(\mathsf{p}, \mathsf{q})$, whence the Turing reduction for the full model-checking problem. □

Note that the reduction in the proof of Lemma 1 preserves the number r of resource types. Below, we show that UCReach restricted to a single counter (denoted by UCReach(1)) is actually as hard as UCReach. Indeed, due to the synchronisation on pseudo-runs with the same plan σ, each dimension can be handled separately thanks to the universal quantification.

Lemma 2. *There is a logarithmic-space reduction from* UCReach *to* UCReach(1).

Thanks to Lemmas 1 and 2, it is enough to show that UCReach(1) is in Ack to prove that $\mathrm{MC}(\mathcal{L}_{kh}(\star))$ is in Ack. Eventually, we get the Ack-completeness of UCReach, UCReach(1), $\mathrm{MC}(\mathcal{L}_{kh}(1))$ and $\mathrm{MC}(\mathcal{L}_{kh}(\star))$.

4 UCReach Is in Ackermannian Time

This section is devoted to showing that UCReach(1), the universal control-state reachability problem restricted to one counter, can be solved in Ackermannian time, following the definition of the complexity class Ack from [42]. To do so, we take advantage of the notion of belief functions from [38] as well as of the well-structured properties of the underlying transition system (WSTS) and the analysis about the length of bad sequences as done in [38, pp. 94-95] (essentially relying on the results from [20], see also [41]). However, unlike what is done in [38] to characterise the complexity of a non-termination problem, we use a backward algorithm along the lines of [7,21,28], since UCReach is essentially a covering problem, as shown below.

Let $\mathcal{V} = (Q, \mathsf{Act}, 1, R), p_0 \in Q, y_0 \in \mathbb{N}, Q_F \subseteq Q$ be an instance of UCReach(1), that is fixed for the rest of this section. A *belief function* is a map $f : Q \to \mathbb{N} \cup \{\infty\}$ [38]. Given $X \subseteq Q \times \mathbb{N}$, we define its belief function f_X as: for all $q \in Q$,

$$f_X(q) \stackrel{\text{def}}{=} \min(\{\infty\} \cup \{v \mid (q, v) \in X\}).$$

Observe that q does not occur in X iff $f_X(q) = \infty$. Moreover, f_X can be understood as a finite abstraction of X. What matters in X are the configurations (q, v) with minimal value v. We write $supp(f) \stackrel{\text{def}}{=} \{q \mid f(q) \in \mathbb{N}\}$ to denote the *support set* of f. Belief functions are introduced in [15, Sect. 3], [38, Sect. 3] and [22, Sect. 5] and though we use such belief functions with a backward algorithm that differs from the forward algorithm in [38, Sect. 3.1], we can still take advantage of results introduced in [38].

Given two belief functions f and g, we define the binary relation $\preceq$ such that $f \preceq g$ iff $supp(f) = supp(g)$ and for all $q \in supp(f)$, we have $f(q) \leq g(q)$ [38, Sect. 3.1]. The relation $\preceq$ on belief functions behaves like the component-wise less than relation on $\mathbb{N}^{|Q|}$. In particular, as a consequence of Dickson's Lemma, the relation $\preceq$ on belief functions can be shown to be a well-quasi-ordering (wqo) too (basic definitions about well-quasi-orderings can be found in [21, Sect. 2.1]). Given a belief function f and a set $\mathcal{X}$ of belief functions, we write $\uparrow f$ to denote the set $\{g \mid f \preceq g\}$ and $\uparrow \mathcal{X}$ to denote the set $\{g \mid f \preceq g, f \in \mathcal{X}\}$. The set $\uparrow f$ (resp. $\uparrow \mathcal{X}$) is known as the *upward closure* of f (resp. $\mathcal{X}$). *Upward closed sets* $\mathcal{X}$ of belief functions are sets of belief functions satisfying $\mathcal{X} = \uparrow \mathcal{X}$.

We define the transition system $\mathcal{T}_\mathcal{V} \stackrel{\text{def}}{=} (\{f \mid f : Q \to \mathbb{N} \cup \{\infty\}\}, (\stackrel{\mathsf{a}}{\to})_{\mathsf{a} \in \mathsf{Act}})$ built over the set of belief functions such that $f \stackrel{\mathsf{a}}{\to} g$ iff the conditions (1)-(2) hold:

(1) for all $q \in supp(f)$ and all $q \stackrel{\mathsf{a},u}{\to} q' \in R$, we have $f(q) + u \geq 0$ ("all the a-transitions are fireable", i.e. no negative budget value is reached),

(2) for all $q' \in Q$,

$$g(q') \stackrel{\text{def}}{=} \min(\{v' \mid \exists\, q \in supp(f) \text{ and } q \stackrel{\mathsf{a},u}{\to} q' \in R \text{ s.t. } v' = f(q) + u\} \cup \{\infty\}).$$

Observe that each relation $\stackrel{\mathsf{a}}{\to}$ is deterministic but not necessarily total. Moreover, for every action $\mathsf{a} \in \mathsf{Act}$, there is a self-loop on the unique belief function with empty support set. We write $\stackrel{*}{\to}$ to denote the reflexive and transitive closure of $(\bigcup_{\mathsf{a} \in \mathsf{Act}} \stackrel{\mathsf{a}}{\to})$. Given a set $\mathcal{X}$ of belief functions, we write $pre_\mathsf{a}(\mathcal{X})$ to denote the set $\{f' \mid \exists\, f \in \mathcal{X} \text{ s.t. } f' \stackrel{\mathsf{a}}{\to} f\}$ and $pre(\mathcal{X})$ to denote $\bigcup_{\mathsf{a} \in \mathsf{Act}} pre_\mathsf{a}(\mathcal{X})$ (sets of predecessors). The set $pre^*(\mathcal{X})$ of belief functions is defined similarly. Formally, $f \in pre^*(\mathcal{X})$ iff there are $f_0, f_1, \ldots, f_n$ and $\mathsf{a}_1, \ldots, \mathsf{a}_n$ ($n \geq 0$) such that $f_n = f$, $f_0 \in \mathcal{X}$ and $f_i \in pre_{\mathsf{a}_i}(\{f_{i-1}\})$ for all $i \in [1, n]$. This means that we have

$$f = f_n \stackrel{\mathsf{a}_n}{\to} f_{n-1} \cdots f_1 \stackrel{\mathsf{a}_1}{\to} f_0 \in \mathcal{X}.$$

In order to establish that $\mathcal{T}_\mathcal{V}$ is a well-structured transition system (WSTS) in the sense of [21, Def. 2.5], we need to establish the monotony property below.

Lemma 3 (Monotony). *If $f \xrightarrow{a} g$ and $f \preceq f'$, then there is $g \preceq g'$ s.t. $f' \xrightarrow{a} g'$.*

Lemma 3 entails that $\mathcal{T}_\mathcal{V}$ is a WSTS, see [21, Def. 2.5] (the main properties being monotony and $\preceq$ is a well-quasi-ordering). As a consequence, we also get (see [21, Prop. 3.1]):

Lemma 4. *For all upward closed sets $\mathcal{X}$ of belief functions, the set $pre^*(\mathcal{X})$ is upward closed, i.e., $pre^*(\mathcal{X}) = {\uparrow}pre^*(\mathcal{X})$.*

Furthermore, Lemma 5 below states that solving UCReach(1) amounts to solving a covering problem on $\mathcal{T}_\mathcal{V}$, which can be decided using [21, Theorem 3.6]. It states that UCReach(1) and the covering problem for the WSTS $\mathcal{T}_\mathcal{V}$ can be used to solved each other. After proving it, it remains to show that the effectivity hypotheses in [21, Theorem 3.6] hold, which requires further developments.

Lemma 5. *The elements $\mathcal{V} = (Q, \mathsf{Act}, 1, R)$, $p_0 \in Q$, $y_0 \in \mathbb{N}$, $Q_F \subseteq Q$ form a positive instance of UCReach(1) iff $f_0 \xrightarrow{*} f$ for some $f_0, f \in \mathcal{T}_\mathcal{V}$ such that $supp(f_0) = \{p_0\}$, $f_0(p_0) = y_0$, and $\varnothing \neq supp(f) \subseteq Q_F$.*

By Lemma 5, in order to solve UCReach(1), we need to design a decision procedure to check whether $f_0 \in pre^*(\{f \mid \varnothing \neq supp(f) \subseteq Q_F\})$, where $\{f \mid \varnothing \neq supp(f) \subseteq Q_F\}$ is an upward closed set whose basis is $\{f \mid \varnothing \neq supp(f) \subseteq Q_F \text{ and } \forall q \in supp(f), f(q) = 0\}$, i.e.,

$$\{f \mid \varnothing \neq supp(f) \subseteq Q_F\} =$$
$${\uparrow}\{f \mid \varnothing \neq supp(f) \subseteq Q_F \text{ and } \forall q \in supp(f), f(q) = 0\}.$$

Given $f : Q \to \mathbb{N} \cup \{\infty\}$ and $\mathsf{a} \in \mathsf{Act}$, we define a finite set $pb(\mathsf{a}, f)$ of belief functions obtained from the belief function $root(\mathsf{a}, f) : Q \to \mathbb{N} \cup \{\infty\}$ defined as follows. Note that 'pb' stands for "pred-basis" by reference to the notation in [21, Sect. 3]. Given $q \in Q$, the value $root(\mathsf{a}, f)(q)$ is ∞ if there is $q \xrightarrow{\mathsf{a},u} q' \in R$ such that $f(q') = \infty$ or if there is no transition $q \xrightarrow{\mathsf{a},u} q'$ in R, for some u, q'; otherwise,

$$root(\mathsf{a}, f)(q) \stackrel{def}{=} \max(\{f(q') - u \mid q \xrightarrow{\mathsf{a},u} q' \in R\} \cup \{0\}).$$

Let $pb(\mathsf{a}, f)$ be the set of belief functions g obtained from $root(\mathsf{a}, f)$ by satisfaction of the three conditions below.

(PB1) $supp(g) \subseteq supp(root(\mathsf{a}, f))$.
(PB2) For all $q \in supp(g)$, we have $g(q) = root(\mathsf{a}, f)(q)$.
(PB3) There is g' such that $g \xrightarrow{\mathsf{a}} g'$ and $f \preceq g'$.

Viewing $root(\mathsf{a}, f)$ as the *seed* from which the elements g in $pb(\mathsf{a}, f)$ are defined, the conditions (PB1) and (PB2) simply state that g must be a restriction of $root(\mathsf{a}, f)$. However, (PB3) enforces that the restriction must not be too limited so that a belief function g' greater than f can be reached from g with $\xrightarrow{\mathsf{a}}$. It is also easy to prove that $pb(\mathsf{a}, f)$ is finite and actually $|pb(\mathsf{a}, f)| \leq 2^{|Q|}$.

In order to illustrate the construction of $pb(\mathsf{a}, f)$, consider the complete VASS with one counter below and the belief function f such that $supp(f) = \{q_3, q_4, q_5\}$ and $f(q_3) = f(q_4) = f(q_5) = 0$.

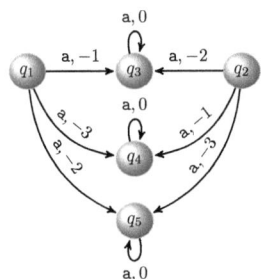

The belief function $root(\mathsf{a}, f)$ is such that $supp(root(\mathsf{a}, f)) = \{q_1, \ldots q_5\}$, $root(\mathsf{a}, f)(q_3) = root(\mathsf{a}, f)(q_4) = root(\mathsf{a}, f)(q_5) = 0$ and both $root(\mathsf{a}, f)(q_1)$ and $root(\mathsf{a}, f)(q_2)$ are equal to 3. We have $\{root(\mathsf{a}, f), f_1, f_2\} \subseteq pb(\mathsf{a}, f)$ with $supp(f_1) = \{q_1\}$, $supp(f_2) = \{q_2\}$ and $f_1(q_1) = f_2(q_2) = 3$. Observe that $root(\mathsf{a}, f)$ and f_i ($i = 1, 2$) are incomparable because the two belief functions have distinct support sets.

We present the main technical lemma to get decidability (and it is also useful to get the Ack upper bound, as explained further below).

Lemma 6. $\uparrow\!pb(\mathsf{a}, f) = \uparrow\!pre_{\mathsf{a}}(\uparrow\!f)$ *for all actions* $\mathsf{a} \in \mathsf{Act}$ *and belief functions* f.

As a consequence of Lemma 6, for all finite sets $\mathcal{Z}$ of belief functions, we have

$$\uparrow\!\bigcup\{pb(\mathsf{a}, f) \mid f \in \mathcal{Z}\} = \uparrow\!pre_{\mathsf{a}}(\uparrow\!\mathcal{Z}),$$

and therefore $\uparrow\!\bigcup\{pb(\mathsf{a}, f) \mid f \in \mathcal{Z}, \mathsf{a} \in \mathsf{Act}\} = pre(\uparrow\!\mathcal{Z})$. To solve the covering problem from Lemma 5, we proceed as follows using the proof method from [21]. We define a family $(\mathcal{Z}_i)_{i \in \mathbb{N}}$ made of finite sets of belief functions such that

- $\mathcal{Z}_0 \stackrel{\text{def}}{=} \{f \mid \emptyset \neq supp(f) \subseteq Q_F \text{ and for all } q \in supp(f), f(q) = 0\}$,
- $\mathcal{Z}_{i+1} \stackrel{\text{def}}{=} \mathcal{Z}_i \cup \bigcup\{pb(\mathsf{a}, f) \mid \mathsf{a} \in \mathsf{Act}, f \in \mathcal{Z}_i\}$.

By [21, Lemma 2.4], there is m such that $\uparrow\!\mathcal{Z}_m = \uparrow\!\mathcal{Z}_{m+1}$ ($\mathcal{T}_\mathcal{V}$ is a WSTS) and therefore the family $(\uparrow\!\mathcal{Z}_i)_{i \in \mathbb{N}}$ stabilises at some point. One can show that a positive instance of the covering problem holds iff $f_0 \in \uparrow\!\mathcal{Z}_m$ (since $\uparrow\!\mathcal{Z}_m = pre^*(\uparrow\!\mathcal{Z}_0)$). Since computing $pb(\mathsf{a}, f)$ is effective and checking whether $\uparrow\!\mathcal{Z} \subseteq \uparrow\!\mathcal{Z}'$ can be done effectively too, this provides a decision procedure. For instance, (A) $\uparrow\!\mathcal{Z} \subseteq \uparrow\!\mathcal{Z}'$ iff (B) for all $f \in \mathcal{Z}$, there is $f' \in \mathcal{Z}'$ such that $f' \preceq f$. First suppose that (A) holds. Let $f \in \mathcal{Z}$ and therefore $f \in {\uparrow}\mathcal{Z}'$ by (A). There is $f' \in \mathcal{Z}'$ such that $f' \preceq f$ and therefore (B) holds. Conversely, suppose that (B) holds and let $f \in {\uparrow}\mathcal{Z}$. There is $f' \in \mathcal{Z}$ such that $f' \preceq f$ and by (B), there is $f'' \in \mathcal{Z}'$ such that $f'' \preceq f'$. Consequently, $f'' \preceq f$ by transitivity of $\preceq$ and therefore $f \in {\uparrow}\mathcal{Z}'$.

Since we can effectively build the sequence $\mathcal{Z}_0, \ldots, \mathcal{Z}_{m+1}$ and each $\mathcal{Z}_{i+1}$ can be built in exponential-time in the respective sizes of the instance of UCReach(1) and of $\mathcal{Z}_i$, UCReach(1) can be solved in Ack as concluded below. Indeed, this holds as soon as we can reasonably bound the length m depending on the size N of the instance. Since ${\uparrow}\mathcal{Z}_0 \subset {\uparrow}\mathcal{Z}_1 \subset \cdots \subset {\uparrow}\mathcal{Z}_m = {\uparrow}\mathcal{Z}_{m+1}$, for all $i \in [1,m]$, there is a belief function f_i such that $f_i \in \mathcal{Z}_i$ and $f_i \notin {\uparrow}\mathcal{Z}_{i-1}$. The sequence $f_1, \ldots, f_m$ is a bad sequence in the sense that for all $i < i'$, we have $f_i \not\preceq f_{i'}$. As we use the notion of belief functions from [38], we can take advantage of the analysis about bad sequences therein, which is itself essentially based on [20]. Hence, we have $m \leq h(N)$ for some function h in $\mathcal{F}_{|Q|+3}$, where $\mathcal{F}_i$ is the ith level of Grzegorczyk hierarchy, see e.g. [42, Sect. 2]. The level $|Q|+3$ in the class $\mathcal{F}_{|Q|+3}$ is shown in [38, page 94]. Thus, $\mathcal{Z}_{m+1}$ can be computed in time $F_\omega(N)$ where F_ω is the fast-growing function at the level ω, see [42, Sect. 2.2.1].

Theorem 2. UCReach(1) *is in* Ack.

By putting together Lemmas 1 and 2 and Theorem 2, we get:

Corollary 1. UCReach *and* MC($\mathcal{L}_{kh}(\star)$) *are in* Ack.

5   Ack-Hardness of the Problem MC($\mathcal{L}_{kh}(\star)$)

Below, we show that MC($\mathcal{L}_{kh}(1)$) is Ack-hard by reducing the control-state reachability problem for VASS with resets, known to be Ack-hard [43, Sect. 6]. By Lemma 1, we get that UCReach(1) is also Ack-hard. Our proof strategy differs from the one for Ack-hardness of the fixed initial credit problem for blind games in [38, Theo. 2] that reduces some halting problem for Minsky machines.

A *VASS with resets* is an extension of the model of VASS by allowing resets on counters. The notions of pseudo-runs, runs, etc. are defined similarly to VASS once the one-step relation is defined (see Sect. 2.2). More precisely, a VASS with resets (RVASS for short) is a structure $\mathcal{V} = (Q, r, R)$, where Q is a finite set of *locations*, $r \in \mathbb{N}$ is its *dimension*, and R is a finite set of *transitions* in $Q \times OP[r] \times Q$ where $OP[r] \stackrel{\text{def}}{=} \mathbb{Z}^r \cup \{\text{reset}(i) \mid i \in [1,r]\}$. Note that there is no finite alphabet Σ in $\mathcal{V}$ as it is of no use in this part. Given a transition $T = q \xrightarrow{\boldsymbol{u}} q'$ with $\boldsymbol{u} \in \mathbb{Z}^r$, we write $(q, \boldsymbol{x}) \xrightarrow{T} (q', \boldsymbol{x}')$ whenever $\boldsymbol{x}' = \boldsymbol{u} + \boldsymbol{x}$ (as for VASS). By contrast, given a transition $T = q \xrightarrow{\text{reset}(i)} q'$, we write $(q, \boldsymbol{x}) \xrightarrow{T} (q', \boldsymbol{x}')$ whenever for all $j \neq i$, we have $\boldsymbol{x}'[j] = \boldsymbol{x}[j]$ and $\boldsymbol{x}'[i] = 0$. Since the operations in $OP[r]$ are deterministic, any pseudo-run $(q_0, \boldsymbol{x}_0) \xrightarrow{T_1} (q_1, \boldsymbol{x}_1) \cdots \xrightarrow{T_N} (q_N, \boldsymbol{x}_N)$ can be represented by the initial configuration $(q_0, \boldsymbol{x}_0)$ and the sequence of transitions $T_1 \cdots T_N$. The *control-state reachability problem for RVASS* (CReach(RVASS)) defined below is Ack-hard [43, Sect. 6]:

Input: an RVASS $\mathcal{V}$ and two locations q_0, q_f.
Question: does there exist a finite run $(q_0, \mathbf{0}) \xrightarrow{*} (q_f, \boldsymbol{x})$ for some $\boldsymbol{x} \in \mathbb{N}^r$?

In order to prepare the proof of Theorem 3 below, we introduce several useful notations. Let $\mathcal{V} = (Q, r, R)$ be an RVASS.

- For each $q \in Q$, we write $R^{src}(q)$ to denote the set of transitions $T \in R$ such that q is the source location (i.e. T is of the form $(q, \cdot, \cdot)$). We also write $R^{tgt}(q)$ to denote the set of transitions $T \in R$ such that q is the target location (i.e. T is of the form $(\cdot, \cdot, q)$).
- For each counter $i \in [1, r]$, we write $R(\mathsf{reset}(i))$ to denote the set of transitions T in R of the form $(\cdot, \mathsf{reset}(i), \cdot)$.

The run witnessing $(q_0, \mathbf{0}) \xrightarrow{*} (q_f, \boldsymbol{x})$ can be represented by the sequence of transitions $T_1 \cdots T_N$ since the initial configuration $(q_0, \mathbf{0})$ is fixed. Observe that $T_1 \in R^{src}(q_0)$, $T_N \in R^{tgt}(q_f)$ and for all $i \in [1, N-1]$ and $q \in Q$, $T_i \in R^{tgt}(q)$ implies $T_{i+1} \in R^{src}(q)$.

Theorem 3. $MC(\mathcal{L}_{kh}(1))$ *is* Ack*-hard.*

Proof (Sketch). We present a reduction from CReach(RVASS) to $MC(\mathcal{L}_{kh}(1))$. Given an RVASS $\mathcal{V} = (Q, r, R)$ and $q_0, q_f \in Q$, we construct a model $\mathcal{S}$ and a formula $\varphi = \mathsf{Kh}^0(\mathsf{init}, \mathsf{acc})$ where init, acc are atomic, and we show that for all states s, we have $\mathcal{S}, s \Vdash \varphi$ iff there is a run of $\mathcal{V}$ from $(q_0, \mathbf{0})$ to some configuration $(q_f, \boldsymbol{x})$ with $\boldsymbol{x} \in \mathbb{N}^r$. W.l.o.g., we can assume that $q_0 \neq q_f$ (otherwise the empty run does the job) and q_f is not the source location of any transition in R (otherwise, we duplicate q_f, one copy having no outgoing transitions).

The model $\mathcal{S} = (\mathsf{S}, (\mathsf{R}_\mathsf{a})_{\mathsf{a} \in \mathsf{Act}}, \mathsf{L})$ has action set $\mathsf{Act} \stackrel{\text{def}}{=} R \cup \{\sharp\}$ ($\sharp$ plays the role of an end marker), and we construct it so that there exists a plan $\sigma \in \mathsf{Act}^*$ that witnesses the satisfaction of φ in $\mathcal{S}$ if and only if $\sigma = \rho\sharp$, where $\rho \in R^*$ corresponds to a run of the RVASS $\mathcal{V}$ from the initial configuration $(q_0, \mathbf{0})$ to a configuration with location q_f. The model $\mathcal{S}$ consists of the union of three types of gadgets ($\bot$ and $\top$ are shared states). In Figs. 2a–c, an edge of the form $s \xrightarrow{\mathsf{a}, u} s'$ denotes $(s, s') \in \mathsf{R}_\mathsf{a}$ with $wf(s, \mathsf{a}, s') = u$. An edge of the form $s \xrightarrow{X, u} s'$ with $X \subseteq \mathsf{Act}$ means that for all $\mathsf{a} \in X$, we have $(s, s') \in \mathsf{R}_\mathsf{a}$ with $wf(s, \mathsf{a}, s') = u$.

In the construction, we wish to guarantee that from any state, any action is fireable (so that strong executability is immediate), which may overcomplicate the definition in a few places, in particular for the structural gadget below.

- The *structural gadget* (Fig. 2a) checks that the first action of the plan is a transition in $R^{src}(q_0)$, the penultimate action is a transition in $R^{tgt}(q_f)$ and the plan is a non-empty sequence followed by $\sharp$.
- The *sequence gadget* (Fig. 2b), parameterised by $q \in Q \setminus \{q_f\}$, checks that consecutive actions in the plan form a path in the control graph of the RVASS: whenever a transition in $\mathrm{R}^{tgt}(q)$ occurs in a plan, it must be followed by a transition in $\mathrm{R}^{src}(q)$.

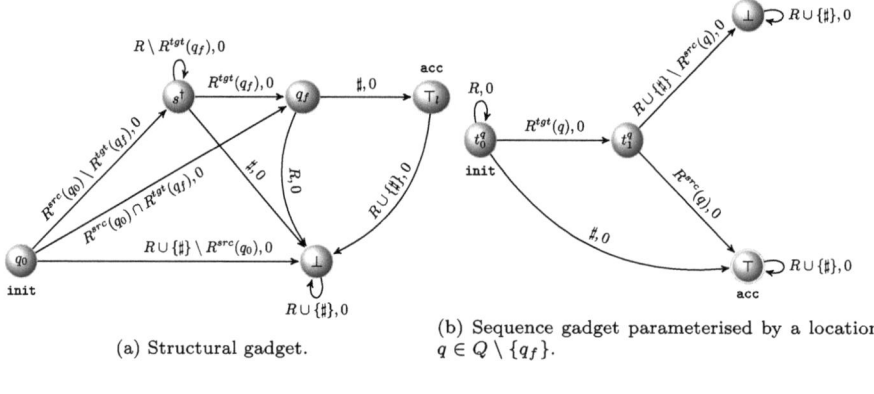

(a) Structural gadget.

(b) Sequence gadget parameterised by a location $q \in Q \setminus \{q_f\}$.

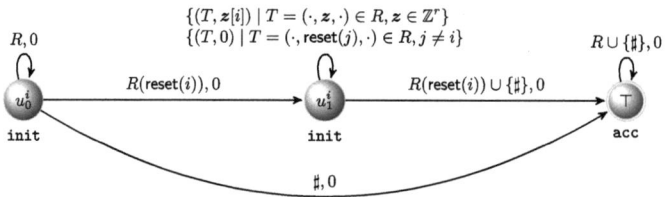

(c) Reset gadget parameterised by a counter $i \in [1, r]$.

Fig. 2. Gadgets.

- The *reset gadget* (Fig. 2c) checks that the i-th counter ($i = 1, \ldots, r$) remains non-negative between two resets: it has two states labelled by init, and guesses either a reset transition (from the leftmost state u_0^i), or the beginning of the run (from the middle state u_1^i) and tracks the value of the i-th counter until another reset transition or the final state is reached (witnessed by $\sharp$).

The model $\mathcal{S}$ consists of the union of (1) the structural gadget, (2) the sequence gadgets instantiated with all $q \in Q \setminus \{q_f\}$ and (3) the reset gadgets instantiated with $i \in [1, r]$. Using the construction above, one can show that $\mathcal{S} \Vdash \mathsf{Kh}^0(\mathsf{init}, \mathsf{acc})$ iff there is a plan $\sigma = \mathsf{a}_1 \cdots \mathsf{a}_L$ such that for all computations $s_0 \xrightarrow{\mathsf{a}_1} s_1 \cdots \xrightarrow{\mathsf{a}_L} s_L$ such that $\mathcal{S}, s_0 \Vdash \mathsf{init}$, we have: (a) $\mathcal{S}, s_L \Vdash \mathsf{acc}$; (b) for all $i \in [1, L]$, $wf(\lambda_{\leq i}) \geq 0$ (the condition about SE trivially holds, since for all states s and actions a, there is a state s' such that $(s, s') \in R_\mathsf{a}$).

It is easy to see that the size of $\mathcal{S}$ is polynomial in the size of the RVASS, and Ack-hardness of the model-checking problem for $\mathcal{L}_{kh}(1)$ follows from the fact that the control-state reachability problem for RVASS is Ack-hard [43]. □

Then, from Corollary 1 and Theorem 3 we conclude that Theorem 1 follows. By Lemma 1, UCReach is Ack-hard too.

6 Final Remarks

We studied the effects of adding budget-like constraints to the knowing how logic from [44], following a fruitful paradigm already used in formalisms like energy games and ATL-like logics. We proved that the model-checking problem for $\mathcal{L}_{kh}(\star)$ is Ack-complete (Theorem 1). The proof relies on the new universal control-state reachability problem over complete VASS (UCReach). To get Ack-membership, we designed an underlying well-structured transition system and showed the effectivity of a backward algorithm for some covering problem, following developments from [21,38]. The Ack-hardness is obtained by reducing the control-state reachability problem for VASS with resets. Hopefully, our work enriches the set of interesting problems on VASS and the scope of their applications. As future work, it would be interesting to find further applications of UCReach.

Acknowledgments. We would like to thank the anonymous reviewers for their comments and suggestions that helped us to improve the quality of the document. R. Fervari is supported by Agencia I+D+i grant PICT 2021-00400, the EU H2020 research and innovation programme under the Marie Skłodowska-Curie grant agreement 101008233 (MISSION), the IRP SINFIN, SeCyT-UNC grant 33620230100178CB, and the France 2030 program ANR-11-IDEX-0003.

References

1. Alechina, N., Bulling, N., Demri, S., Logan, B.: On the complexity of resource-bounded logics. TCS **750**, 69–100 (2018). https://doi.org/10.1016/j.tcs.2018.01.019
2. Alechina, N., Bulling, N., Logan, B., Nguyen, H.: The virtues of idleness: a decidable fragment of resource agent logic. Artif. Intell. **245**, 56–85 (2017). https://doi.org/10.1016/j.artint.2016.12.005
3. Areces, C., Cassano, V., Castro, P., Fervari, R., Saravia, A.R.: How easy it is to know how: an upper bound for the satisfiability problem. In: JELIA 2023. LNCS, vol. 14281, pp. 405–419. Springer (2023). https://doi.org/10.1007/978-3-031-43619-2_28
4. Blockelet, M., Schmitz, S.: Model-checking coverability graphs of vector addition systems. In: MFCS 2011, LNCS, vol. 6907, pp. 108–119. Springer (2011)
5. Bojańczyk, M., David, C., Muscholl, A., Schwentick, T., Segoufin, L.: Two-variable logic on data words. ACM Trans. Comput. Log. **12**(4), 27 (2011)
6. Bouyer, P., Fahrenberg, U., Larsen, K.G., Markey, N., Srba, J.: Infinite runs in weighted timed automata with energy constraints. In: Cassez, F., Jard, C. (eds.) FORMATS 2008. LNCS, vol. 5215, pp. 33–47. Springer, Heidelberg (2008). https://doi.org/10.1007/978-3-540-85778-5_4
7. Bozzelli, L., Ganty, P.: Complexity analysis of the backward coverability algorithm for VASS. In: Delzanno, G., Potapov, I. (eds.) RP 2011. LNCS, vol. 6945, pp. 96–109. Springer, Heidelberg (2011). https://doi.org/10.1007/978-3-642-24288-5_10
8. Bulling, N., Goranko, V.: Combining quantitative and qualitative reasoning in concurrent multi-player games. Auton. Agent. Multi-Agent Syst. **36**(1), 1–33 (2021). https://doi.org/10.1007/s10458-021-09531-9

9. Cao, R., Naumov, P.: Budget-constrained dynamics in multiagent systems. In: IJCAI 2017, pp. 915–921. ijcai.org (2017). https://www.ijcai.org/proceedings/2017/0127.pdf
10. Cardoza, E., Lipton, R., Meyer, A.: Exponential space complete problems for petri nets and commutative semigroups: Preliminary report. In: STOC 1976, pp. 50–54. ACM (1976)
11. Chatterjee, K., Doyen, L., Henzinger, T.A.: The cost of exactness in quantitative reachability. In: Aceto, L., Bacci, G., Bacci, G., Ingólfsdóttir, A., Legay, A., Mardare, R. (eds.) Models, Algorithms, Logics and Tools. LNCS, vol. 10460, pp. 367–381. Springer, Cham (2017). https://doi.org/10.1007/978-3-319-63121-9_18
12. Cimatti, A., Pistore, M., Roveri, M., Traverso, P.: Weak, strong, and strong cyclic planning via symbolic model checking. Artif. Intell. **147**(1–2), 35–84 (2003)
13. Czerwinski, W., Orlikowski, L.: Reachability in vector addition systems is Ackermann-complete. In: STOC 2021, pp. 1229–1240. IEEE (2021)
14. David, C.: Analyse de XML avec données non-bornées. Ph.D. thesis, LIAFA, Université Paris VII (2009)
15. Degorre, A., Doyen, L., Gentilini, R., Raskin, J.-F., Toruńczyk, S.: Energy and mean-payoff games with imperfect information. In: Dawar, A., Veith, H. (eds.) CSL 2010. LNCS, vol. 6247, pp. 260–274. Springer, Heidelberg (2010). https://doi.org/10.1007/978-3-642-15205-4_22
16. Delzanno, G., Sangnier, A., Zavattaro, G.: Parameterized verification of ad hoc networks. In: Gastin, P., Laroussinie, F. (eds.) CONCUR 2010. LNCS, vol. 6269, pp. 313–327. Springer, Heidelberg (2010). https://doi.org/10.1007/978-3-642-15375-4_22
17. Demri, S.: On selective unboundedness of VASS. JCSS **79**(5), 689–713 (2013)
18. Demri, S., Fervari, R.: Model-checking for ability-based logics with constrained plans. In: AAAI 2023, pp. 6305–6312. AAAI Press (2023). https://ojs.aaai.org/index.php/AAAI/article/view/25776
19. Esparza, J.: Decidability and complexity of petri net problems — an introduction. In: Reisig, W., Rozenberg, G. (eds.) ACPN 1996. LNCS, vol. 1491, pp. 374–428. Springer, Heidelberg (1998). https://doi.org/10.1007/3-540-65306-6_20
20. Figueira, D., Figueira, S., Schmitz, S., Schnoebelen, P.: Ackermannian and primitive-recursive bounds with dickson's lemma. In: LiCS 2011, pp. 269–278 (2011). https://arxiv.org/abs/1007.2989
21. Finkel, A., Schnoebelen, P.: Well-structured transitions systems everywhere! TCS **256**(1–2), 63–92 (2001). https://doi.org/10.1016/S0304-3975(00)00102-X
22. Hofman, P., Totzke, P.: Trace inclusion for one-counter nets revisited. TCS **735**, 50–63 (2018). https://doi.org/10.1016/j.tcs.2017.05.009
23. Hopcroft, J., Pansiot, J.: On the reachability problem for 5-dimensional vector addition systems. TCS **8**, 135–159 (1979). https://doi.org/10.1016/0304-3975(79)90041-0
24. Kaiser, A., Kroening, D., Wahl, T.: Dynamic cutoff detection in parameterized concurrent programs. In: Touili, T., Cook, B., Jackson, P. (eds.) CAV 2010. LNCS, vol. 6174, pp. 645–659. Springer, Heidelberg (2010). https://doi.org/10.1007/978-3-642-14295-6_55
25. Karp, R.M., Miller, R.E.: Parallel program schemata. JCSS **3**(2), 147–195 (1969). https://doi.org/10.1016/S0022-0000(69)80011-5
26. Kosaraju, R.: Decidability of reachability in vector addition systems. In: STOC 1982, pp. 267–281 (1982)
27. Lambert, J.: A structure to decide reachability in petri nets. TCS **99**, 79–104 (1992)

28. Lazić, R., Schmitz, S.: The ideal view on Rackoff's coverability technique. Inf. Comput. **277**, 104582 (2021). https://doi.org/10.1016/j.ic.2020.104582
29. Leroux, J.: The general vector addition system reachability problem by Presburger inductive invariants. In: LiCS 2009, pp. 4–13. IEEE (2009)
30. Leroux, J.: Vector addition system reachability problem (A short self-contained proof). In: POPL 2011, pp. 307–316 (2011)
31. Leroux, J.: The reachability problem for petri nets is not primitive recursive. In: FOCS 2021, pp. 1241–1252. IEEE (2021). https://arxiv.org/abs/2104.12695
32. Leroux, J., Schmitz, S.: Reachability in vector addition systems is primitive-recursive in fixed dimension. In: LiCS 2019, pp. 1–13. IEEE (2019)
33. Li, Y.: Knowing what to do: a logical approach to planning and knowing how. Ph.D. thesis, University of Groningen (2017). https://pure.rug.nl/ws/portalfiles/portal/47919164/Complete_thesis.pdf
34. Li, Y., Wang, Y.: Achieving while maintaining: - a logic of knowing how with intermediate constraints. In: ICLA 2017, LNCS, vol. 10119, pp. 154–167. Springer (2017). https://doi.org/10.1007/978-3-662-54069-5_12
35. Lipton, R.: The reachability problem requires exponential space. Technical Report 62, Department of Computer Science, Yale University (1976). http://www.cs.yale.edu/publications/techreports/tr63.pdf
36. Mayr, E.: An algorithm for the general petri net reachability problem. SIAM J. Comput. **13**(3), 441–460 (1984)
37. Minsky, M.: Computation, Finite and Infinite Machines. Prentice Hall (1967)
38. Pérez, G.: The fixed initial credit problem for partial-observation energy games is ack-complete. IPL **118**, 91–99 (2017). https://doi.org/10.1016/j.ipl.2016.10.005
39. Rackoff, C.: The covering and boundedness problems for vector addition systems. TCS **6**(2), 223–231 (1978). https://doi.org/10.1016/0304-3975(78)90036-1
40. Reutenauer, C.: The mathematics of Petri nets. Masson and Prentice (1990)
41. Schmitz, S.: Algorithmic complexity of well-quasi-orders, November 2017, habilitation thesis
42. Schmitz, S.: Complexity hierarchies beyond elementary. ACM Trans. Comput. Theor. **8**(1), 3:1–3:36 (2016). https://doi.org/10.1145/2858784
43. Schnoebelen, P.: Revisiting ackermann-hardness for lossy counter machines and reset petri nets. In: Hliněný, P., Kučera, A. (eds.) MFCS 2010. LNCS, vol. 6281, pp. 616–628. Springer, Heidelberg (2010). https://doi.org/10.1007/978-3-642-15155-2_54
44. Wang, Y.: A logic of knowing how. In: LORI 2015, LNCS, vol. 9394, pp. 392–405. Springer (2015). https://doi.org/10.1007/978-3-662-48561-3_32
45. Wang, Y.: A logic of goal-directed knowing how. Synthese **195**(10), 4419–4439 (2018). https://doi.org/10.1007/s11229-016-1272-0
46. Wang, Y.: Beyond knowing that: a new generation of epistemic logics. In: van Ditmarsch, H., Sandu, G. (eds.) Jaakko Hintikka on Knowledge and Game-Theoretical Semantics. OCL, vol. 12, pp. 499–533. Springer, Cham (2018). https://doi.org/10.1007/978-3-319-62864-6_21

Nets-Within-Nets Through the Lens of Data Nets

Francesco Di Cosmo[1], Soumodev Mal[2(✉)], and Tephilla Prince[3]

[1] Free University of Bozen-Bolzano, Bolzano, Italy
frdicosmo@unibz.it
[2] Chennai Mathematical Institute, Kelambakkam, India
soumodevmal@cmi.ac.in
[3] IIT Dharwad, Dharwad, India
tephilla.prince.18@iitdh.ac.in

Abstract. Elementary Object Systems (EOSs) are a model in the nets-within-nets (NWNs) paradigm, where tokens in turn can host standard Petri nets. We study the complexity of the reachability problem of EOSs when subjected to non-deterministic token losses. It is known that this problem is equivalent to the coverability problem with no lossiness of conservative EOSs (cEOSs). We precisely characterize cEOS coverability into the framework of data nets, whose tokens carry data from an infinite domain. Specifically, we show that cEOS coverability is equivalent to the coverability of an interesting fragment of data nets that extends beyond νPN (featuring globally fresh name creation), yet remains less expressive than Unordered Data Nets (featuring lossy name creation as well as powerful forms of whole-place operations and broadcasts). This insight bridges two apparently orthogonal approaches to PN extensions, namely data nets and NWNs. At the same time, it enables us to analyze cEOS coverability taking advantage of known results on data nets. As a byproduct, we immediately get that the complexity of cEOS coverability lies between $\mathbf{F}_{\omega^2}$ and $\mathbf{F}_{\omega^\omega}$, two classes beyond Primitive Recursive.

Keywords: Data nets · Nets-within-Nets · Coverability · Hyper-Ackermannian problems · Fast-growing complexity hierarchy

1 Introduction

Recent works have studied the Nets Within Nets (NWN) paradigm [1,7,18], i.e., Petri Nets (PNs) whose tokens in turn carry PNs, as a model for the robustness of multiagent systems against agent breakdowns and, more generally, agent imperfections, modeled as token losses. These works focus on the reachability/coverability problems of Elementary Object Systems (EOS), i.e., NWNs where there is only one level of nesting. Other forms of NWNs can be found in [6,8,11,12,18]. Out of the several combinations of problem type (reachability and coverability), lossiness degree (none, finite, unbounded number of token losses), and level (at the outer, nested, or both levels), only reachability/coverability of EOSs under an unbounded amount of lossiness at both levels is decidable [1]. The picture is

© The Author(s), under exclusive license to Springer Nature Switzerland AG 2026
P. Ganty and A. Mansutti (Eds.): RP 2025, LNCS 16230, pp. 156–170, 2026.
https://doi.org/10.1007/978-3-032-09524-4_11

moderately more optimistic when the constraint of conservativity is applied. In EOSs, each place can host tokens of a fixed type. In a conservative EOS (cEOS), if a transition consumes an object of a given type, at least one object of the same type must be produced, i.e., the set of types available in the net is conserved. When subjected to lossiness at both levels, the cEOS reachability/coverability problem is equivalent to perfect cEOS coverability (see [2]), which is decidable. Instead, perfect cEOS reachability is known to be undecidable. Hence, the decidability boundary of reachability/coverability of lossy/perfect EOS/cEOS is fully charted [1]. However, the precise complexity class of lossy EOS reachability and perfect cEOS coverability is unknown, besides an $F_{\omega 2}$ lower bound [2].

The complexity of verification problems for NWNs have not been well-studied. In contrast, many results about the complexity of coverability for several data extensions of PNs [9], whose tokens carry data from an infinite domain, are available in the literature (see Fig. 1 in [10]). For νPN [15] the coverability problem is double-Ackermannian, $F_{\omega 2}$-complete [10]. For unordered data nets (UDNs), the coverability is hyper-Ackermannian, F_{ω^ω}-complete [14]. Ordered data nets and ordered data Petri nets are both $F_{\omega^{\omega^\omega}}$-complete [5].

In this paper, we pave the way to the study of cEOS coverability and its equivalent forms (such as lossy EOS reachability). Instead of directly attacking this problem, for example by exploiting techniques for complexity over Well-Structured Transition Systems [3, 14, 16, 17], we bridge the nesting paradigm with the apparently orthogonal extension of Petri nets with data. We show that cEOS coverability can be characterized as the coverability of a type of net with data, sitting in between νPN and UDNs. Our technical contributions are as follows:

1. We introduce channel-νPN (c-νPN), a clean minimal extension of νPNs suitable to characterize the complexity power of cEOS coverability.
2. We show that cEOS coverability and c-νPN coverability are inter-reducible. As a consequence, we obtain inter-reducibility also between c-νPN coverability and lossy EOS reachability under an unbounded number of losses as well as cEOS reachability with any non-zero number of token losses.
3. On top of the $\mathcal{F}_{\omega 2}$ lower bound from [2], we obtain a novel $\mathcal{F}_{\omega^\omega}$ upper bound from the literature on Unordered Data Nets (UDNs).

Overall, we provide a novel connection between the nested and data paradigm, useful to study the complexity of perfect/imperfect EOS/cEOS reachability/coverability problems under the lens of data nets.

This paper is an abridged version of the technical report [2]. Section 2 provides preliminaries on PNs, νPNs, and cEOSs. Section 3 introduces c-νPNs and rename-νPN (r-νPN). Section 4 (Sect. 5) reduces r-νPN (cEOS) coverability to cEOS (c-νPN) coverability. Section 6 discusses the results.

2 Preliminaries

Multisets. Given $n \in \mathbb{N}$, $[n] = \{1, 2, \cdots, n\}$. A *multiset* m on a set D is a mapping $m : D \to \mathbb{N}$. The *support* of m is the set $\text{Supp}(m) = \{i \mid m(i) > 0\}$.

The multiset m is finite if $\text{Supp}(m)$ is finite. The family of all multisets over D is denoted by $D^\oplus$. We denote a finite multiset m by enumerating the elements $d \in \text{Supp}(m)$ exactly $m(d)$ times in between $\{\!\{$ and $\}\!\}$. The empty multiset $\{\!\{\}\!\}$ is also denoted by $\emptyset$. The empty multiset on the empty domain is denoted by ε. Given two multisets m_1 and m_2 on D, we define the multisets $m_1 + m_2$ and $m_1 - m_2$ on D as follows: $(m_1 + m_2)(d) = m_1(d) + m_2(d)$ and $(m_1 - m_2)(d) = \max(m_1(d) - m_2(d), 0)$, for each $d \in D$. We write $m_1 \sqsubseteq m_2$ if, for each $d \in D$, we have $m_1(d) \le m_2(d)$.

Petri Nets. A PN [13] is a tuple $N = (P, T, F)$, where P is a finite *place set*, T is a finite *transition set*, and $F : (P \times T) \cup (T \times P) \longrightarrow \mathbb{N}$ is a *flow function*. We set the functions of *pre-* and *post-conditions* $\text{pre}_N, \text{post}_N : T \to (P \to \mathbb{N})$ where $\text{pre}_N(t)(p) = F(p, t)$ and $\text{post}_N(t)(p) = F(t, p)$. A *marking* μ is a finite multiset on P. A $t \in T$ is enabled on μ if, for each $p \in P$, we have $\text{pre}_N(t)(p) \le \mu(p)$. Its firing results in the marking μ' such that $\mu'(p) = \mu(p) - \text{pre}_N(t)(p) + \text{post}_N(t)(p)$, for each $p \in P$. We always assume that P is ordered. Thus, we can denote a marking m as a multiset or as a vector $\langle m(p) \rangle_{p \in P}$. We also work with the special *empty PN* $\blacktriangle = (\emptyset, \emptyset, \emptyset)$, whose only marking is ε. Places are depicted by circles, transitions by rectangles, markings by multisets of black tokens $\bullet$ in the respective places, and the flow by labeled arrows (see Fig. 1a and Fig. 1b).

νPN. A νPN is a PN where each token is associated with a data value that comes from a countable domain. The flow function is extended to check equality/inequality of data values using a finite set of variables as labels. We recall νPN as in [10]. Let Υ and $\mathcal{X}$ be disjoint sets of *fresh* and *standard variables*, denoted by x_i and ν_i for $i \in \mathbb{N}$, respectively. Let $Vars \stackrel{\text{def}}{=} \mathcal{X} \bigcup \Upsilon$.

Definition 1. *A νPN is a PN $\mathcal{D} = \langle P, T, F \rangle$ with the provision that $F : (P \times T) \bigcup (T \times P) \to Vars^\oplus$ and, for each $t \in T$, $\Upsilon \cap \text{pre}(t) = \emptyset$ and $\text{post}(t) \setminus \Upsilon \subseteq \text{pre}(t)$, where $\text{pre}(t) = \bigcup_{p \in P} supp(F(p, t))$ and $\text{post}(t) = \bigcup_{p \in P} supp(F(p, t))$.*

For each $t \in T$, we set $\text{Var}(t) = \text{pre}(t) \cup \text{post}(t)$. In this section, we work with a fixed arbitrary νPN $\mathcal{D} = \langle P, T, F \rangle$ where $P = \{p_1, \ldots, p_\ell\}$. The flow F_x of a variable $x \in \text{Var}$ is $F_x : (P \times T) \bigcup (T \times P) \to \mathbb{N}$ where $F_x(p, t) \stackrel{\text{def}}{=} F(p, t)(x)$ and $F_x(t, p) \stackrel{\text{def}}{=} F(t, p)(x)$. We denote $\langle F_x(p_1, t), \ldots, F_x(p_\ell, t) \rangle \in \mathbb{N}^\ell$ by $F_x(P, t)$ and $\langle F_x(t, p_1), \ldots, F_x(t, p_\ell) \rangle \in \mathbb{N}^\ell$ by $F_x(t, P)$. The set of *configurations* of $\mathcal{D}$ is the set $(\mathbb{N}^P)^\oplus$. For each $t \in T$, let $out_\Upsilon(t) \stackrel{\text{def}}{=} \sum_{\nu \in \Upsilon(t)} \{\!\{F_\nu(t, P)\}\!\}$. Given a configuration $M = \{\!\{m_1, \ldots, m_{|M|}\}\!\}$, a transition t is fireable from M if there is a function $e : \mathcal{X}(t) \longrightarrow \{1, \ldots, |M|\}$, called *mode*, such that, for each $x \in \mathcal{X}(t)$, $F_x(P, t) \le m_{e(x)}$. We write $M \to^{t,e} M'$ if, for some configuration M'', $M = M'' + \sum_{x \in \mathcal{X}(t)} \{\!\{m_{e(x)}\}\!\}$ and $M' = M'' + out_\Upsilon(t) + \sum_{x \in \mathcal{X}(t)} \{\!\{m'_{e(x)}\}\!\}$ where, for $x \in \mathcal{X}(t)$, $m'_{e(x)} = m_{e(x)} - F_x(P, t) + F_x(t, P)$. The firing of t with mode e over M applies F_x, for each $x \in \mathcal{X}(t)$, to a distinct tuple $m \in M$ such that $m_{e(x)} \ge F_x(P, t)$ and replaces it with $m'_{e(x)}$. It also adds the new markings $F_\nu(t, P)$ for $\nu \in \Upsilon$. Each tuple m in a configuration M is depicted similarly to PN, but using a dedicated symbol in place of $\bullet$ (see Fig. 1c and Fig. 1d).

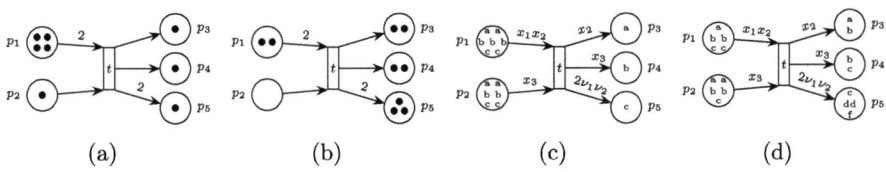

Fig. 1. (a) a simple Petri net and (b) the resulting marking on firing t. (c) A ν-net and (d) the resulting configuration on firing t with mode e instantiating x_1 to (the tuple identified by) a, x_2 to b, x_3 to c, ν_1 to d, and ν_2 to f.

Elementary Object Systems. An EOS [6,18] is a PN where each token is, in turn, a PN. The underlying PN is known as *system net*, and its tokens, which are PNs, are known as *object net*. A firing can happen asynchronously in either the system net or the object net or synchronously in both.

Definition 2. *An EOS $\mathfrak{E}$ is a tuple $\mathfrak{E} = \langle \hat{N}, \mathcal{N}, d, \Theta \rangle$ where:*

1. $\hat{N} = \langle \hat{P}, \hat{T}, \hat{F} \rangle$ *is a PN called* system net; $\hat{T}$ *contains a special set* $ID_{\hat{P}} = \{id_p \mid p \in \hat{P}\} \subseteq \hat{T}$ *of idle transitions such that, for each distinct $p, q \in \hat{P}$, we have* $\hat{F}(p, id_p) = \hat{F}(id_p, p) = 1$ *and* $\hat{F}(q, id_p) = \hat{F}(id_p, q) = 0$.
2. $\mathcal{N}$ *is a finite set of PNs, called* object PNs, *such that* $\blacktriangle \in \mathcal{N}$ *and if* $(P_1, T_1, F_1), (P_2, T_2, F_2) \in \mathcal{N} \cup \{\hat{N}\}$, *then* $P_1 \cap P_2 = \emptyset$ *and* $T_1 \cap T_2 = \emptyset$.
3. $d : \hat{P} \to \mathcal{N}$ *is called the* typing function.
4. Θ *is a finite set of events where each event is a pair $\langle \hat{\tau}, \theta \rangle$, where $\hat{\tau} \in \hat{T}$ and $\theta : \mathcal{N} \to \bigcup_{(P,T,F) \in \mathcal{N}} T^{\oplus}$, such that $\theta((P, T, F)) \in T^{\oplus}$ for each $(P, T, F) \in \mathcal{N}$ and, if $\hat{\tau} = id_p$, then $\theta(d(p)) \neq \emptyset$.*

A *nested token* is a system net token carrying an internal marking.

Definition 3. *Let $\mathfrak{E} = \langle \hat{N}, \mathcal{N}, d, \Theta \rangle$ be an EOS. The set of* nested tokens $\mathcal{T}(\mathfrak{E})$ *of $\mathfrak{E}$ is the set* $\bigcup_{(P,T,F) \in \mathcal{N}} (d^{-1}(P, T, F) \times P^{\oplus})$. *The set of* nested markings $\mathcal{M}(\mathfrak{E})$ *of $\mathfrak{E}$ is $\mathcal{T}(\mathfrak{E})^{\oplus}$. Given $\lambda, \rho \in \mathcal{M}(\mathfrak{E})$, we say that λ is a sub-marking of μ if $\lambda \sqsubseteq \mu$.*

Nested tokens are depicted via a dashed line from the token in the system net place to an instance of the object type where the internal marking is represented in the standard PN way. However, if the nested token is $\langle p, \varepsilon \rangle$ where p is of type $\blacktriangle$, we represent it with a black-token $\blacktriangle$ on p. Events $\langle \hat{\tau}, \theta \rangle$ are depicted by labeling the system net transition $\hat{\tau}$ by the multiset θ of object net transitions (see Fig. 2). If there are several events involving $\hat{\tau}$, then $\hat{\tau}$ has several labels.

Example 1. Fig. 2a depicts the EOS $\mathfrak{E} = (\hat{N}, \mathcal{N}, d, \Theta)$ where we have

- the system net $\hat{N} = (\{p_1, p_2, p_3, p_4, p_5\}, \{\hat{t}\}, \hat{F})$ where $\hat{F}(p_1, \hat{t}) = 2$ and $\hat{F}(p_2, \hat{t}) = \hat{F}(\hat{t}, p_3) = \hat{F}(\hat{t}, p_4) = \hat{F}(\hat{t}, p_5) = 1$. Otherwise, $\hat{F}$ is the constant function ε.
- the set of object nets $\mathcal{N} = \{N_1, N_2, \blacktriangle\}$, where $N_1 = (\{q_1, q_2\}, \{t_1\}, F_1)$, $N_2 = (\{r_1\}, \{t_2\}, F_2)$ where $F_1(q_1, t_1) = F_1(t_1, q_2) = 1$ and $F_2(t_2, r_1) = 1$.

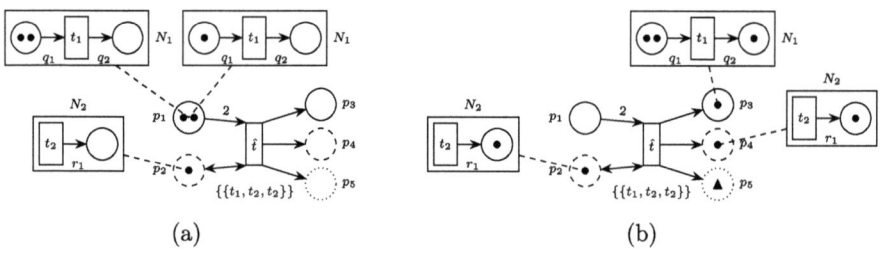

Fig. 2. An EOS depicting the firing of synchronized event $\langle \hat{t}, \{\{t_1, t_2, t_2\}\}\rangle$ for a given mode (λ, ρ), where $\lambda = \langle p_1, \{\{q_1, q_1\}\}\rangle + \langle p_1, \{\{q_1\}\}\rangle + \langle p_2, \{\{\}\}\rangle$, $\rho = \langle p_2, \{\{r_1\}\}\rangle + \langle p_3, \{\{q_1, q_1, q_2\}\}\rangle + \langle p_4, \{\{r_1\}\}\rangle + \langle p_5, \varepsilon\rangle$, $d(p_1) = d(p_3) = N_1$, $d(p_2) = d(p_4) = N_2$, $d(p_5) = \blacktriangle$.

- the typing function d, where $d(p_1) = d(p_3) = N_1, d(p_2) = d(p_4) = N_2$, and $d(p_5) = \blacktriangle$.
- the set of events $\Theta = \{\langle \hat{t}, \{\{t_1, t_2, t_2\}\}\rangle\}$.

Fig. 2a also depicts the nested marking $\langle p_1, \{\{q_1, q_1\}\}\rangle + \langle p_1, \{\{q_1\}\}\rangle + \langle p_2, \{\{\}\}\rangle$.

Intuitively, the firing of an event $e = \langle \tau, \theta \rangle$ can be characterized as follows: first, merge, type by type, all objects handled by the preconditions of τ (identified by a mode, as in νPNs), obtaining merged tuples; second, fire, at the same time, all transitions in θ on the merged tuples; third, non-deterministically distribute the tokens in the updated tuples among new objects in the places indicated by the post-conditions of τ. Possibly, some of the new objects have empty marking. This process is formalized by the next definitions.

Definition 4. *Let $\mathfrak{E}$ be an EOS $\langle \hat{N}, \mathcal{N}, d, \Theta \rangle$. The projection operator Π^1 maps each nested marking $\mu = \sum_{i \in I} \langle \hat{p}_i, M_i \rangle$ for $\mathfrak{E}$ to the PN marking $\sum_{i \in I} \hat{p}_i$ for $\hat{N}$. Given an object net $N \in \mathcal{N}$, the projection operator Π_N^2 maps each nested marking $\mu = \sum_{i \in I} \langle \hat{p}_i, M_i \rangle$ for $\mathfrak{E}$ to the PN marking $\sum_{j \in J} M_j$ for N where $J = \{i \in I \mid d(\hat{p}_i) = N\}$.*

Example 2. Let $M = \langle p_1, \{\{q_1, q_1\}\}\rangle + \langle p_1, \{\{q_1\}\}\rangle + \langle p_2, \varepsilon\rangle$ be the nested marking in Fig. 2a. We have: $\Pi^1(M) = \{\{p_1, p_1, p_2\}\}$; $\Pi_{N_1}^2(M) = \{\{q_1, q_1\}\} + \{\{q_1\}\}$; $\Pi_{N_2}^2 = \{\{\}\}$; $\Pi_{\blacktriangle}^2 = \varepsilon$.

We now define the *enabledness condition*. Set $\mathtt{pre}_N(\theta(N)) = \sum_{i \in I} \mathtt{pre}_N(t_i)$ and $\mathtt{post}_N(\theta(N)) = \sum_{i \in I} \mathtt{post}_N(t_i)$ where $(t_i)_{i \in I}$ is an enumeration of $\theta(N)$.

Definition 5. *Let $\mathfrak{E}$ be an EOS $\langle \hat{N}, \mathcal{N}, d, \Theta \rangle$. Given an event $e = \langle \hat{\tau}, \theta \rangle \in \Theta$ and markings $\lambda, \rho \in \mathcal{M}(\mathfrak{E})$, the enabledness condition $\Phi(\langle \hat{\tau}, \theta \rangle, \lambda, \rho)$ holds iff*

$$\Pi^1(\lambda) = \mathtt{pre}_{\hat{N}}(\hat{\tau}) \wedge \Pi^1(\rho) = \mathtt{post}_{\hat{N}}(\hat{\tau}) \wedge \forall N \in \mathcal{N}, \Pi_N^2(\lambda) \geq \mathtt{pre}_N(\theta(N)) \wedge$$
$$\forall N \in \mathcal{N}, \Pi_N^2(\rho) = \Pi_N^2(\lambda) - \mathtt{pre}_N(\theta(N)) + \mathtt{post}_N(\theta(N))$$

The event e is enabled with mode (λ, ρ) on a marking μ iff $\Phi(e, \lambda, \rho)$ holds and $\lambda \sqsubseteq \mu$. Its firing results in the step $\mu \xrightarrow{(e, \lambda, \rho)} \mu - \lambda + \rho$.

Example 3. Let $\lambda = \langle p_1, \{\{q_1, q_1\}\}\rangle + \langle p_1, \{\{q_1\}\}\rangle + \langle p_2, \{\{\}\}\rangle$ and $\rho = \langle p_2, \{\{r_1\}\}\rangle + \langle p_3, \{\{q_1, q_1, q_2\}\}\rangle + \langle p_4, \{\{r_1\}\}\rangle + \langle p_5, \varepsilon\rangle$. In Fig. 2, the event $e = \langle \hat{t}, \{\{t_1, t_2, t_2\}\}\rangle$ is enabled under mode (λ, ρ). Figure 2b depicts the nested marking reached after firing e under mode (λ, ρ). When compared to the intuitive presentation (provided above) of the semantics, we have, for each type $N \in \mathcal{N}$:

- $\Pi^1(\lambda)$ amounts to the consumed system net tokens.
- $\Pi_N^2(\lambda)$ is the resultant marking obtained by merging the consumed system net tokens of type N.
- $\Pi_N^2(\rho)$ amounts to the firing, on the merged marking above, of the transitions in $\{\{t_1, t_2, t_2\}\}$ from net N.
- $\Pi^1(\rho)$ amounts to the produced system net tokens.
- $\rho - \lambda$ amounts to the distributed marking.

Note that the non-determinism of the final distribution is captured by the choice of one of the many possible enabling modes.

An EOS is *conservative* if, for each system net transition t, if t consumes a nested token on a place of type N, then it produces at least one token on a place of that same type.

Definition 6. *A cEOS is an EOS* $\mathfrak{E} = \langle \hat{N}, \mathcal{N}, d, \Theta\rangle$ *with* $\hat{N} = \langle \hat{P}, \hat{T}, \hat{F}\rangle$ *where, for all* $\hat{t} \in \hat{T}$, $d(\text{Supp}(\text{pre}_{\hat{N}}(\hat{t}))) \subseteq d(\text{Supp}(\text{post}_{\hat{N}}(\hat{t})))$.

Let $\leq_f$ be the order among configurations such that $\mu \leq_f \mu'$ if μ' is obtained from μ by adding 1. tokens in the inner markings of objects in μ and/or 2. nested tokens at the system net places. The *EOS coverability problem* asks, given an EOS $\mathfrak{E}$ and configurations μ_f and μ_1, whether there is a *run* (sequence of event firings) from μ_0 to a configuration $\mu_1 \geq_f \mu_f$. We may denote $\leq_f$ also by $\leq$.

Definition 7. *EOS (cEOS) coverability is the following decision problem:*
Input: *An EOS (cEOS)* $\mathfrak{E}$, *an initial and target configuration* M_0 *and* M_t.
Output: *Whether there is a configuration* M_1 *reachable from* M_0 *and* $M_t \leq M_1$.

EOS (cEOS) coverability is undecidable (decidable; Th. 4.3 and Th. 5.2 in [6]).

3 Extended νPN

We introduce c-νPN, an extension of νPN by restricted forms of whole-place operations in the style of UDNs. Afterwards, we identify r-νPN as a simpler, yet equivalent fragment. They prove themselves valuable in Sect. 4.

3.1 Channel-νPN

c-νPN extend νPN by *special* transitions that perform transfers with renaming. The firing of special transitions is similar to UDNs (and affine nets [4]): 1. the preconditions are fired, 2. the transfers (defined by matrices $G_t(x_i, x_j)$ for $x_i, x_j \in \mathcal{X}(t)$) are performed 3. the post-conditions are fired.

Technically, $G_t(x_i, x_j) \in \{0,1\}^{P \times P}$. $G_t(x_i, x_j)(p, q) = n$ means that, after firing the pre-conditions, the transition puts on q, in the tuple instantiating x_j, n times the number of tokens on p from the tuple instantiating x_i. We use the following notation: if, for some set P of places, $M \in \{0,1\}^{P \times P}$ is a matrix and $p \in P$ is a place, $M[p]$ denotes the row of M indexed by p. Moreover, given a place $p \in P$, we denote by $\delta_p \in \{0,1\}^{|P|}$ the vector such that, for each $q \in P$, $\delta_p(q) = 1$ if and only if $p = q$. For example, given a tuple m and a place p, the tuple $m(p)\delta_p$ is the projection of m to p.

Definition 8. *A channel νPN (c-νPN) is a tuple $N = (P, T, F, G)$ where (P, T, F) is a νPN and G maps each $t \in T$ to a function $G_t : \mathcal{X}(t) \times \mathcal{X}(t) \to \{0,1\}^{P \times P}$ such that, for each variable $x_i \in \mathcal{X}(t)$ and place $p \in P$:*

1. *for each variable $x_j \in \mathcal{X}(t)$ either $G_t(x_i, x_j)[p] = 0$ or $\exists q \in P$ $G_t(x_i, x_j) = \delta_q$, i.e., $G_t(x_i, x_j)[p]$ contains at most a single 1.*
2. *if $\exists q \in P \exists x_j \in \mathcal{X}(t) \setminus \{x_i\}$ $G_t(x_i, x_j)[p] = \delta_q$, then $\forall x_k \in \mathcal{X}(t) \setminus \{x_j\}$ $G_t(x_i, x_k)[p] = 0^{|P|}$, else $G_t(x_i, x_i)[p] = \delta_p$.*

The graphical representation of a c-νPN $N' = (P, T, F, G)$ adds the representation of G_t, for each $t \in T$, on top of the representation of the νPN $N = (P, T, F)$. This is achieved by special arrows called *channels*. A channel c is a pair of decorated arrows (e.g., double or triple arrows in Fig. 3), one to t, called pre-channel of c, and one from t, called post-channel of c. Different channels use different decorations. The pre- and post-channels of c are labeled by two non-empty sequences σ_{pre} and σ_{post} of variables, respectively, such that $|\sigma_{\text{pre}}| = |\sigma_{\text{post}}|$. Moreover, for each place p, each variable $x \in \mathcal{X}(t)$ can occur at most once in the label of at most one pre-channel from p.[1] Intuitively, c represents $|\sigma_{\text{pre}}|$ transfers from p to q while renaming the tokens taken from the tuple instantiating $\sigma_{\text{pre}}(i)$ as tokens of the tuple instantiating $\sigma_{\text{post}}(i)$. A transition is *special* if it has at least one channel.

Each PN $N = (P, T, F)$ corresponds (cf. c-νPN semantics in Def. 9 below) to the c-νPN $N' = (P, T, F, G)$ where, for each $t \in T$ and $x \in \mathcal{X}(t)$, $y \in \mathcal{X}(t) \setminus \{y\}$, $G_t(x, x) = \text{Id}$, and $G_t(x, y) = 0$. Both N and N' have the same graphical representation. N' has no special transition.

Example 4. Fig. 3 depicts a c-νPN $N = (P, T, F, G)$ with several configurations (cf. Definition 9 below) where $P = \{p_1, \cdots, p_5\}$, $T = \{t\}$, $\mathcal{X}(t) = \{x_1, x_2, x_3\}$, $\Upsilon(t) = \{\nu_1, \nu_2\}$, and for each place $p \in P$ and $x \in \text{Var}(t)$, $G_t(x_1, x)[p]$ is δ_p if

---

[1] This condition is necessary to forbid the representation of transfers with duplication.

$x = x_1$ and 0 otherwise and:

$$F_x(p,t) = \begin{cases} 1 & \text{if } p = p_2, x \in \{x_2, x_3\} \\ 1 & \text{if } p = p_1, x = x_1 \\ 0 & \text{otherwise} \end{cases} \quad F_x(t,p) = \begin{cases} 1 & \text{if } p = p_3, x = x_3 \\ 2 & \text{if } p = p_3, x = x_2 \\ 2 & \text{if } p = p_5, x \in \{\nu_1, \nu_2\} \\ 0 & \text{otherwise} \end{cases}$$

$$G_t(x_2, x)[p] = \begin{cases} \delta_{p_3} & \text{if } p = p_1, x = x_3 \\ \delta_p & \text{if } p \neq p_1, x = x_2 \\ 0 & \text{otherwise} \end{cases} \quad G_t(x_3, x)[p] = \begin{cases} \delta_{p_4} & \text{if } p = p_2, x = x_1 \\ \delta_p & \text{if } p \neq p_2, x = x_3 \\ 0 & \text{otherwise} \end{cases}$$

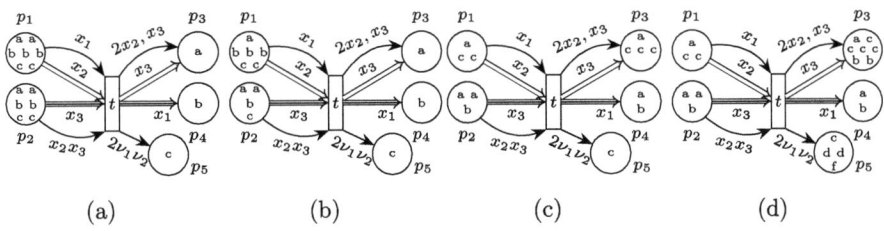

Fig. 3. The c-νPN discussed in Example 4 and Example 5, with configurations.

The firing of a special transition t serializes its standard pre-conditions, the transfers, and, finally, the post-conditions. In Definition 9 below, M'' captures the tokens uninvolved with the firing, $out_\Upsilon(t)$ accounts for the creation of fresh tuples via ν variables, m'' accounts for the firing of the standard preconditions, the summation in $m'_{e(x_j)}$ formalizes the transfers to the tuple instantiating x_j, while m' takes into account also the final firing of the standard post-conditions.

Definition 9. *A configuration M of a c-νPN $N = (P, T, F, G)$ is a configuration of (P, T, F), and a transition $t \in T$ is enabled in N if it is enabled in (P, T, F). A configuration M' is reached from M by firing a transition $t \in T$ with mode e, denoted by $M \to^{t,e} M'$, if*

$$M = M'' + \sum_{x \in \mathcal{X}(t)} \{\!\{m_{e(x)}\}\!\} \quad M' = M'' + out_\Upsilon(t) + \sum_{x \in \mathcal{X}(t)} \{\!\{m'_{e(x)}\}\!\}$$

where $m'_{e(x_j)} = \sum_{x_i \in \mathcal{X}(t)} \left(m''_{e(x_i)} * G_t(x_i, x_j)\right) + F_{x_j}(t, P)$ and $m''_{e(x_i)} = m_{e(x_i)} - F_{x_i}(P, t)$. We define $\to^*$ as the reflexive and transitive closure of $\to^{t,e}$.

Example 5. Figure 3d depicts the firing of transition t from Fig. 3a with mode $e(x_1) = a$, $e(x_2) = b$, $e(x_3) = c$, $e(\nu_1) = d$, and $e(\nu_2) = f$. Fig 3b and Fig. 3c depict the intermediate steps.

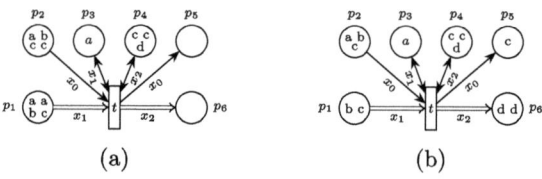

Fig. 4. Firing of a special r-νPN transition. x_0 maps to c, x_1 to a, and x_2 to d.

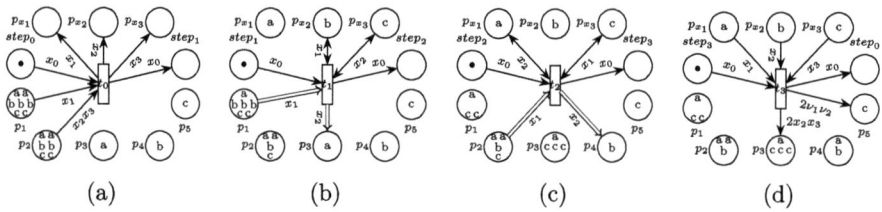

Fig. 5. r-νPN transitions to simulate the special c-νPN transition in Fig. 3.

3.2 Rename-νPN

A r-νPN is a c-νPN where all transitions are either νPN transitions, or c-νPN transitions as in Fig. 4a (up to place names), with a single channel and few extra pre- and post-conditions.

Its firing results in the configuration in Fig. 4b. If p_1 did not contain any token bidden to x_1, then the transition could still fire, but no transfer happens. c-νPN coverability can be reduced to r-νPN coverability: special c-νPN transitions can be simulated by chains of r-νPN special transitions (see [2]). Consequently, in the following, we can show reductions regarding c-νPN or r-νPN coverability interchangeably.

Lemma 1. *Each c-νPN is simulated by a polynomial-time constructible r-νPN.*

Example 6. Fig. 5 depicts a r-νPN N consisting of four r-νPN transitions $t_0, \ldots, t_3$ that simulate the c-νPN N' in Fig. 3. Figure 5 depicts also the configurations as t_0, t_1, t_2 fire. t_0 simulates the preconditions, t_1 the channel from p_1 to p_3, t_2 the channel from p_2 to p_4, and t_3 the postconditions. A token in step_0 signals that t is not currently simulated. A firing of t is simulated by the sequential firing of $t_0, \ldots, t_3$. This is assured by the passage of a token among the step_i places, for $i \in \{0, \ldots, 3\}$. When t_3 fires, the token is moved back to step_0, signaling that the simulation round has finished. The initial configuration of N (Fig. 5a) is that of N' plus a fresh token on step_0. The coverability in the c-νPN in Fig. 3 of a target C is equivalent to the coverability in the r-νPN in Fig. 5 of the the extension of C by the same fresh token on step_0.

4 From r-νPN to cEOS

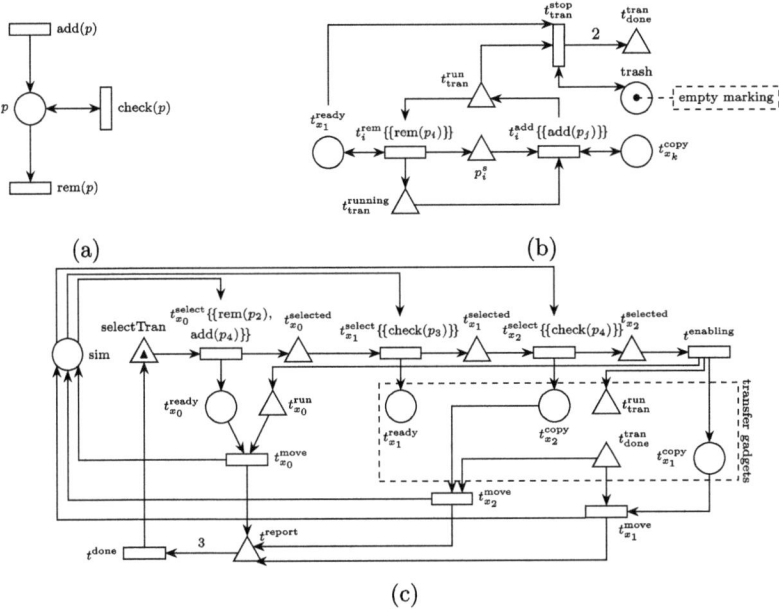

Fig. 6. (a) Transitions $add(p)$, $check(p)$, and $rem(p)$ assumed to be in the νPN $N = (P, T, F)$, for each place $p \in P$. (b) Gadget transfer$_{p_i}$, with initial marking, for each place p_i in Fig. 4: $j = 6$ if $i = 1$ and $j = i$ otherwise; $k = 2$ if $i = 1$ and $k = 1$ otherwise. Note that all the places except t_i^{rem}, p_i^s, and t_i^{add} are shared by all transfer gadgets. (c) The part of $\mathfrak{E}$ dedicated to the simulation of the special transition t in Fig. 4. The dashed area represents the gadget transfer$_{p_i}$. Only interface places are shown. Circles and triangles denote places of type N and $\blacktriangle$.

We show how to reduce r-νPN coverability to cEOS coverability (proofs can be found in the full version [2]). Lemma 2 in [2] shows that arbitrary standard νPN transitions can be simulated by cEOSs. Thus, taking advantage of a similar strategy, we show how to simulate special transitions of a r-νPN $N' = (P, T, F, G)$. We encode a configuration $M = \{\!\{m_1, \ldots, m_n\}\!\}$ of N' in a cEOS place sim: each m_i corresponds to an object at sim with internal marking m_i. Since all special transitions have the same shape (up to the name of the involved places), it is sufficient to show the simulation of the special transition t in Fig. 4. We assume, without loss of generality, that the net N' contains, for each $p \in P$, the standard transitions $add(p), check(p), rem(p) \in T$ depicted in Fig. 6a. We encode the arbitrary special r-νPN transition t in Fig. 4 into a cEOS $\mathfrak{E}$ (depicted in Fig. 6c) that employs the two object net types $N = (P, T, F)$ and $\blacktriangle$. The synchronization structure contains no object autonomous event.

The key idea of this construction is that, by using the gadgets transfer$_{p_i}$ in Fig. 6b, we simulate t up to lossiness, i.e., the simulation of t may return either the encoding of the configuration reached by firing t or a sub-configuration w.r.t. to $\leq_f$. The transfer gadget destroys the object encoding the tuple instantiating x_1 and redirects its tokens into two new objects, one that captures the token not affected by the renaming (those in places $p_2, \ldots, p_6$) and one that captures the tokens that have to be renamed and transferred (those from places p_1 to p_6). These new objects are stored respectively in $t_{x_1}^{copy}$ and $t_{x_2}^{copy}$. Since EOS transitions do not enjoy whole-place operations, specifically transfer, the redirection moves the tokens one by one, via the transition t_i^{rem} and t_i^{add} (and their synchronization with $rem(p_i)$ and $add(p_j)$). However, this way, it is not possible to detect termination of this procedure. To solve this, the gadget non-deterministically (possibly too early) forces termination by trashing the object (and its remaining internal marking) into a place $trash$ via a transition t_{tran}^{stop}, which disables further transfer steps. The internal tokens of the object in $trash$ are those that, by the end of the simulation of t, were not moved to $t_{x_1}^{copy}$ or $t_{x_2}^{copy}$, but essentially lost. Lossiness does not impair the reduction for coverability.

Observe that the runs of the cEOS in Fig. 6c simulate, up to lossiness, the firing of the transition in Fig. 4. This is done in three steps. First, an array of transitions $t_{x_i}^{select}$, for $i \in \{0, 1, 2\}$, select objects from sim (each capturing, by their internal marking, a tuple of the νPN), amounting to a νPN mode: $t_{x_0}^{select}$ simulates t restricted to the variable x_0, producing a single object into $t_{x_0}^{ready}$; the other transitions of the array initialize the transfer gadgets. Note that the so encoded mode enables the transition t of the r-νPN because of the synchronization structure. Second, the transfer gadgets are executed, simulating the selective transfers up to lossiness, returning two objects, one in $t_{x_1}^{move}$ and one in $t_{x_2}^{move}$. Third, the so obtained objects are moved back to sim.

Theorem 1. *There is a polynomial reduction from r-νPN to cEOS coverability.*

Note that the simulation of standard νPN transitions in [2] is obtained by gadgets that serialize several copies of transitions like $t_{x_0}^{select}$ and $t_{x_0}^{move}$.

5 From cEOS to c-νPN

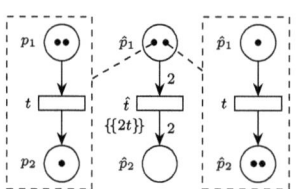

Fig. 7. A cEOS Example.

In this section, we show how cEOS $\mathfrak{E} = (\hat{N}, \mathcal{N}, d, \Theta)$ can be simulated by polynomial-time constructible c-νPN $W = (P, T, F, G)$. We assume, without loss of generality, that each $\hat{t} \in \hat{T}$ is involved in only one, synchronous event $e = \langle \hat{t}, \theta \rangle$. The c-$\nu$PN W combines several modules, to be fired sequentially: 1. $\hat{p}$-**Block** modules, used to encode the configurations of $\mathfrak{E}$; 2. N-**merged** and N-**updated** modules, auxiliary places to store the result of the merging phase and conduct the internal firing phase of the event e; 3. e-**merging** modules, to capture the dynamics of the merging stage; 4. *e*-**firing module**, to capture the dynamics of the internal firing stage; 5. *e*-**distributing module**, to capture the dynamics of the distribution stage.

Theorem 2. *Given a cEOS $\mathfrak{E} = (\hat{N}, \mathcal{N}, d, \Theta)$, there is a polynomial time constructible c-νPN $W = (P, T, F, G)$ that simulates $\mathfrak{E}$.*

The proof is in [2] and is exemplified below on the cEOS in Fig. 7.

Module $\hat{p}$-Block. For each $\hat{p} \in \hat{P}$, $\hat{p}$-**Block** encodes in W the objects hosted by $\hat{p}$: it contains a disjoint copy of the places of N as well as a place $\mathrm{Id}_{\hat{p}}$. The latter stores identifiers dedicated to each encoded object. A configuration $\langle \hat{p}, m \rangle$ of $\mathfrak{E}$ is captured by the tuple m' of $\hat{p}$-**Block** that extends m with a token on $\mathrm{Id}_{\hat{p}}$. If $m = \emptyset$, then m is captured by just a token on $\mathrm{Id}_{\hat{p}}$.

Modules N-Merged and N-Updated. These blocks store the objects after the merging and internal firing stages and are analogous to Fig. 8. They contain a disjoint copy of the places of N and the place Id_N^m and Id_N^u, respectively. Their encoding follows from Fig. 8.

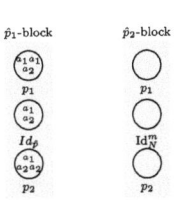

Fig. 8. Config in Fig. 7 captured by W.

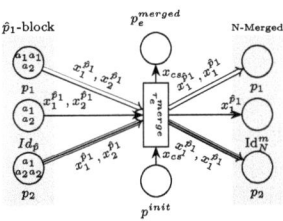

Fig. 9. Module *e*-merging.

Module *e*-merging. This module captures the dynamics of the merging phase. It contains the places p_e^{init} and p_e^{merged} as well as the transition τ_e^{merge}. The two places behave as enable and acknowledgment for the transition. Initially, p^{init} is populated by a *control sequence* token (moved around by the variable x_{cs}). In $\mathfrak{E}$, for each place $\hat{p}$ from which τ consumes, there is a channel from each non-Id place of $\hat{p}$-**Block** to the corresponding place in N-merged. The pre-channels from $\hat{p}$ are labeled by $x_1^{\hat{p}}, \ldots, x_n^{\hat{p}}$, where n is the number of objects consumed by τ from $\hat{p}$ in $\mathfrak{E}$. The corresponding post-channel is labeled by the constant sequence $x_1^{\hat{p}}, \ldots, x_1^{\hat{p}}$ of length n. Moreover, to take care of the identifiers of the encoded objects consumed by τ, τ_e^{merge} consumes from $\mathrm{Id}_{\hat{p}}$ one token for each $x_1^{\hat{p}}, \ldots, x_n^{\hat{p}}$ and produces one token x_1^p in Id_N^m, where p is a fixed place among the places from which τ consumes (Fig. 9).

Module *e*-updating. This module captures the dynamics of the internal firing phase applied to the N-merged modules after the merging phase, storing its result in the module N-updated. It contains the places p_e^{select} as well as a set $\{p_{e,t}^x \mid x \in \{fire, fired\}, t \in supp(\theta)\}$ of new places, and the transitions τ_e^{init} and τ_e^{fin} as well as a copy t_e of the transitions $t \in supp(\theta)$. Places p_e^{init}, p_e^{select}, and p_e^{fin} are used to fire τ_e^{init} and τ_e^{fin} in sequence. The

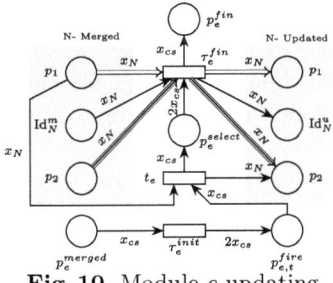

Fig. 10. Module *e*-updating.

places $p_{e,t}^{fire}$ and $p_{e,t}^{fired}$ as enable and acknowledgment for t. For each $t \in supp(\theta)$, τ_e^{init} puts $\theta(t)$ tokens in $p_{e,t}^{fire}$. Thus, the transition t_e gets enabled $\theta(t)$ times. When firing, t_e behaves as t, except that its preconditions stem from places in N-**merged**, while its post-conditions are directed towards places in N-**updated**. The movement of tokens from N-**merged** to N-**updated** ensures that either the transitions in θ are all enabled at the start of this phase, or a deadlock is reached, i.e., the firing of a $t' \in \theta$ is not responsible for the enabling of another transition $t'' \in \theta$. Finally, all tokens remaining in N-**merged** are moved to the respective places in N-**updated** (also using channels; see Fig. 10).

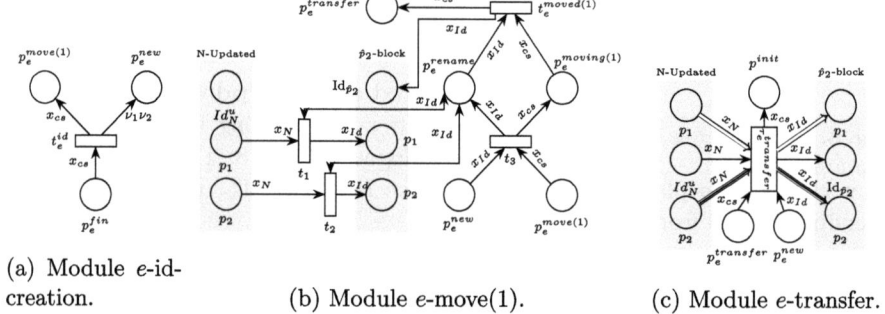

(a) Module e-id-creation.

(b) Module e-move(1).

(c) Module e-transfer.

Fig. 11. Module e-distributing.

Module e-Distributing. This module captures the dynamics of the final distribution of updated tokens. Let n be the number of objects created by τ and let $\hat{p}_1, \ldots, \hat{p}_n$ be an enumeration (possibly with repetition) of the places in the post-conditions of τ. The module concatenates the sub-module e-**id-creation**, a sequence of sub-modules e-**move**(i) for $i \in \{1, \ldots, n-1\}$, and the sub-module e-**transfer**. The sub-module e-**id-creation** generates n new identifiers for the encoding of the objects to be created. Then, each e-**move**(i) selects one such identifier, say a, and moves while renaming to a, one by one, some token not in Id_N^u from N-**updated** to the corresponding place in $\hat{p}_i$-**block**. By taking advantage of concurrency and of a couple of enable/acknowledge places as in the previous modules, each e-**move**(i) passes the turn to the next sub-module after a non-deterministic number of movements. While passing the turn, e-**move**(i) finally moves the identifier a to $\mathrm{Id}_{\hat{p}_i}$, completing the encoding of the next object created by τ at place $\hat{p}_i$. Finally, the e-**transfer** sub-module creates the last encoding by moving, via channels, all remaining tokens in N-**updated** to $\hat{p}_n$-**block** while renaming to the last new identifier (Fig. 11).

6 Conclusions

We have detected c-νPN as the data-aware counterpart of cEOS, when restricted to coverability. νPNs are captured by c-νPN which, in turn, are captured by

UDNs. This immediately yields three results: 1. the computational power of cEOS coverability corresponds to anonymous unordered data (as νPN) plus selective transfer with renaming 2. an $F_{\omega 2}$ lower bound from standard νPN coverability and 3. an F_{ω^ω} upper bound from UDN coverability. Arguably, both these bounds are sub-optimal. Consequently, we conjecture that cEOS coverability is F_α-complete for some ordinal α such that $\omega 2 < \alpha < \omega^\omega$. As future works, we aim at the precise complexity class of cEOS coverability. This endevour is now easier, since we can directly take advantage of the wide literature on coverability of various forms of data nets. Furthermore, we plan to study the impact of several iterations of nesting, as in general NWNs. Results of this kind would enable a deep understanding of the computational power of PN nesting.

References

1. Di Cosmo, F., Mal, S., Prince, T.: Deciding reachability and coverability in lossy EOS. In: International Workshop on Petri Nets and Software Engineering. CEUR Workshop Proceedings, vol. 3730, pp. 74–95 (2024), https://ceur-ws.org/Vol-3730/paper04.pdf
2. Di Cosmo, F., Mal, S., Prince, T.: Nets-within-nets through the lens of data nets (2025). https://arxiv.org/abs/2506.22344
3. Finkel, A., Schnoebelen, P.: Well-structured transition systems everywhere! Theoretical Comput. Sci. **256**(1-2), 63–92 (2001). https://linkinghub.elsevier.com/retrieve/pii/S030439750000102X
4. Finkel, A., McKenzie, P., Picaronny, C.: A well-structured framework for analysing petri net extensions. Inf. Comput. **195**(1–2), 1–29 (2004). https://doi.org/10.1016/J.IC.2004.01.005
5. Haddad, S., Schmitz, S., Schnoebelen, P.: The ordinal-recursive complexity of timed-arc petri nets, data nets, and other enriched nets. In: 2012 27th Annual IEEE Symposium on Logic in Computer Science. pp. 355–364 (2012). https://doi.org/10.1109/LICS.2012.46
6. Köhler-Bußmeier, M.: A survey of decidability results for elementary object systems. Fundamenta Informaticae **130**(1), 99–123 (2014). https://doi.org/10.3233/FI-2014-983
7. Köhler-Bussmeier, M., Capra, L.: Robustness: a natural definition based on nets-within-nets. In: Proceedings of the 2023 International Workshop on Petri Nets and Software Engineering. CEUR Workshop Proceedings, vol. 3430, pp. 70–87 (2023). https://ceur-ws.org/Vol-3430/paper5.pdf
8. Köhler-Bußmeier, M., Rölke, H.: Analysing adaption processes of hornets. Trans. Petri Nets Other Model. Concurr. **17**, 84–107 (2023). https://doi.org/10.1007/978-3-662-68191-6_4
9. Lazic, R., Newcomb, T.C., Ouaknine, J., Roscoe, A.W., Worrell, J.: Nets with tokens which carry data. Fundam. Informaticae **88**(3), 251–274 (2008). https://doi.org/10.5555/1497079.1497082
10. Lazic, R., Schmitz, S.: The complexity of coverability in ν-petri nets. In: Proceedings of the 31st Annual ACM/IEEE Symposium on Logic in Computer Science, LICS '16, July 5-8, 2016, pp. 467–476. ACM, New York (2016)
11. Lomazova, I.A.: Nested petri nets - a formalism for specification and verification of multi-agent distributed systems. Fundam. Informaticae **43**(1-4), 195–214 (2000). https://doi.org/10.3233/FI-2000-43123410

12. Lomazova, I.A., Schnoebelen, P.: Some decidability results for nested petri nets. In: Perspectives of System Informatics, Third International Andrei Ershov Memorial Conference. LNCS, vol. 1755, pp. 208–220. Springer (1999), https://doi.org/10.1007/3-540-46562-6_18
13. Murata, T.: Petri nets: Properties, analysis and applications. IEEE **77**(4), 541–580 (1989). https://doi.org/10.1109/5.24143
14. Rosa-Velardo, F.: Ordinal recursive complexity of unordered data nets. Inf. Comput. **254**, 41–58 (2017)
15. Rosa-Velardo, F., de Frutos-Escrig, D.: Decidability and complexity of petri nets with unordered data. Theor. Comput. Sci. **412**(34), 4439–4451 (2011). https://doi.org/10.1016/j.tcs.2011.05.007
16. Schmitz, S., Schnoebelen, P.: Multiply-recursive upper bounds with higman's lemma. In: Automata, Languages and Programming - 38th International Colloquium, ICALP 2011, Zurich, Switzerland, July 4-8, 2011, Proceedings, Part II. LNCS, vol. 6756, pp. 441–452. Springer (2011). https://doi.org/10.1007/978-3-642-22012-8_35
17. Schmitz, S., Schnoebelen, P.: Algorithmic aspects of wqo theory (2012)
18. Valk, R.: Object petri nets: using the nets-within-nets paradigm. In: Lectures on Concurrency and Petri Nets, Advances in Petri Nets. LNCS, vol. 3098, pp. 819–848. Springer (2003). https://doi.org/10.1007/978-3-540-27755-2_23

Compositional Verification of Almost-Sure Büchi Objectives in MDPs

Marck van der Vegt[1](✉)[iD], Kazuki Watanabe[2][iD], Ichiro Hasuo[2][iD], and Sebastian Junges[1][iD]

[1] Radboud University, Nijmegen, The Netherlands
{marck.vandervegt,sebastian.junges}@ru.nl
[2] National Institute of Informatics, Tokyo, Japan
{kazukiwatanabe,hasuo}@nii.ac.jp

Abstract. This paper studies the verification of almost-sure Büchi objectives in MDPs with a known, compositional structure based on string diagrams. In particular, we ask whether there is a strategy that ensures that a Büchi objective is almost-surely satisfied. We first show that proper exit sets—the sets of exits that can be reached within a component without losing locally—together with the reachability of a Büchi state are a sufficient and necessary statistic for the compositional verification of almost-sure Büchi objectives. The number of proper exit sets may grow exponentially in the number of exits. We define two algorithms: (1) A straightforward bottom-up algorithm that computes this statistic in a recursive manner to obtain the verification result of the entire string diagram and (2) a polynomial-time iterative algorithm which avoids computing all proper exit sets by performing iterative strategy refinement.

1 Introduction

Markov decision processes (MDPs) are a ubiquitous model to describe systems with uncertain action outcomes. Their efficient verification is hindered by typical state space explosion problems, see e.g., [1,12]. To overcome this state space explosion, a promising research direction to overcome scalability concerns is to explicitly capture and utilize the (compositional) structure of the MDP [5,13,16].

Concretely, we consider sequentially composed MDPs (potentially with loops), that is, we consider MDPs whose state space is partitioned into *components*. Specifically, we study the setting in which these components and their interaction is specified as string diagrams [20]: That is, components are MDPs with entrance and exit states and their composition is defined by a number of operations that glue these components together to obtain the *monolithic MDP* (see e.g., Fig. 1). The key concept in compositional verification is then to verify objectives without constructing the monolithic MDP, but rather, by (iteratively

---

I.H. was supported by ERATO HASUO Metamathematics for Systems Design Project (JST, No. JPMJER1603). I.H. and M.V. were supported by the ASPIRE grant (JST, No. JPMJAP2301). K.W. was supported by ACT-X grant (JST, No. JPMJAX23CU).

© The Author(s), under exclusive license to Springer Nature Switzerland AG 2026
P. Ganty and A. Mansutti (Eds.): RP 2025, LNCS 16230, pp. 171–185, 2026.
https://doi.org/10.1007/978-3-032-09524-4_12

or exhaustively) analyzing the components and combining the result of their analysis. This has been done for quantitative reachability in [20,23], but not for almost-sure Büchi objectives. Although the analysis of Büchi objectives on a monolithic MDP can be done in polynomial time in the number of states of the MDP, the size of the MDPs is often prohibitively large. Compositional approaches avoid the necessity to handle the monolithic MDP.

The key question this paper answers affirmatively is whether such a compositional approach can work for almost-sure Büchi objectives. We study MDPs with a set of marked states, called the *Büchi states*, and algorithmically answer whether given an entrance it is possible to *win* the almost-sure Büchi objective, that is, can we visit a Büchi state infinitely often with probability one.

In this paper, we first discuss what information is necessary to sufficiently describe the behavior of the components *exhaustively*. We call this information the *solution*. The solution should include for each entrance the set of exits that can be reached using strategies that we call *no-lose*, that is, a strategy which avoid surely violating the Büchi objective in the component. We call the set of reachable exits the *proper exit set*. Intuitively, after playing a no-lose strategy, we can reach a proper exit which is connected to an entrance of another component, after which we play another no-lose strategy. However, constantly playing no-lose strategies is not sufficient to satisfy the Büchi objective. In the solution, beside the proper exit set we must therefore also include whether there is a positive probability of reaching a Büchi state. If for a given entrance, there exists a no-lose strategy that can reach a Büchi state, then reaching this entrance infinitely often implies satisfaction of the Büchi objective. We call the combination of the proper exit set and the positive reachability of a Büchi state the *effect* of the strategy and it is *the* essential information for our *compositional solution*.

The characterization of the problem in terms of a compositional solution gives rise to a natural (exhaustive) algorithm in the spirit of [20,22]: Consider the abstract syntax tree of the string diagram, first analyze the leafs of this tree exhaustively by computing its solution, and then compose the analysis results. The exhaustive analysis is exponential in the number of exits, but polynomial in the number of components. Compared to the quantitative reachability case [20, 22], this is already good news, because in the quantitative case, the computation can be superpolynomial even with a constant number of exits per component.

Our second (iterative) algorithm avoids computing all possible effects by taking the context of each component into account, inspired by a compositional formulation of value iteration in [23]. In contrast to existing work, this algorithm can limit the amount of iterations necessary by determining that certain entrances may not be reached by any winning strategy. In particular, we start with a strategy that reaches as many entrances as possible, and refine this strategy by pruning entrances from which we cannot avoid losing. Our iterative algorithm mimics the *classical Büchi algorithm* [4,6,7] on the level of components. We show that the sets of possible effects computed by the exhaustive algorithm exhibit a join semilattice structure, which the iterative algorithm uses to prevent computing all effects, yielding a polynomial time algorithm.

In conclusion, the contributions of this work are a characterization of the necessary and sufficient information about the behavior of a component in order to prove satisfaction of a Büchi objective globally, an exponential-time bottom-up algorithm that computes this information on individual components and composes them according to the structure of the string diagram, and an iterative approach that avoids the exponential influence of the number of exits.

2 Preliminaries

For $n \in \mathbb{N}$, $[n]$ denotes the set $\{1, 2, \ldots, n\}$. We denote the disjoint union by $\uplus$. We use ν and μ to denote the greatest and least fixpoint operators, respectively.

A (directed) *graph* $\mathcal{G} = \langle V, E \rangle$ is a tuple with finitely many vertices V and edges $E \subseteq V \times V$. The *successors* of $x \in V$ are $\mathrm{Post}(x) := \{y \in V \mid \langle x, y \rangle \in E\}$. Similarly, $\mathrm{Pre}(y) := \{x \in V \mid \langle x, y \rangle \in E\}$ are the *predecessors* of $y \in V$. We lift Post and Pre to sets: $\mathrm{Post}(X) := \bigcup_{x \in X} \mathrm{Post}(x)$, $\mathrm{Pre}(X) := \bigcup_{x \in X} \mathrm{Pre}(x)$.

We are interested in *almost-sure acceptance* in *Markov decision processes* [2, 11] (MDPs). Then, the amplitude of probabilities is irrelevant—a non-zero probability, however small it is, can break almost-sure acceptance. Therefore, we study the following abstraction of MDPs (see, e.g. [3]), which preserves all results.

Definition 1 (MDP graph and strategy). An *MDP graph* $M := \langle V_1, V_P, E \rangle$ is a graph $\langle V, E \rangle$ where $V := V_1 \uplus V_P$. V_1 are the *player-1 vertices* and V_P are the *probabilistic vertices*. A *strategy* $\sigma \colon V_1 \to \mathcal{P}(V_P)$ maps player-1 vertices to successor vertices, such that for all vertices $v \in V_1$: $\sigma(v) \subseteq \mathrm{Post}(v)$.

In the following we simply refer to MDP graphs as MDPs. We denote the pointwise union of strategies σ_1, σ_2 by $\sigma_1 \cup \sigma_2$, with $(\sigma_1 \cup \sigma_2)(v) := \sigma_1(v) \cup \sigma_2(v)$ for all $v \in V_1$. We note that our strategies are memoryless randomized strategies. These are sufficient for satisfying almost-sure Büchi objectives [7]. We assume w.l.o.g. that V_1 and V_P vertices *alternate*, i.e., V_1 vertices are only reachable from V_P vertices and vice versa.

Definition 2 (reachable vertices). Let $\mathrm{Post}_\sigma(x)$ be equal to $\sigma(x)$ if $x \in V_1$ and $\mathrm{Post}(x)$ otherwise. We define the *vertices reachable from* $X \subseteq V$ *under* σ as $\mathrm{Reach}_\sigma(X) := \bigcup_{n=0}^\infty X_n$, where $X_0 := X$ and $X_{n+1} := X_n \cup \mathrm{Post}_\sigma(X_n)$.

The general notion of Büchi acceptance is over *infinite traces*. When one restricts to almost-sure acceptance (as we do), there is an alternative characterization [2, Theorem 10.29, Cor. 10.30], namely that a vertex i is *almost-sure accepting* iff there exists a strategy σ such that for every vertex v reachable from i under σ, a Büchi vertex v' is reachable from v under σ.

This characterization justifies the following definition.

Definition 3 (Büchi objective). A strategy σ *satisfies the Büchi objective* $B \subseteq V_P$ for vertex $i \in V$ if $\mathrm{Reach}_\sigma(v) \cap B \neq \emptyset$ for all v in $\mathrm{Reach}_\sigma(i)$. We denote $i, \sigma \models \Box \Diamond B$ or simply $i \models \Box \Diamond B$ if such a strategy exists. We denote the *winning region* by $\mathrm{Win}_{\Box \Diamond B} := \{v \in V_1 \mid v \models \Box \Diamond B\}$.

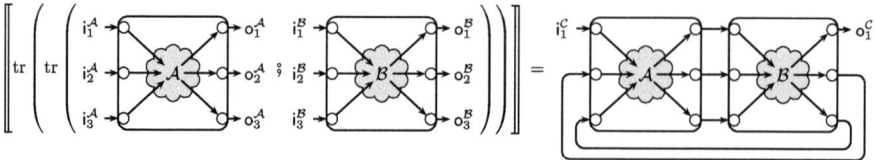

Fig. 1. $\mathcal{C} := \mathrm{tr}(\mathrm{tr}(\mathcal{A} \,\S\, \mathcal{B}))$ and its operational semantics.

For ease of presentation, we assume w.l.o.g. that all Büchi vertices are probabilistic vertices. We briefly recall the *classical Büchi algorithm* [4,6,7].

Definition 4 (Büchi operator). Let $F(X,Y) := \{v \in V_1 \mid \exists v' \in \mathrm{Post}(v)\colon \mathrm{Post}(v') \subseteq Y \text{ and } v' \in B \vee v' \in \mathrm{Pre}(X)\}$, the *Büchi operator*.

We can compute the winning region $\mathrm{Win}_{\Box\Diamond B}$ using the Büchi operator:

Lemma 1. $\nu Y. \mu X. F(X,Y) = \mathrm{Win}_{\Box\Diamond B}$

For a fixed Y, a vertex v is in the 'inner fixpoint' $\mu X. F(X,Y)$ iff it can reach a Büchi vertex while staying in Y. Then, a vertex v is in the 'outer fixpoint' $\nu Y. \mu X. F(X,Y)$ iff all vertices reachable from v can reach a Büchi vertex, which is exactly the winning region for Büchi acceptance. The classical Büchi algorithm computes the above fixpoint using Kleene iteration. Initially, we fix Y to V_1 and compute the least fixpoint of F by iteratively applying F to $\emptyset$, i.e., we compute the sequence $F(\emptyset, Y), F(F(\emptyset, Y), Y), \cdots$. The obtained fixpoint is the new value of Y. The process is repeated until we obtain a fixpoint of Y, in which case Y is equal to the winning region $\mathrm{Win}_{\Box\Diamond B}$.

All proofs are given in the appendix of the technical report [17].

2.1 Compositional MDPs

We define *rightward open MDP* which have *open ends* via which they compose.

Definition 5 (roMDP). A *rightward open MDP* $\mathcal{A} = \langle M, \mathrm{IO} \rangle$ is a pair consisting of an MDP M with *open ends* $\mathrm{IO} = \langle \mathrm{I}, \mathrm{O} \rangle$, where $\mathrm{I}, \mathrm{O} \subseteq V_1$ are pairwise disjoint and totally ordered. The vertices in I and O are the *entrances* and *exits* respectively. We require that $\mathrm{Pre}(\mathrm{I}) = \mathrm{Post}(\mathrm{O}) = \emptyset$.

We denote the *arity* of $\mathcal{A}$ as $\mathrm{arity}(\mathcal{A})\colon m \to n$ meaning that $|\mathrm{I}| = m$ and $|\mathrm{O}| = n$. In the following, we will fix arbitrary roMDPs $\mathcal{A}$ and $\mathcal{B}$. We often use superscript to refer to elements of an roMDP $\mathcal{A}$, e.g., $\mathrm{I}^\mathcal{A}$ denotes the entrances of $\mathcal{A}$.

Definition 6 (string diagram). A *string diagram* $\mathbb{D}$ of roMDPs is a term adhering to $\mathbb{D} := \mathsf{c}_\mathcal{A} \mid \mathbb{D} \,\S\, \mathbb{D} \mid \mathbb{D} \oplus \mathbb{D} \mid \mathrm{tr}(\mathbb{D})$, where $\mathsf{c}_\mathcal{A}$ designates an roMDP $\mathcal{A}$.

In the definition of a string diagram, $\S$ denotes *sequential composition*, $\oplus$ denotes *sum composition* and tr denotes *trace*. In the following we fix $\mathbb{D}$ to be an arbitrary string diagram of roMDPs. Figure 1 demonstrates the graphical intuition behind trace and sequential composition. In the figure, the clouds represent a large number of vertices and transitions. We now define these operations.

Definition 7 ($\mathcal{A} \mathbin{\raisebox{0.5ex}{,}} \mathcal{B}$). Let $\mathcal{A}, \mathcal{B}$ be roMDPs such that arity($\mathcal{A}$): $m \to l$ and arity($\mathcal{B}$): $l \to n$. Their *sequential composition* $\mathcal{A} \mathbin{\raisebox{0.5ex}{,}} \mathcal{B}$ is the roMDP $\langle M, \mathsf{IO} \rangle$ where $\mathsf{IO} := \langle \mathsf{I}^\mathcal{A}, \mathsf{O}^\mathcal{B} \rangle$, $M := \langle (V_1^\mathcal{A} \uplus V_1^\mathcal{B}) \setminus \mathsf{O}^\mathcal{A}, V_P^\mathcal{A} \uplus V_P^\mathcal{B}, E \rangle$ and E satisfies the following: $\langle v, v' \rangle \in E$ iff either: (1) $\langle v, v' \rangle \in E^\mathcal{A} \wedge v' \in V^\mathcal{A} \setminus \mathsf{O}^\mathcal{A}$, (2) $\langle v, \mathsf{o}_i^\mathcal{A} \rangle \in E^\mathcal{A} \wedge v' = \mathsf{i}_i^\mathcal{B}$ for some $i \in [l]$, (3) $\langle v, v' \rangle \in E^\mathcal{B}$.

Definition 8 ($\mathcal{A} \oplus \mathcal{B}$). The *sum* $\mathcal{A} \oplus \mathcal{B}$ is the roMDP $\langle M, \mathsf{IO} \rangle$ where $\mathsf{IO} := \langle \mathsf{I}^\mathcal{A} \uplus \mathsf{I}^\mathcal{B}, \mathsf{O}^\mathcal{A} \uplus \mathsf{O}^\mathcal{B} \rangle$, $M := \langle V_1^\mathcal{A} \uplus V_1^\mathcal{B}, V_P^\mathcal{A} \uplus V_P^\mathcal{B}, E \rangle$, and E satisfies $\mathrm{Post}(v) = \mathrm{Post}^\mathcal{D}(v)$ for $\mathcal{D} \in \{\mathcal{A}, \mathcal{B}\}$ if $v \in V^\mathcal{D}$.

Definition 9 (tr($\mathcal{A}$)). Let $\mathcal{A}$ be an roMDP such that arity($\mathcal{A}$): $m+1 \to n+1$. The *trace* $\mathrm{tr}(\mathcal{A})$ is the roMDP $\langle M, \mathsf{IO} \rangle$ where $\mathsf{IO} := \langle \mathsf{I}^\mathcal{A} \setminus \{\mathsf{i}_{m+1}^\mathcal{A}\}, \mathsf{O}^\mathcal{A} \setminus \{\mathsf{o}_{n+1}^\mathcal{A}\} \rangle$, $M := \langle V_1^\mathcal{A}, V_P^\mathcal{A} \uplus \{*\}, E \rangle$ and $E := E^\mathcal{A} \cup \{\langle \mathsf{o}_{n+1}^\mathcal{A}, * \rangle, \langle *, \mathsf{i}_{m+1}^\mathcal{A} \rangle\}$.

The trace operation introduces two edges that create a loop from the last exit to the last entrance. The '*' vertex is added to maintain vertex alternation.

Definition 10 (operational semantics $[\![\mathbb{D}]\!]$). The *operational semantics* $[\![\mathbb{D}]\!]$ is the roMDP inductively defined by Definitions 7 to 9, with the base case $[\![c_\mathcal{A}]\!] = \mathcal{A}$. We assume that every string diagram $\mathbb{D}$ has matching arities, w.r.t. the definitions of the compositional operators.

We call $[\![\mathbb{D}]\!]$ the *monolithic roMDP*. We will now formally state the main problem.

Main Problem Statement: Given entrance $i \in \mathsf{I}^{[\![\mathbb{D}]\!]}$, does $i \models^{[\![\mathbb{D}]\!]} \Box \Diamond B$?

---

3 Compositional Solution

A natural idea, following [20,22], is to define a compositional algorithm which takes the string diagram, computes a solution for every leaf of the string diagram, and then composes these solutions into an answer on the complete string diagram. This section develops exactly such an approach. We first define a (local) solution: This solution captures the necessary and sufficient information that needs to be extracted from an roMDP. Next, we define the compositional operations on these solutions so that they can be composed into a final answer.

We continue by introducing the necessary definitions for our compositional solution, which are *no-lose strategies* and their *effect*. Intuitively, no-lose strategies represent *local strategies* that can be part of a *globally winning strategy* in $[\![\mathbb{D}]\!]$.

Definition 11 (no-lose strategy). Let σ be a strategy in $\mathcal{A}$. We say that σ is a *no-lose strategy* from $i \in V$ if for every $v \in \mathrm{Reach}_\sigma(i)$ either $\mathrm{Reach}_\sigma(v) \cap \mathsf{O} \neq \emptyset$ or $v, \sigma \models \Box \Diamond B$. Additionally we require that σ is i-local, that is, $\sigma(v) = \emptyset$ for $v \notin \mathrm{Reach}_\sigma(i)$. We denote the set of no-lose strategies from i by $\Sigma_{NL}[i]$.

The intuition of a no-lose strategy is as follows. When playing such a strategy, we almost-surely either (1) reach some exit or (2) satisfy the Büchi condition. A strategy that is not no-lose is *losing*. We will use the following characterization:

Lemma 2. *Strategy σ is no-lose from $i \in V$ iff for all v in $\mathrm{Reach}_\sigma(i)$ we have that $\mathrm{Reach}_\sigma(v) \cap (\mathsf{O} \cup B) \neq \emptyset$.*

The lemma implies that no-lose strategies can be computed as strategies that satisfy the Büchi objective $\square\lozenge(\mathsf{O}\cup B)$, using off-the-shelf algorithms (e.g., [4,7]).

Towards determining whether a no-lose strategy σ is indeed part of a globally winning strategy, we consider its *proper exit set* and *effect*.

Definition 12 (proper exit sets and effects). Let σ be a no-lose strategy from vertex $v \in V$. We denote the *proper exit set* of σ as $\mathrm{Exits}_\sigma(v) := \mathrm{Reach}_\sigma(v) \cap \mathsf{O}$. We define the *effect* of σ as $\mathrm{Effect}(v, \sigma) := \langle \mathrm{Exits}_\sigma(v), \mathrm{Reach}_\sigma(v) \cap B \neq \emptyset \rangle$.

We use $\langle T_1, b_1 \rangle \sqcup \langle T_2, b_2 \rangle$ to denote the pointwise join of effects, that is, $\langle T_1 \cup T_2, b_1 \vee b_2 \rangle$. By definition, $\langle \emptyset, \bot \rangle$ cannot be the effect of a no-lose strategy σ: it would imply that for all vertices v we have that $\mathrm{Reach}_\sigma(v) \cap (\mathsf{O} \cup B) = \emptyset$, contradicting that σ is no-lose.

3.1 Local Solution

We first introduce the solution for a given roMDP and then define the composition of such solutions. Finally, we prove the compositionality in Theorem 1. Our solution contains all possible effects of no-lose strategies.

Definition 13 (local solution). The *local solution* of roMDP $\mathcal{A}$ is the function $\mathsf{S}_\mathcal{A} \colon \mathsf{I}^\mathcal{A} \to \mathcal{P}(\mathcal{P}(\mathsf{O}^\mathcal{A}) \times \{\top, \bot\})$ s.t. $\mathsf{S}_\mathcal{A}(\mathsf{i}_i^\mathcal{A}) := \{\mathrm{Effect}^\mathcal{A}(\mathsf{i}_i^\mathcal{A}, \sigma) \mid \sigma \in \Sigma_{NL}[\mathsf{i}_i^\mathcal{A}]\}$.

We can replace every entrance and exit in the above definition by their index to get $\mathsf{S}_\mathcal{A} \colon [m] \to \mathcal{P}(\mathcal{P}([n]) \times \{\top, \bot\})$, with the advantage that it no longer depends on the identity of the vertices. For brevity, we (sometimes) implicitly convert between entrances/exits and their index, e.g., we denote i for $\mathsf{i}_i^\mathcal{A}$ and vice versa. We define $\mathrm{arity}(\mathsf{S}_\mathcal{A})$ as $\mathrm{arity}(\mathcal{A})$. As a special case of local solutions, we can derive whether one can win within an roMDP without leaving it.

Lemma 3. *For all $i \in \mathsf{I}^\mathcal{A}$, we have $\langle \emptyset, \top \rangle \in \mathsf{S}_\mathcal{A}(i)$ iff $i \models^{[\![\mathcal{A}]\!]} \square\lozenge B$.*

See Fig. 2 for a graphical intuition of local solutions.

3.2 Compositionality of the Solution

The information in a local solution suffices to compositionally compute the solution of a string diagram. We first state the theorem and then clarify the semantics of the operators $\mathbin{\raisebox{0.2ex}{\scriptsize$\circ$}}, \oplus$, and tr on the solutions in the following definitions.

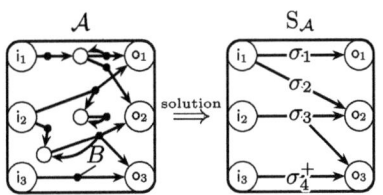

Fig. 2. Computing the local solution.

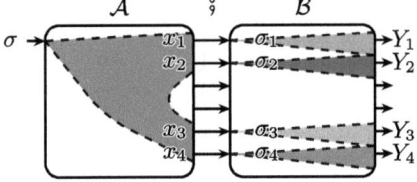

Fig. 3. $S_\mathcal{A} \mathbin{\text{\textcolon}} S_\mathcal{B}$.

Theorem 1 (solution is compositional).

$$S_{\mathcal{A}\mathbin{\text{\textcolon}}\mathcal{B}} = S_\mathcal{A} \mathbin{\text{\textcolon}} S_\mathcal{B}, \qquad S_{\mathcal{A}\oplus\mathcal{B}} = S_\mathcal{A} \oplus S_\mathcal{B}, \qquad \text{and} \qquad S_{\mathrm{tr}(\mathcal{A})} = \mathrm{tr}(S_\mathcal{A}).$$

Sequential Composition. As stated by Theorem 1, we want the definition of $S_\mathcal{A} \mathbin{\text{\textcolon}} S_\mathcal{B}$ to be equal to $S_{\mathcal{A}\mathbin{\text{\textcolon}}\mathcal{B}}$. We can combine the effects of $S_\mathcal{A}$ and $S_\mathcal{B}$ to obtain an effect in $S_{\mathcal{A}\mathbin{\text{\textcolon}}\mathcal{B}}$ by considering the underlying no-lose strategies.

We explain the intuition using Fig. 3. First, we choose a no-lose strategy σ in $\mathcal{A}$ for some entrance i, with the goal of constructing a no-lose strategy in $\mathcal{A}\mathbin{\text{\textcolon}}\mathcal{B}$. Let $\langle \{x_1, \ldots, x_N\}, b \rangle$ be the effect of σ such that $N > 0$ (the case where $N = 0$ is trivial, as it implies that the strategy is also a no-lose strategy in $\mathcal{A}\mathbin{\text{\textcolon}}\mathcal{B}$). The strategy σ is not defined on the vertices of $\mathcal{B}$ and is therefore losing in $\mathcal{A}\mathbin{\text{\textcolon}}\mathcal{B}$. Thus, to construct a no-lose strategy in $\mathcal{A}\mathbin{\text{\textcolon}}\mathcal{B}$, we need to ensure that the continuation in $\mathcal{B}$ is also no-lose. If there exist no-lose strategies $\sigma_1, \ldots, \sigma_N$ in $\mathcal{B}$ such that σ_k is no-lose from x_k, we can compute their union with σ to obtain a no-lose strategy in $\mathcal{A}\mathbin{\text{\textcolon}}\mathcal{B}$.

We formalize the above intuition using the following *decomposition lemma*.

Lemma 4 (sequential decomposition). *Let σ be a no-lose strategy in $\mathcal{A}$ and let $\langle \{x_1, \ldots, x_N\}, b \rangle := \mathrm{Effect}^\mathcal{A}(i, \sigma)$. For $1 \leq k \leq N$, let σ_k be a no-lose strategy from x_k in $\mathcal{B}$. We have that*

$$\mathrm{Effect}^{\mathcal{A}\mathbin{\text{\textcolon}}\mathcal{B}}(i, \sigma \cup \sigma_1 \cup \cdots \cup \sigma_N) = \langle \emptyset, b \rangle \sqcup \bigsqcup_{1 \leq k \leq N} \mathrm{Effect}^\mathcal{B}(x_k, \sigma_k).$$

Based on Lemma 4, we define the sequential composition of solutions.

Definition 14 ($S_1 \mathbin{\text{\textcolon}} S_2$). Let S_1 and S_2 be solutions such that $\mathrm{arity}(S_1)\colon m \to l$ and $\mathrm{arity}(S_2)\colon l \to n$. We define $S_1 \mathbin{\text{\textcolon}} S_2$ for entrance i as

$$(S_1 \mathbin{\text{\textcolon}} S_2)(i) := \left\{ \langle \emptyset, b \rangle \sqcup \bigsqcup_{1 \leq k \leq N} Y_k \;\middle|\; \begin{array}{l} \langle \{x_1, \ldots, x_N\}, b \rangle \in S_1(i), \\ Y_1 \in S_2(x_1), \ldots, Y_N \in S_2(x_N) \end{array} \right\}.$$

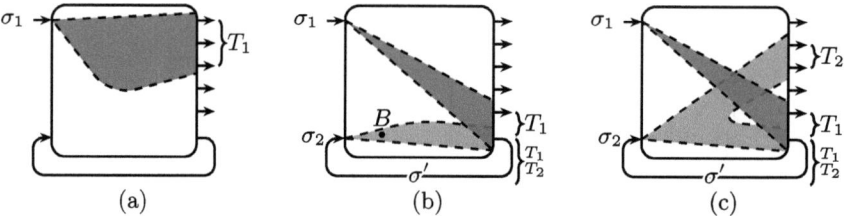

Fig. 4. Three ways of obtaining an effect in tr(S).

We conclude by outlining a proof of compositionality.

Lemma 5 (compositionality of $\mathring{,}$). $S_{\mathcal{A}\mathring{,}\mathcal{B}} = S_{\mathcal{A}} \mathring{,} S_{\mathcal{B}}$.

proof sketch. The $\supseteq$ direction follows from Lemma 4. For the $\subseteq$ direction we show that we can always decompose a no-lose strategy in $\mathcal{A} \mathring{,} \mathcal{B}$ into a no-lose strategy σ in $\mathcal{A}$ and no-lose strategies in $\mathcal{B}$ such that we can again apply Lemma 4.

Sum Composition. The sum of two solutions is simply their disjoint union.

Definition 15 ($S_1 \oplus S_2$). Let S_1 and S_2 be solutions such that arity(S_1): $m_1 \to n_1$ and arity(S_2): $m_2 \to n_2$. We define $(S_1 \oplus S_2)(i)$ as $S_1(i)$ if $1 \leq i \leq m_1$ and $S_2(i - m_1)$ otherwise.

The following statement follows straightforwardly.

Lemma 6 (compositionality of $\oplus$). $S_{\mathcal{A} \oplus \mathcal{B}} = S_{\mathcal{A}} \oplus S_{\mathcal{B}}$.

Trace. We explain the intuition of the trace of a solution using Fig. 4.

First, we pick a no-lose strategy σ_1 in $\mathcal{A}$ and let $\langle T_1, b_1 \rangle$ be its effect. Depending on the values of T_1 and b_1, we can distinguish three cases.

(a): If o_{n+1} (the *trace exit*) is not in T_1, then σ_1 is also a no-lose strategy in tr($\mathcal{A}$) with the same effect. Otherwise, for the remaining two cases, we need a no-lose strategy σ_2 from entrance $m + 1$. Let $\langle T_2, b_2 \rangle$ be the effect of σ_2.

(b): If $T_2 = \{o_{n+1}\}$, then b_2 must be true. Otherwise it would imply that σ_2 forms a *losing cycle* in tr($\mathcal{A}$).

(c): If $T_2 \neq \{o_{n+1}\}$, then the union of σ_1 and σ_2 forms a no-lose strategy in tr($\mathcal{A}$). If additionally we have that $o_{n+1} \in T_2$, then, intuitively, the formed cycle ensures that we eventually reach an exit.

We formalize the above intuition using the following *decomposition lemma*.

Lemma 7 (trace decomposition). *Let*

- *$\mathcal{A}$ be an roMDP such that* arity($\mathcal{A}$): $m + 1 \to n + 1$, *and* $i \in [m]$,
- *σ_1, σ_2 be no-lose strategies in $\mathcal{A}$ for entrances i and $m + 1$, respectively,*

- $\langle T_1, b_1 \rangle := \text{Effect}^{\mathcal{A}}(i, \sigma_1)$ such that $o_{n+1} \in T_1$,
- $\langle T_2, b_2 \rangle := \text{Effect}^{\mathcal{A}}(m+1, \sigma_2)$ such that $\langle T_2, b_2 \rangle \neq \langle \{o_{n+1}\}, \bot \rangle$,
- $\sigma' := \{o_{n+1} \mapsto \{*\}\} \cup \{v \mapsto \emptyset \mid v \in V^{\text{tr}(\mathcal{A})}, v \neq o_{n+1}\}$.

Then, $\text{Effect}^{\text{tr}(\mathcal{A})}(i, \sigma_1 \cup \sigma_2 \cup \sigma') = \langle (T_1 \cup T_2) \setminus \{o_{n+1}\}, b_1 \vee b_2 \rangle$.

Based on Lemma 7 we define the trace of a solution.

Definition 16 (tr(S)). Let S be a solution such that $\text{arity}(S): m+1 \to n+1$. We define tr(S) with $\text{arity}(\text{tr}(S)): m \to n$ as

$$\big(\text{tr}(S)\big)(i) := \{\langle T, b \rangle \in S(i) \mid n+1 \notin T\} \cup$$

$$\bigcup_{\substack{\langle T_1, b_1 \rangle \in S(i) \\ \text{s.t.} \\ n+1 \in T_1}} \left\{ \langle (T_1 \cup T_2) \setminus \{n+1\}, b_1 \vee b_2 \rangle \;\middle|\; \begin{array}{l} \langle T_2, b_2 \rangle \in S(m+1) \text{ s.t.} \\ \langle T_2, b_2 \rangle \neq \langle \{n+1\}, \bot \rangle \end{array} \right\}.$$

We conclude by outlining a proof of compositionality.

Lemma 8 (compositionality of tr). $S_{\text{tr}(\mathcal{A})} = \text{tr}(S_{\mathcal{A}})$.

proof sketch. The $\supseteq$ direction of the proof follows from Lemma 7. For the $\subseteq$ direction of the proof, we take a no-lose strategy in $\text{tr}(\mathcal{A})$ and show that we can decompose it into $\sigma_1, \sigma_2, \sigma'$, and again apply Lemma 7. We show that taking the loop of the traced roMDP multiple times does not change the reachable states.

3.3 One-Shot Bottom-up Algorithm

Theorem 1 gives rise to a natural, bottom-up computation. We provide pseudocode for the bottom-up algorithm in Appendix B.3 of the technical report [17]. Its correctness follows directly from Theorem 1 and Lemma 3. Because our solution is compositional, we can make use of *solution sharing*.

Corollary 1 (solution sharing). *Let $\mathcal{A}, \mathcal{B}$ and $\mathcal{C}$ be roMDPs. If $S_{\mathcal{A}} = S_{\mathcal{B}}$ then*

$$S_{\mathcal{A} \S \mathcal{C}} = S_{\mathcal{B} \S \mathcal{C}}, \qquad S_{\mathcal{A} \oplus \mathcal{C}} = S_{\mathcal{B} \oplus \mathcal{C}}, \qquad S_{\text{tr}(\mathcal{A})} = S_{\text{tr}(\mathcal{B})}.$$
$$S_{\mathcal{C} \S \mathcal{A}} = S_{\mathcal{C} \S \mathcal{B}}, \qquad S_{\mathcal{C} \oplus \mathcal{A}} = S_{\mathcal{C} \oplus \mathcal{B}},$$

Thus, an efficient implementation of the bottom-up algorithm only needs to compute the solution of each unique subtree of $\mathbb{D}$ (up to solution equality). If we limit the number of exits each roMDP can have to a constant, the maximum number of effects that a solution can have is also a constant, yielding a *polynomial-time* algorithm. However, in general, the running time of the algorithm grows exponentially in the number of exits.

4 Strategy Refinement Algorithm

The bottom-up algorithm above computes the set of all effects, which can grow exponentially in the number of exits. Below, we present an algorithm for solving almost-sure Büchi by iteratively reasoning about at most one effect per entrance.

First, we introduce the *shortcut graph*, adapting the shortcut MDP from [22]. This graph contains all entrances of $\mathbb{D}$, connected by transitions which represent the effect of some local strategy (the *shortcuts*). We show that an entrance in this shortcut graph is winning iff it is winning in $[\![\mathbb{D}]\!]$.

Next, we show that for each entrance the set of all effects forms a *join semi-lattice*. This allows us to define the *maximum effect* given some subset of 'allowed exits' (akin to the role of Y in the classical Büchi operator). We define a specialized *refinement Büchi operator* that uses maximum effects to prevent constructing the entire shortcut graph, yielding a polynomial time algorithm.

4.1 Shortcut Graph

We need some additional definitions to reason about string diagrams.

Definition 17. The set of *component entrances* is inductively defined by the following: $\text{CPI}(\mathcal{A}) := \{I^{\mathcal{A}}\}$, $\text{CPI}(\mathcal{A} * \mathcal{B}) := \text{CPI}(\mathcal{A}) \uplus \text{CPI}(\mathcal{B})$ for $* \in \{\fatsemi, \oplus\}$, and $\text{CPI}(\text{tr}(\mathcal{A})) := \text{CPI}(\mathcal{A})$.

We define a *connection mapping* $\text{Conn}_{\mathcal{A}}^{\mathbb{D}} \colon \mathcal{P}(O^{\mathcal{A}}) \to \mathcal{P}(\text{CPI}(\mathbb{D}))$ for each roMDP $\mathcal{A}$, which maps a set of (local) exits $X \subseteq O^{\mathcal{A}}$ to the connected component entrances in $\mathbb{D}$ (exits that are not connected to any entrance do not map to any entrance). A precise definition of $\text{Conn}_{\mathcal{A}}^{\mathbb{D}}$ straightforwardly follows by induction on the construction of $[\![\mathbb{D}]\!]$.

Definition 18 (shortcut graph). The *shortcut graph* $\mathcal{G}_{\mathbb{D}}$ is a graph $\langle V_1, V_P, E \rangle$ where $V_1 := \text{CPI}(\mathbb{D})$, $V_P := \bigcup_{i \in \text{CPI}(\mathbb{D})} \{i\} \times S_{\mathcal{A}}(i)$ and E satisfies the following equations for all $\langle i_i^{\mathcal{A}}, \langle X, b \rangle \rangle \in V_P$:

$$\text{Pre}(\langle i_i^{\mathcal{A}}, \langle X, b \rangle \rangle) = \{i_i^{\mathcal{A}}\}, \quad \text{Post}(\langle i_i^{\mathcal{A}}, \langle X, b \rangle \rangle) = \begin{cases} \text{Conn}_{\mathcal{A}}^{\mathbb{D}}(X) & \text{if } X \neq \emptyset \\ \{i_i^{\mathcal{A}}\} & \text{if } X = \emptyset \end{cases}$$

The Büchi vertices (of $\mathcal{G}_{\mathbb{D}}$) are $B_{\mathbb{D}} := \{\langle i, \langle X, b \rangle \rangle \in V_P \mid b = \top\}$.

Intuitively, the shortcut graph transforms each leaf roMDP $\mathcal{A}$ of the string diagram into a graph representing the effects present in $S_{\mathcal{A}}$. These 'leaf graphs' are then combined using the usual operational semantics of the string diagram. Note that the size of the shortcut graph, like the number of effects in the solution of an roMDP, is exponential in the number of exits.

Example 1. Figure 5a depicts the shortcut graph of $\mathcal{C}$ in Fig. 1. Each arrow represents a no-lose strategy σ that can be played. The vertices reached by the arrow depict the proper exit set of σ. In the figure, we have omitted no-lose

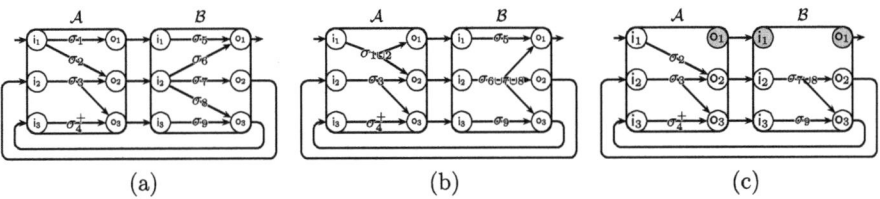

Fig. 5. Example refinement algorithm execution.

strategies that can be obtained by taking the union of two no-lose strategies. Beside a strategy we write a + to indicate that a Büchi state is reachable. We see that $\text{Effect}^{\mathcal{A}}(i_2, \sigma_3) = \langle\{o_2, o_3\}, \bot\rangle$ and $\text{Effect}^{\mathcal{A}}(i_3, \sigma_4) = \langle\{o_3\}, \top\rangle$. To satisfy the Büchi condition, we must take the transition from $i_3^{\mathcal{A}}$ to $o_3^{\mathcal{A}}$ infinitely often. ∎

The shortcut graph preserves the winning region of the entrances:

Theorem 2. *For all* $i \in \text{CPI}(\mathbb{D})$ *we have* $i \models^{\llbracket \mathbb{D} \rrbracket} \Box\Diamond B$ *iff* $i \models^{\mathcal{G}_\mathbb{D}} \Box\Diamond B_\mathbb{D}$.

The proof is given in the appendix of the technical report [17].

4.2 Strategy Refinement Algorithm

In our novel strategy refinement algorithm, we exploit the fact that the effects of a given entrance are a *join semilattice*. In short, the reason is that no-lose strategies are closed under union and $S_{\mathcal{A}}(i)$ is finite.

Lemma 9 (no-lose strategy union). *Let* σ_1, σ_2 *be no-lose strategies from entrance* i. *We have that* $\text{Effect}(i, \sigma_1 \cup \sigma_2) = \text{Effect}(i, \sigma_1) \sqcup \text{Effect}(i, \sigma_2)$.

Lemma 10. $\langle S_{\mathcal{A}}(i), \sqsubseteq \rangle$ *is a join semilattice where* $\sqsubseteq$ *is the lexicographical order, that is,* $\langle T_1, b_1 \rangle \sqsubseteq \langle T_2, b_2 \rangle$ *iff* $T_1 \subset T_2 \vee [T_1 \subseteq T_2 \wedge b_1 \leq b_2]$.

Using the join semilattice structure of effects, we now define *maximum effects*.

Definition 19 (maximum effect). *The maximum effect for* i *restricted to* $E \subseteq [n]$ *is* $S_{\mathcal{A}}(i, E) := \max\{\langle E', b\rangle \in S_{\mathcal{A}}(i) \mid E' \subseteq E\}$ *where* $\max \emptyset := \langle \emptyset, \bot \rangle$.

Note that the maximum effect restricted to E is equal to the join of all effects that have a proper exit set that is a subset of E, and therefore always exists. Using this notion of maximum effect, we define the *refinement Büchi operator* which is specialized for shortcut graphs.

Definition 20 (refinement Büchi operator). *Let* $X, Y \subseteq V_1^{\mathcal{G}_\mathbb{D}}$, *we define*

$$F'(X, Y) := \left\{ i_i^{\mathcal{A}} \in \text{CPI}(\mathbb{D}) \;\middle|\; \begin{array}{l} \text{let } \langle T, b\rangle := S_{\mathcal{A}}(i, Y) \\ \text{s.t. } \langle T, b\rangle \neq \langle \emptyset, \bot\rangle \text{ and } b \vee \text{Conn}_{\mathcal{A}}^{\mathbb{D}}(T) \cap X \neq \emptyset \end{array} \right\}.$$

This new operator closely follows the classic Büchi operator defined in Sect. 2. We will now show that the new refinement Büchi operator matches the classic Büchi operator when applied to the shortcut graph.

Theorem 3 *On $\mathcal{G}_\mathbb{D}$, $F'(X,Y) = F(X,Y)$ for all $X, Y \subseteq V_1$.*

Proof Starting with the definition of F (Definition 4), we expand its definition in a shortcut graph and work towards the definition of F'.

$$F(X,Y) := \{v \in V_1 \mid \exists v' \in \text{Post}(v) \colon \text{Post}(v') \subseteq Y \text{ and } v' \in B \vee v' \in \text{Pre}(X)\}$$

$(*\ \textit{expand shortcut graph definition}\ *)$

$$= \left\{ i_i^\mathcal{A} \in V_1 \mid \exists \langle T, b \rangle \in S_\mathcal{A}(i) \colon \text{Conn}_\mathcal{A}^\mathbb{D}(T) \subseteq Y \text{ and } b \vee \text{Conn}_\mathcal{A}^\mathbb{D}(T) \cap X \neq \emptyset \right\}$$

$(*\ \textit{an effect satisfies the previous condition iff the maximum effect does}\ *)$

$$= \left\{ i_i^\mathcal{A} \in V_1 \,\middle|\, \begin{array}{l} \text{let } \langle T, b \rangle := \max\{\langle T', b' \rangle \in S_\mathcal{A}(i) \mid T \subseteq Y\} \\ \text{s.t. } \langle T, b \rangle \neq \langle \emptyset, \bot \rangle \text{ and } b \vee \text{Conn}_\mathcal{A}^\mathbb{D}(T) \cap X \neq \emptyset \end{array} \right\}$$

$(*\ \textit{use Definition 19 of maximum effect}\ *)$

$$= \left\{ i_i^\mathcal{A} \in V_1 \,\middle|\, \begin{array}{l} \text{let } \langle T, b \rangle := S_\mathcal{A}(i, Y) \\ \text{s.t. } \langle T, b \rangle \neq \langle \emptyset, \bot \rangle \text{ and } b \vee \text{Conn}_\mathcal{A}^\mathbb{D}(T) \cap X \neq \emptyset \end{array} \right\} = F'(X,Y)$$

□

As a consequence of Theorem 3, F can be replaced by F' in the fixpoint computation of Lemma 1. Our new strategy refinement algorithm is then simply the Kleene iteration of F'. Crucially, this algorithm does *not* require an explicit representation of the shortcut graph.

Example 2. We again consider Fig. 5 and follow the Kleene iteration steps as described below Lemma 1, but using the refinement Büchi operator F' instead. Initially, Y is equal to V_1. The maximum effects considered by F' for each entrance are depicted in (b). The next value of Y (after computing the 'inner fixpoint', i.e., $\mu X. F'(X, Y)$) is equal to all vertices that can reach a Büchi vertex by playing the depicted effects. Entrances that cannot reach a Büchi vertex by playing the maximum effects must be losing; in this example, this is only $i_1^\mathcal{B}$. Therefore, $i_1^\mathcal{B}$ is not in the updated value of Y. In the next iteration of computing the inner fixpoint, we use the updated value for Y, see (c). For $i_1^\mathcal{A}$, we compute the maximum effect not reaching o_1, which is $\langle \{o_2\}, \bot \rangle$. Similarly, for $i_2^\mathcal{B}$, we compute the maximum effect not reaching o_1, which is $\langle \{o_2, o_3\}, \bot \rangle$. Finally, all entrances in Y can reach a Büchi vertex and Y is the greatest fixpoint of F'. Consequently, Y is equal to the winning region of the shortcut graph. ∎

In the classical Büchi algorithm, the number of times F is computed (and therefore F' in the refinement algorithm) is polynomial in the size of the MDP [4]. Thus, the strategy refinement algorithm runs in polynomial time if we can compute maximum effects in polynomial time, which we show below. We also outline how intermediate results of the refinement algorithm can be cached.

Computation of Maximum Effect. We briefly sketch how to efficiently compute the maximum effect for entrance i restricted to some exit set $E \subseteq O^{\mathcal{A}}$ in roMDP $\mathcal{A}$. First, we compute the winning region W of the almost-sure Büchi objective $\Box\Diamond(E\cup B)$, which can be done in polynomial time [4]. These are exactly the vertices that have a no-lose strategy that does not reach exits $O^{\mathcal{A}} \setminus E$. If W includes i, then the strategy σ_W that randomizes over vertices that remain in W reaches the biggest set of vertices and is no-lose. Hence, $S_{\mathcal{A}}(i, E) = \text{Effect}(i, \sigma_W)$.

Caching. The main computation of the refinement algorithm is that of maximum effects, which we try to cache. As an initial cache, we simply store the maximum effects and the inputs that were used to obtain it. For example, if at some point during the refinement algorithm we compute that $S_{\mathcal{A}_1}(i, E) = E'$, then for some other occurrence of $\mathcal{A}$ in the string diagram, say, $\mathcal{A}_2$, we use the cache to obtain that $S_{\mathcal{A}_2}(i, E) = E'$. Here we assume that B is defined equally for repeated occurrences of a component (otherwise, the cache may not be applied as is).

We can improve our caching by observing that the maximum effect is *monotone* for a fixed entrance i:

Lemma 11. *If $E \subseteq E'$ then $S_{\mathcal{A}}(i, E) \sqsubseteq S_{\mathcal{A}}(i, E')$.*

We can use this monotonicity to cache the maximum effect of some exit set that is 'in between' two previous computations, formalized by the following lemma:

Lemma 12. *Let $\langle Y', b \rangle := S_{\mathcal{A}}(i, Y)$. If $Y' \subseteq Y'' \subseteq Y$ then $S_{\mathcal{A}}(i, Y'') = \langle Y', b \rangle$.*

5 Related Work

Compositional model checking has been an active area of research for several decades, originating with the seminal work of Clarke et al. [5]. In this context, we focus on two closely related lines of inquiry: compositional probabilistic model checking and compositional algorithms for two-player infinite games. Other forms of compositions, such as parallel compositions, have also been studied [9,10].

Sequential Compositional Probabilistic Model Checking. Compositional probabilistic model checking using string diagrams of MDPs has been developed in a series of works [20,22,23]. These studies focus on quantitative properties such as reachability probabilities and rewards. Likewise, model checking of hierarchical MDPs [13] is also closely related to our approach. In particular, recent works [15,16] incorporate parameter synthesis techniques into hierarchical model checking algorithms. None of the works above considers repeated reachability. Sequentially composed MDPs have also been studied in the context of learning-based approaches, such as in [8,14].

Compositional Algorithms for Two-Player Infinite Games. The works [19,21] introduce string diagrammatic frameworks for parity games and mean-payoff games, respectively. The compositional algorithm proposed in [21], applicable to both types of games, relies on enumerating positional strategies for each subsystem. In contrast, [18] presents a compositional algorithm for string diagrams of parity games that avoids such enumeration, in the spirit of [22]. While its worst-case complexity is exponential in the number of exits, our strategy refinement algorithm operates in polynomial time.

6 Conclusion

In this paper, we have presented two approaches for the verification of almost-sure Büchi objectives in sequentially composed MDPs, expressed using string diagrams. The first approach is a bottom-up algorithm which computes a *compositional solution* for each leaf of the string diagram and combines them to obtain a solution of the entire string diagram. The second approach is an iterative algorithm that closely resembles the classical Büchi algorithm, but reasons locally wherever possible. Natural directions for future work are extensions to parity objectives and to quantitative variants of the problem.

References

1. Andriushchenko, R., et al.: Tools at the frontiers of quantitative verification. CoRR, abs/2405.13583 (2024)
2. Baier, C., Katoen, J.-P.: Principles of model checking. MIT Press (2008)
3. Chatterjee, K., Henzinger, M.: Faster and dynamic algorithms for maximal end-component decomposition and related graph problems in probabilistic verification. In: SODA, pp. 1318–1336. SIAM (2011)
4. Chatterjee, K., Jurdziński, M., Henzinger, T.A.: Simple stochastic parity games. In: Baaz, M., Makowsky, J.A. (eds.) CSL 2003. LNCS, vol. 2803, pp. 100–113. Springer, Heidelberg (2003). https://doi.org/10.1007/978-3-540-45220-1_11
5. Clarke, E.M., Long, D.E., McMillan, K.L.: Compositional model checking. In: LICS, pp. 353–362. IEEE Computer Society (1989)
6. de Alfaro, L.: Formal verification of probabilistic systems. PhD thesis, Stanford University, USA (1997)
7. de Alfaro, L., Henzinger, T.A.: Concurrent omega-regular games. In: LICS, pp. 141–154. IEEE Computer Society (2000)
8. Delgrange, F., Avni, G., Lukina, A., Schilling, G., Nowé, A., Pérez, G.A.: Composing reinforcement learning policies, with formal guarantees. In: AAMAS, pp. 574–583. International Foundation for Autonomous Agents and Multiagent Systems/ACM (2025)
9. Feng, L., Kwiatkowska, M.Z., Parker, D.: Compositional verification of probabilistic systems using learning. In: QEST, pp. 133–142. IEEE Computer Society (2010)
10. Feng, L., Kwiatkowska, M., Parker, D.: Automated learning of probabilistic assumptions for compositional reasoning. In: Giannakopoulou, D., Orejas, F. (eds.) FASE 2011. LNCS, vol. 6603, pp. 2–17. Springer, Heidelberg (2011). https://doi.org/10.1007/978-3-642-19811-3_2
11. Forejt, V., Kwiatkowska, M., Norman, G., Parker, D.: Automated verification techniques for probabilistic systems. In: Bernardo, M., Issarny, V. (eds.) SFM 2011. LNCS, vol. 6659, pp. 53–113. Springer, Heidelberg (2011). https://doi.org/10.1007/978-3-642-21455-4_3

12. Hartmanns, A., Junges, S., Quatmann, T., Weininger, M.: A practitioner's guide to MDP model checking algorithms. In: TACAS (1), vol. 13993. LNCS, pp. 469–488. Springer (2023)
13. Hauskrecht, M., Meuleau, N., Kaelbling, L.P., Dean, T.L., Boutilier, C.: Hierarchical solution of Markov decision processes using macro-actions. In: UAI, pp. 220–229. Morgan Kaufmann (1998)
14. Jothimurugan, K., Bansal, S., Bastani, O., Alur, R.: Compositional reinforcement learning from logical specifications. In: NeurIPS, pp. 10026–10039 (2021)
15. Junges, S., Spaan, M.T.J.: Abstraction-refinement for hierarchical probabilistic models. In: CAV (1), vol. 13371. LNCS, pp. 102–123. Springer (2022)
16. Neary, C., Verginis, C.K., Cubuktepe, M., Topcu, U.: Verifiable and compositional reinforcement learning systems. In: ICAPS, pp. 615–623. AAAI Press (2022)
17. van der Vegt, M., Watanabe, K., Hasuo, I., Junges, S.: Compositional verification of almost-sure Büchi objectives in MDPs. CoRR, abs/2508.13087 (2025)
18. Watanabe, K.: Pareto fronts for compositionally solving string diagrams of parity games. In: CALCO, vol. 342. LIPIcs, pp. 14:1–14:20. Schloss Dagstuhl - Leibniz-Zentrum für Informatik (2025)
19. Watanabe, K., Eberhart, C., Asada, K., Hasuo, I.: A compositional approach to parity games. In: MFPS, vol. 351. EPTCS, pp. 278–295 (2021)
20. Watanabe, K., Eberhart, C., Asada, K., Hasuo, I.: Compositional probabilistic model checking with string diagrams of MDPs. In: CAV (3), vol. 13966. LNCS, pp. 40–61. Springer (2023)
21. Watanabe, K., Eberhart, C., Asada, K., Hasuo, I.: Compositional solution of mean payoff games by string diagrams. In: Principles of Verification (3), vol. 15262. LNCS, pp. 423–445. Springer (2024)
22. Watanabe, K., van der Vegt, M., Hasuo, I., Rot, J., Junges, S.: Pareto curves for compositionally model checking string diagrams of MDPs. In: TACAS (2), vol. 14571. LNCS, pp. 279–298. Springer (2024)
23. Watanabe, K., van der Vegt, M., Junges, S., Hasuo, I.: Compositional value iteration with Pareto caching. In: CAV (3), vol. 14683. LNCS, pp. 467–491. Springer (2024)

DTMC Model Checking by Path Abstraction Revisited

Arnd Hartmanns and Robert Modderman(✉)

University of Twente, Enschede, The Netherlands
r.modderman@utwente.nl

Abstract. Computing the probability of reaching a set of goal states G in a discrete-time Markov chain (DTMC) is a core task of probabilistic model checking. We can do so by directly computing the probability mass of the set of *all* finite paths from the initial state to G; however, when refining counterexamples, it is also interesting to compute the probability mass of subsets of paths. This can be achieved by splitting the computation into *path abstractions* that calculate "local" reachability probabilities as shown by Ábrahám et al. in 2010. In this paper, we complete and extend their work: We prove that splitting the computation into path abstractions indeed yields the same result as the direct approach, and that the splitting does not need to follow the SCC structure. In particular, we prove that path abstraction can be performed along *any* finite sequence of sets of non-goal states. Our proofs proceed in a novel way by interpreting the DTMC as a structure on the free monoid on its state space, which makes them clean and concise. Additionally, we provide a compact reference implementation of path abstraction in PARI/GP.

1 Introduction

In this paper, we study methods to split computations of total reachability probabilities on discrete-time Markov chains (DTMCs) into computing *local* reachability probabilities: Given a DTMC with finite state space S and initial state a, instead of computing the total probability $\mathbf{P}^a(\Diamond G)$ of reaching the set of goal states G from a (where all $s \in G$ are absorbing) by directly computing the probability mass of the set of paths from a to G, we select a sequence of subsets of the set of non-goal states, compute local reachability probabilities *over* those state subsets—i.e. the probabilities of the paths within the subset that start from a transition entering the subset and end in a transition leaving the subset—and in the end compute the total probabilities from there.

Computing these local probabilities can be done by *path abstraction* [1], which is an operation that, given a subset S_1 of the set of all non-goal states, moves the probability masses of the *paths* that pass through S_1 onto new *transitions* in

---

This work was supported by the EU's Horizon 2020 research and innovation programme under MSCA grant agreement 101008233 (MISSION), by the Interreg North Sea project STORM_SAFE, and by NWO VIDI grant VI.Vidi.223.110 (TruSTy).

© The Author(s), under exclusive license to Springer Nature Switzerland AG 2026
P. Ganty and A. Mansutti (Eds.): RP 2025, LNCS 16230, pp. 186–201, 2026.
https://doi.org/10.1007/978-3-032-09524-4_13

a new DTMC: the path abstraction of the original DTMC *over* S_1. Most states in S_1 become unreachable in the new DTMC and could be removed. The path abstraction operation can be used to split the computation of total reachability probabilities into multiple steps of computing local reachability probabilities on subsets of states (and thus of paths). It also provides a recipe for probabilistic *counterexample refinement* [2]: each separate computation yields a candidate for the violation of a certain specified reachability constraint, e.g. those of the form $\mathbf{P}^a(\Diamond s) \leqslant \lambda$ for some goal state $s \in G$ and some given $\lambda \in [0,1]$.

Ábrahám et al. in [1] give a procedure to split the computation of total reachability probabilities by path abstraction based on a recursive decomposition into *strongly connected components* (SCCs) of the DTMC's underlying digraph up to trivial SCCs. Furthermore, they give a concrete algorithm for computing path abstractions. However, in [1], no proof is supplied for the statement that the direct approach and the one that proceeds by splitting the computation over SCCs yield the same result. Furthermore, the algorithm as given in [1, Sec. III] can be used only for subsets $S_1 \subseteq S$ for which there is a path from every state within S_1 to a state outside of S_1, i.e. for S_1 that do not contain bottom-SCCs (BSCCs), i.e., SCCs with no transitions leading outwards. This can be ensured by additional preprocessing of the DTMC (e.g. finding and collapsing all states that reach s with probability zero [7]), but limits the algorithm's generality.

Our Contribution. In Sect. 3, we prove that path abstraction is "monotonically absorbing" and thus both the direct "global" approach (of path abstracting a DTMC over the set of all non-goal states straight away) and the "local" approach (of path abstracting along *any* finite sequence of subsets of non-goal states and *then* over the set of all non-goal states) yield the exact same result. We achieve this in a novel, elegant way by interpreting a DTMC as a structure on the free monoid on its state space (in Sect. 2). From our proof follows correctness of the approach of [1], which we show in Sect. 4, along with coupling our findings to counterexample refinement. Furthermore, in Sect. 5, we give a numerical algorithm to compute the path abstraction of a DTMC over *any* subset of the state space, taking into account that, when abstracting over a subset $S_1 \subseteq S$, S_1 may contain states that do not reach outside of S_1. Finally, in Sect. 6, we give a high-level reference implementation in the computer algebra system PARI/GP [25] to compute path abstractions, which in addition allows for the input of parametric DTMCs. This reference implementation is close to our abstract formulation of the algorithm, providing some confidence in its correctness. It is a first step towards formalizing and machine-checking our algorithm and proofs using an interactive theorem prover, like recently done for the iterative interval iteration algorithm that computes reachability probabilities in a global manner [18].

Related Approaches. Computing reachability probabilities in DTMCs, mathematically, means solving a linear equation system [3, Sec. 10.1.1]; research lies in doing so efficiently under different constraints and with a view towards different purposes. Our work extends and completes that of Ábrahám et al. [1], which uses *path abstraction* over the SCC structure, with a view towards counterexample refinement. We briefly compare path abstraction to *state elimination*, which is prominently used for checking parametric DTMCs, in Sect. 7. The idea of exploiting the SCC structure was already part of the model reduction techniques for Markov decision processes (MDPs), of which DTMCs are a special case, proposed by Ciesinski et al. [4] in 2008. It also underpins the iterative-numeric *topological value iteration* algorithm [5] as well as the incremental approach by Kwiatkowska et al. [19], and improves parametric DTMC model checking [17]. Gui et al. [23] then use the DTMC's SCC structure for computing reachability probabilities via Gaussian elimination, with a view towards improved scalability and performance by eliminating cycles. Their work was later extended to MDPs [9]. The idea of reducing to acyclic DTMCs was in fact used earlier by Andrés et al. [2] for finding and describing probabilistic counterexamples.

2 Background

$\mathbb{N} = \{0, 1, \dots\}$ is the set of natural numbers, and $\mathbb{N}^+ \stackrel{\text{def}}{=} \mathbb{N} \setminus \{0\}$. Given $n \in \mathbb{N}^+$, let $[n] \stackrel{\text{def}}{=} \{1, \dots, n\}$. Given a set X, we write 2^X for the powerset of X. Given a propositional formula φ and objects A and B, let $[\varphi, A, B]$ be the object A if φ holds and B otherwise. In addition, let $[\varphi] \stackrel{\text{def}}{=} [\varphi, 1, 0]$ (the *Iverson bracket* of φ).

We index matrices by finite sets, which is more general than indexing them by positive integers. For a matrix T indexed by $A \times B$, we let $T(a, b)$ be the entry of T indexed by the pair $(a, b) \in A \times B$. If T is an $A \times B$ matrix, and $A' \subseteq A$ and $B' \subseteq B$, then let $T(A', B')$ denote the submatrix of T whose rows and columns are those indexed by A' and B', respectively. If $A' = B'$ then write $T(A')$ for $T(A', A')$. Let $\mathbf{1}(A, B)$ denote the matrix with $\mathbf{1}(a, b) = [a = b]$. If A and B are clear from the context, then we write $\mathbf{1} = \mathbf{1}(A, B)$.

2.1 Combinatorics on Words: The Free Monoid

Let us formalize and introduce some notions on free monoids on finite sets (or, equivalently, combinatorics on words), which is a field within mathematics that studies words and formal languages. We borrow most of the terminology from [22].

Let Σ be a finite set, the *alphabet*, and let Σ^* be the set of all finite sequences over Σ, the *words* over Σ. Given $x \in \Sigma^*$, let $|x| \in \mathbb{N}$ be the *length* of x; if in addition x is non-empty, then let x_i denote the i-th entry of x for all $i \in [|x|]$. Entries of words are called *letters*. The *empty word* is denoted by ε. Given $n \in \mathbb{N}$, let $\Sigma^{\bowtie n} \stackrel{\text{def}}{=} \{x \in \Sigma^* \mid |x| \bowtie n\}$ where $\bowtie \in \{<, \leq, \geq, >\}$, and let $\Sigma^n \stackrel{\text{def}}{=} \{x \in \Sigma^* \mid |x| = n\}$. Whenever appropriate, we identify Σ^1 with Σ. Furthermore, conforming to [22], we write $\Sigma^+ = \Sigma^* \setminus \{\varepsilon\}$.

Given $x, y \in \Sigma^*$, let xy denote the *concatenation* of x and y. Σ^* together with the concatenation operation and the empty sequence ε forms a *monoid*, as outlined in [22, Sec. 1.2.1], known as the *free monoid* on Σ. Concatenation extends to sets of words: Given $X, Y \subseteq \Sigma^*$, we can form $XY \stackrel{\text{def}}{=} \{xy \mid x \in$

$X \wedge y \in Y\}$. Furthermore, for $x, y \in \Sigma^*$ and $a \in \Sigma$, set $(xa) \star (ay) \stackrel{\text{def}}{=} xay$.[1] If $X, Y \subseteq \Sigma^*$ are such that there exists an $a \in \Sigma$ with $X \subseteq \Sigma^* a$ and $Y \subseteq a\Sigma^*$, then we may form $X \star Y \stackrel{\text{def}}{=} \{x \star y \mid x \in X \wedge y \in Y\}$.

If $x \in \Sigma^*$, then $x' \in \Sigma^*$ is called a *factor* of x if there exist $y, z \in \Sigma^*$ such that $x = yx'z$, denoted by $x' \sqsubseteq x$. If y can be chosen empty, then x' is called a *prefix* of x, denoted by $x' \leqslant x$. Note that $(\Sigma^*, \leqslant)$ in this fashion becomes a partially ordered set. If $x' \leqslant x$ but $x' \neq x$, then this is denoted by $x' < x$. If z can be chosen empty, then x' is called a *suffix* of x. Given $L \subseteq \Sigma^*$, we let $L^{\leqslant} \stackrel{\text{def}}{=} \{x \in L \mid x' < x \Rightarrow x' \notin L\}$ denote the subset of L of minimal elements w.r.t. $\leqslant$.[2] Note that the operator $(\cdot)^{\leqslant}$ on 2^{Σ^*} is idempotent, i.e., $(L^{\leqslant})^{\leqslant} = L^{\leqslant}$.

Now, given $\Sigma_1 \subseteq \Sigma$, we define the operation $-\Sigma_1 : \Sigma^* \to \Sigma^*$ on Σ^* as follows, where for ease of notation we write $x - \Sigma_1$ instead of $(-\Sigma_1)(x)$: Given a word $x \in \Sigma^*$, we consider x as the unique minimal-length[3] word over the alphabet $\Sigma_1^+ \uplus (\Sigma \setminus \Sigma_1)^+$ and let $x - \Sigma_1$ denote the word obtained from x by replacing each letter $a \in \Sigma_1^+$ of x by the first letter a_1 of a when a is considered as a word over Σ. In other words, from every maximal factor $a \sqsubseteq x$ of x with $a \in \Sigma_1^+$, we only keep its first entry. To expand this, if $\Sigma_1, \ldots, \Sigma_n \subseteq \Sigma$, then we write $-(\Sigma_1, \ldots, \Sigma_n)$ for the operation $(-\Sigma_n) \circ \cdots \circ (-\Sigma_1) : \Sigma^* \to \Sigma^*$. (Note that function composition order is read right-to-left.) Furthermore, write, for $x \in \Sigma^*$, $x - (\Sigma_1, \ldots, \Sigma_n)$ instead of $(-(\Sigma_1, \ldots, \Sigma_n))(x)$. Given $x \in \Sigma^*$, write $x + (\Sigma_1, \ldots, \Sigma_n)$ for the pre-image $(-(\Sigma_1, \ldots, \Sigma_n))^{-1}(x) = \{x' \in \Sigma^* \mid x' - (\Sigma_1, \ldots, \Sigma_n) = x\}$. For $x + (\Sigma_1)$ we simply write $x + \Sigma_1$—thus, $x + \Sigma_1 = \{x' \in \Sigma^* \mid x' - \Sigma_1 = x\}$.

Example 1. Consider sequences over the Latin alphabet $\Sigma = \{a, b, \ldots, z\}$. We illustrate the $-\Sigma_1$-operation by a few cases of $x \in \Sigma^*$ and $\Sigma_1 = \{b, r, e, a, k\}$.

1. Let $x = error$. Then, x as the unique minimal-length word over the alphabet $\Sigma_1^+ \uplus (\Sigma \setminus \Sigma_1)^+$ is written as $x = (err)(o)(r)$. To obtain $x - \Sigma_1$, we replace the occurrence of err by e and the occurrence of r by r, so $x - \Sigma_1 = eor$.
2. Since $x - \Sigma_1 = eor$ we have $x \in eor + \Sigma_1$.
3. For $x = spacebar$ we have $x - \Sigma_1 = space$, since x considered as the minimal-length word over $\Sigma_1^+ \uplus (\Sigma \setminus \Sigma_1)^+$ is written as $x = (sp)(a)(c)(ebar)$.
4. For $x = coffee$ we have $x + \Sigma_1 = \emptyset$, as no $y \in \Sigma^*$ satisfies $y - \Sigma_1 = x$ because $ee \sqsubseteq x$ is a factor of x containing e twice while $e \in \Sigma_1$.

2.2 Discrete-Time Markov Chains

Let us give a first formal definition of *discrete-time Markov chains* (DTMCs).

Definition 1. *Let S be a finite set and $a \in S$. A* **discrete-time Markov chain** *(DTMC) with* state space S *and* initial state $a \in S$ *is a triple (S, a, T) where $T \in \mathbb{R}^{S \times S}$ is a stochastic (substochastic) matrix, i.e. $T(s, t) \geqslant 0$ for all $s, t \in S$, and for all $s \in S$, $\sum_{t \in S} T(s, t) = 1$ ($\leqslant 1$).*

---

[1] In this fashion, we obtain a function $\star : \biguplus_{a \in \Sigma} \Sigma^* a \times a\Sigma^* \to \Sigma^+$.
[2] In [1], the prime symbol $'$ is used for this operation; we use $\leqslant$ instead for readability.
[3] I.e., the letters of x as a word over $\Sigma_1^+ \uplus (\Sigma \setminus \Sigma_1)^+$ alternate over Σ_1 and $\Sigma \setminus \Sigma_1$.

Since for the mechanism of path abstraction we need to reason about probabilities of paths and sets thereof, a convenient alternative interpretation of a DTMC is as a structure on S^+. The key observation is that a (sub)stochastic matrix $T \in \mathbb{R}^{S \times S}$ corresponds to a unique function $\mathbf{P}\colon S^+ \to [0,1]$ such that $\mathbf{P}(x) = 1$ for all $x \in S^1$, $\mathbf{P}(st) = T(s,t)$ for all $s,t \in S$, and $\mathbf{P}(xsy) = \mathbf{P}(xs)\mathbf{P}(sy)$ for all $x,y \in S^*$ and $s \in S$.

Definition 2. *Let S be a finite set. Then, let $\mathcal{T}(S)$ denote the set of all functions $\mathbf{P}\colon S^+ \to [0,1]$ satisfying*

1. *$\mathbf{P}(x) = 1$ for all $x \in S^1$;*
2. *for all $s \in S$, $\sum_{t \in S} \mathbf{P}(st) \leq 1$;*
3. *for all $x, y \in S^*$ and $s \in S$, $\mathbf{P}(xsy) = \mathbf{P}(xs)\mathbf{P}(sy)$—or, equivalently, for all $s \in S$ and $u \in S^*s$ and $v \in sS^*$, $\mathbf{P}(u \star v) = \mathbf{P}(u)\mathbf{P}(v)$.*

Let $\mathcal{T}_1(S) \stackrel{\text{def}}{=} \{\,\mathbf{P} \in \mathcal{T}(S) \mid \forall s \in S \colon \sum_{t \in S} \mathbf{P}(st) = 1\,\}$. Functions $\mathbf{P} \in \mathcal{T}(S)$ are called **substochastic transition probability functions** on S^+, and functions $\mathbf{P} \in \mathcal{T}_1(S)$ are simply called **transition probability functions** on S^+.

Note that any $\mathbf{P} \in \mathcal{T}(S)$ can be extended to a function $2^{S^+} \to [0,\infty]$ by setting, for $R \subseteq S^+$, $\mathbf{P}(R) \stackrel{\text{def}}{=} \sum_{x \in R^{\leq}} \mathbf{P}(x)$, which is guaranteed by [24, Theorem 0.0.2] to exist as either a nonnegative real number[4] or $+\infty$. Note that we do not sum over R but over $R^{\leq}$ instead. This way of defining probability masses of sets of paths is standard in the realm of probabilistic model checking on DTMCs, see e.g. [1, Sec. II]. We can now give the following alternative definition of DTMCs, which we call the "free monoid interpretation":

Definition 3. *Let S be a finite set and let $a \in S$. Then, a **discrete-time Markov chain** (DTMC) with state space S and initial state a is a pair[5] $M = (\mathbf{P}, a)$ where $\mathbf{P} \in \mathcal{T}_1(S)$. Pairs $M = (\mathbf{P}, a)$ with $\mathbf{P} \in \mathcal{T}(S)$ are called **substochastic discrete-time Markov chains**. We write $\mathcal{M}_1(S, a)$ and $\mathcal{M}(S, a)$ for the two classes of Markov chains, respectively.*

From now on, let $M = (\mathbf{P}, a) \in \mathcal{M}(S, a)$ always be given implicitly. We are ultimately interested in computing **reachability probabilities**, i.e. the probability $\mathbf{P}^a(\lozenge G)$ of reaching some set of goal states $G \subseteq S \setminus \{\,a\,\}$. We write $\mathbf{P}^a(\lozenge s)$ for $\mathbf{P}^a(\lozenge \{\,s\,\})$. W.l.o.g. we assume that each goal state is absorbing and a is not.

Definition 4. *Let $G \subseteq \{\,s \in S \mid \mathbf{P}(ss) = 1\,\}$ be a set of absorbing states with $a \notin G$. Then the probability of reaching G from a is $\mathbf{P}^a(\lozenge G) \stackrel{\text{def}}{=} \mathbf{P}(aS^*G)$.*

Example 2. Figure 1 visualizes DTMC $M_e \in \mathcal{M}_1(S_e, s_1)$ with $S_e = \{\,s_1, \ldots, s_8\,\}$. We have $\mathbf{P}(s_2) = 1$, $\mathbf{P}(s_2 s_2) = 0$, $\mathbf{P}(s_2 s_5) = \frac{1}{3}$, and $\mathbf{P}(s_2 (s_5 s_6)^+ s_2) = \frac{1}{6}$. Some reachability probabilities are $\mathbf{P}^{s_1}(\lozenge \{\,s_7, s_8\,\}) = 1$ and $\mathbf{P}^{s_1}(\lozenge s_7) = \frac{5}{9}$ (see Example 4).

---

[4] By default, to the empty sum we assign the value 0, so we obtain $\mathbf{P}(\varnothing) = 0$.
[5] The state space S is implicitly given by the domain of $\mathbf{P}$ hence omitted.

Definition 5. A transition *is a pair* $(s,t) \in S \times S$ *such that* $\mathbf{P}(st) > 0$. *The* underlying digraph *of M is the graph with vertex set S whose edges are the transitions in M. A* finite path *in M is a sequence* $x \in S^+$ *such that* $\mathbf{P}(x) > 0$. *State t is* reachable *from state s in M if* $\mathbf{P}(sS^*t) > 0$. *Given $K \subseteq S$, let* $\mathrm{Comp}\,(M,K)$ *denote the coarsest partition of K such that each $C \in \mathrm{Comp}\,(M,K)$ satisfies* $\mathbf{P}(sC^*t) > 0$ *for all distinct* $s,t \in C$. *That is,* $\mathrm{Comp}\,(M,K)$ *are the* strongly connected components *(SCCs) of the underlying digraph of M restricted to K.*

3 Path Abstraction via the Free Monoid

We define path abstraction using our new free monoid interpretation of DTMCs, and in contrast to [1], without altering the state space (we simply do not draw states that become isolated). We then prove our main results on the monotonic absorption property of path abstraction that equates the global and local approaches to DTMC model checking. For that, we need two fundamental results:

Lemma 1. *Let Σ be a finite set, and let $\Sigma_1 \subseteq \Sigma$. Then, for all $x \in \Sigma^*$, we have* $x' \leqslant x \Rightarrow x' - \Sigma_1 \leqslant x - \Sigma_1$.

Proof (sketch). By induction on $|x| - |x'|$. The full proof is included in [15, App. A].

Theorem 1. *Let Σ be a finite set. Then, the following hold:*

1. *Let $\Sigma_1 \subseteq \Sigma$, let $x, y \in \Sigma^*$, and let $a \in \Sigma$. Suppose that $xa \notin \Sigma^* \Sigma_1^2$. Then, we have $xay - \Sigma_1 = (xa - \Sigma_1) \star (ay - \Sigma_1)$.*
2. *Again let $\Sigma_1 \subseteq \Sigma$, let $x, y \in \Sigma^*$, and let $a \in \Sigma$. Then, we have $(xa + \Sigma_1)^{\leqslant} \star (ay + \Sigma_1)^{\leqslant} = (xay + \Sigma_1)^{\leqslant}$.*
3. *If $\Sigma_1 \subseteq \Sigma_2 \subseteq \Sigma$, then for all $x \in \Sigma^*$ we have $x - (\Sigma_1, \Sigma_2) = x - \Sigma_2$.*
4. *Let $\Sigma_1 \subseteq \Sigma_2 \subseteq \Sigma$. Then, for all $x \in \Sigma^*$ we have $(x + \Sigma_2)^{\leqslant} = \biguplus \{ (y + \Sigma_1)^{\leqslant} \mid y \in (x + \Sigma_2)^{\leqslant} \wedge y - \Sigma_1 = y \}$.*

Proof (sketch). Part 1: by case distinction $a \in \Sigma_1$ vs. $a \notin \Sigma_1$. Part 2: by part 1. Part 3: by considering x as a word of minimal length over the alphabet $\Sigma_2^+ \uplus (\Sigma \setminus \Sigma_2)^+$. Part 4: by part 3 and Lemma 1. The full proofs are included in [15, App. A].

We also need some initial results on transition probability functions. We note that parts 2 and 3 below may appear trivial, but are later needed.

Theorem 2. *Let S be a finite set, and let $\mathbf{P} \in \mathcal{T}(S)$. Then,*

1. *for all $R \subseteq S^+$, we have $\mathbf{P}(R) = \lim_{k \to \infty} \mathbf{P}(R^{\leqslant} \cap S^{\leqslant k})$;*
2. *if $R \subset sS^*$ for some $s \in S$ with R finite, then $\mathbf{P}(R) \leqslant 1$;*
3. *if $R \subseteq sS^*$ for some $s \in S$ with R of any cardinality, then still $\mathbf{P}(R) \leqslant 1$;*
4. *if $R \subseteq T \subseteq S^+$, then $\mathbf{P}(R) \leqslant \mathbf{P}(T)$.*

Proof (sketch). Part 1: by [24, Theorem 0.0.2]. Part 2: by induction on the maximum length of a sequence in $R^{\leqslant}$. Part 3: by combining parts 1 and 2. Part 4: via the function $f\colon R^{\leqslant} \to T^{\leqslant}$ where $f(x)$ is the unique nonempty prefix of x in T (hence is in $T^{\leqslant}$). The full proofs are included in [15, App. A].

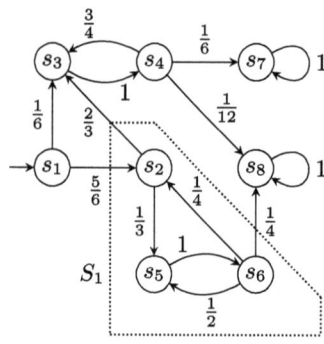
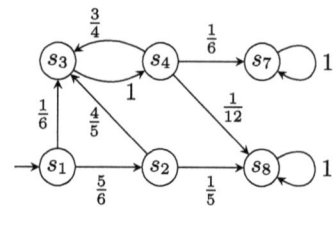

Fig. 1. DTMC M_e and set S_1 **Fig. 2.** Abstracted DTMC $M_e - S_1$

3.1 Path Abstraction

The path abstraction operation on (substochastic) DTMCs intuitively collapses (all paths through) the abstraction set into (the paths crossing) the set's border states, while preserving the DTMC's overall reachability probabilities.

Definition 6. *For $S_1 \subseteq S$, the* **path abstraction** *of M over S_1 is $M - S_1 \overset{\text{def}}{=} (\mathbf{P}^a_{S_1}, a) \in \mathcal{M}(S, a)^6$ where $\mathbf{P}^a_{S_1} \colon S^+ \to [0, \infty]$ is defined as follows:*

$$\mathbf{P}^a_{S_1}(x) = \begin{cases} 1 & \text{if } x \in S^1 \\ 0 & \text{if } x \in S^{\geqslant 2} \wedge \exists\, i \colon x_i \in (S_1)^M_0 \\ \mathbf{P}(x + S_1) & \text{otherwise} \end{cases}$$

with $(S_1)^M_0 \overset{\text{def}}{=} \{\, s \in S_1 \setminus \{\, a\,\} \mid \mathbf{P}((S \setminus S_1)s) = 0\,\}$.

The set $(S_1)^M_0$ collects the states in the "interior" of S_1, i.e. those with no incoming transition from outside S_1. Note that $M - S_1$ has at most as many transitions as M; the second case removes all paths (and thus transitions) that pass through $(S_1)^M_0$. For $S_1, \ldots, S_n \subseteq S$, we again write $-(S_1, \ldots, S_n)$ instead of $(-S_n) \circ \cdots \circ (-S_1)$, and $M - (S_1, \ldots, S_n)$ instead of $(-(S_1, \ldots, S_n))(M)$.

Example 3. The path abstraction $M_e - S_1$ of M_e over $S_1 = \{\, s_2, s_5, s_6\,\}$ is visualized in Fig. 2. In the abstraction from M_e to $M - S_1$, the transitions between states from S_1 are removed, the states from $(S_1)^{M_e}_0 = \{\, s_5, s_6\,\}$ are removed from the diagram (as their incoming and outgoing probabilities are set to zero), and the probabilities of the transitions $s_2 \to s_3$ and $s_2 \to s_8$ are replaced with the probability masses of the sets $s_2 S_1^* s_3$ and $s_2 S_1^* s_8$, respectively.

For Definition 6 to be well-defined, we need to confirm that $(\mathbf{P}^a_{S_1}, a) \in \mathcal{M}(S, a)$:

---

[6] Formally, $M - S_1$ again denotes $(-S_1)(M)$ for a function $-S_1 \colon \mathcal{M}(S, a) \to \mathcal{M}(S, a)$.

Lemma 2. *In Definition 6, we have* $\mathbf{P}^a_{S_1} \in \mathcal{T}(S)$.

Proof (sketch). We show that $\mathbf{P}^a_{S_1}$ satisfies the three axioms of Definition 2 as follows. Axiom 1: by definition of $\mathbf{P}^a_{S_1}$, case 1. Axiom 2: by Theorem 2, part 3. Axiom 3: by Theorem 1, part 2. The full proof is included in [15, App. B].

The reason why we consider substochastic DTMCs (see Definition 3) is because there exist DTMCs $M \in \mathcal{M}_1(S, a)$ and $S_1 \subseteq S$ such that $M - S_1 \notin \mathcal{M}_1(S, a)$, e.g. when S_1 contains a BSCC, and we want to keep the option open to abstract over sets that overlap with BSCCs.

3.2 Monotonic Absorption of Path Abstraction

We are now in position to prove the two main results of this paper, which state that path abstraction is "monotonically absorbing" in the sense that applying it first to subsets of later abstraction sets does not change the final result.[7]

Theorem 3. *If* $S_1 \subseteq S_2 \subseteq S$, *then* $\forall M \in \mathcal{M}(S, a) \colon M - (S_1, S_2) = M - S_2$.

Proof (sketch). By Theorem 1, part 3, Theorem 1, part 4, and Theorem 2, part 4. The full proof is included in [15, App. B].

The preceding theorem provides structure in the analysis of path abstracting DTMCs over (many) more than simply two subsets of the state space.

Corollary 1. *Let* $S_1, \ldots, S_t \subseteq K \subseteq S$. *Then* $M - (S_1, \ldots, S_t, K) = M - K$.

Proof (sketch). By induction on t and by Theorem 3; the full proof is in [15, App. B].

4 DTMC Model Checking by Path Abstraction

The main goal of DTMC model checking by path abstraction is to obtain the *result* of abstracting a DTMC M over the set $K \stackrel{\text{def}}{=} \{\, s \in S \mid \mathbf{P}(ss) < 1 \,\}$ of its non-absorbing states, i.e. the DTMC $M - K$. Clearly, this preserves the self-loops with probability 1 of all absorbing states, and the only other transitions remaining are the transitions $a \to s$ with $s \in S \setminus K$ absorbing. They carry precisely the reachability probabilities $\mathbf{P}^a_K(as) = \mathbf{P}(aK^*s) = \mathbf{P}(aS^*s) = \mathbf{P}^a(\Diamond s)$.[8] Then, for any goal set $G \subseteq S \setminus K$ we can simply compute $\mathbf{P}^a(\Diamond G) = \sum_{s \in G} \mathbf{P}^a(\Diamond s)$.

---

[7] This could also be seen as a form of *confluence*, but we prefer the term *absorption* to emphasize the invariance of the result w.r.t. abstracting over *subsets*.
[8] $\mathbf{P}(aS^*s) = \mathbf{P}^a(\Diamond s)$ is by Definition 4. The fact that $\mathbf{P}(aK^*s) = \mathbf{P}(aS^*s)$ is seen as follows: We have $(aS^*s)^{\leqslant} = a(S \setminus \{s\})^*s$ and $(aK^*s)^{\leqslant} = aK^*s$, so any $x \in (aS^*s)^{\leqslant} \setminus (aK^*s)^{\leqslant}$ must contain a state $x_i \neq s$ that is absorbing, hence, as x ends in s and x_i has outgoing probability zero, has probability zero.

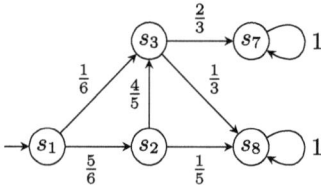

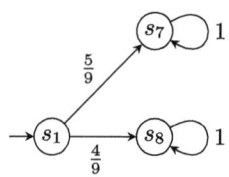

Fig. 3. DTMC $M_e - (S_1, S_2)$ **Fig. 4.** The final DTMC $M_e - K$

Ábrahám et al. in [1] give two methods to perform DTMC model checking by path abstraction: The first method just abstracts over each SCC (leaving the algorithm to abstract, i.e. solve, each SCC open), while the second method also recursively applies path abstraction within SCCs. We argue that both methods implicitly generate a finite sequence $S_1, \ldots, S_t \subseteq K$ of subsets of K along which the original DTMC is abstracted, before it is eventually abstracted over K. This means that they compute $M - (S_1, \ldots, S_t, K)$, which by Corollary 1 is precisely $M - K$: the desired output (substochastic) DTMC. Let us fix $S^{(0)} = \{\,\{s\} \mid s \in S \land \mathbf{P}(ss) = 0\,\}$ for the set of singleton sets of states that do not have self-loops.

SCC abstraction. The first method, given by [1, Algorithm 1 alone], is functionally equivalent to transforming M into $M - (U_1, \ldots, U_t, K)$ where $U_1, \ldots, U_t$ is any enumeration of $\mathrm{Comp}\,(M, K) \setminus S^{(0)}$. Since every U_i is a subset of K, by Corollary 1 this in turn is functionally equivalent to transforming M into $M - K$.

Example 4. If we apply the first method to M_e from Fig. 1, then we abstract over $S_1 = \{s_2, s_5, s_6\}$, $S_2 = \{s_3, s_4\}$, and finally $K = \{s_1, \ldots, s_6\}$. We obtain the sequence M_e, $M_e - S_1$, $M_e - (S_1, S_2)$, and $M_e - (S_1, S_2, K) = M_e - K$, depicted in Figs. 1, 2, 3, and 4, respectively. From Fig. 4 it is clear that $\mathbf{P}^{s_1}(\Diamond s_7) = \frac{5}{9}$ and $\mathbf{P}^{s_1}(\Diamond s_8) = \frac{4}{9}$, where $\mathbf{P}$ is the transition probability function of the first DTMC M_e as depicted in Fig. 1, but, due to Corollary 1, in expressing the reachability probabilities of $\Diamond s_7$ and $\Diamond s_8$, can be replaced with the transition probability function of *any* of the abstracted DTMCs we computed along the way.

Recursive Abstraction. The second method, described by [1, Algorithm 1 using Algorithm 2], uses a recursive approach: To abstract an SCC S_1, it abstracts each SCC of $(S_1)_0^M$ (which is the interior of S_1 from Definition 6), and so on. We capture this by a new operator $\ominus$:

$$M \ominus S_1 := (M \ominus (U_1, \ldots, U_t)) - S_1$$

where (i) $S_1 \subseteq S$ is strongly connected (i.e. $\mathbf{P}(sS_1^*t) > 0$ for all distinct $s, t \in S_1$), (ii) $(S_1)_0^M \subsetneq S_1$, and (iii) $U_1, \ldots, U_t$ is an enumeration of $\mathrm{Comp}\,(M, (S_1)_0^M) \setminus S^{(0)}$. If $U_1, \ldots, U_t$ in condition (iii) is empty, $\ominus$ returns $M - S_1$ as a base case.

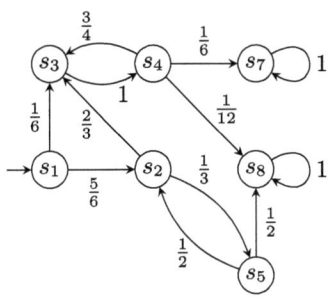
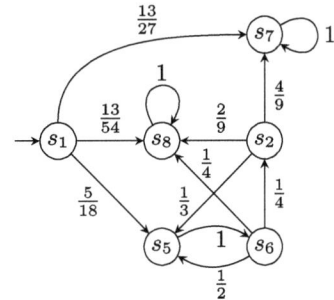

Fig. 5. DTMC $M_e - S_0$

Fig. 6. DTMC $M_e - \{s_1, s_2, s_3, s_4\}$

At some point, $\ominus$ will be applied to an S_1 for which $(S_1)_0^M$ does not contain non-trivial SCCs, and then regular "$-$"-abstractions are computed bottom-up.[9] In this way, the input DTMC M is, again, "$-$"-abstracted along a sequence of subsets of K and finally over K itself, which by Corollary 1 yields $M - K$ again.

Example 5. Consider M_e from Fig. 1 again. Now following the recursive structure of "$\ominus$", one additional path abstraction is computed compared to Ex. 4: over $S_0 = \{s_5, s_6\}$ at the beginning. Our sequence of subsets of K becomes (S_0, S_1, S_2), and our sequence of DTMCs becomes M_e, $M_e - S_0$, $M_e - (S_0, S_1) = M_e - S_1$,[10] $M_e - (S_1, S_2)$, and $M_e - K$, depicted in Figs. 1, 5, 2, 3, and 4, respectively.

Our formulation of the two methods is slightly more general, and fixes a small mistake, compared to [1]. We provide detailed notes on the differences in [15, App. C].

Counterexample refinement. Suppose we choose a threshold $\lambda \in [0, 1]$ and wish to determine whether $\mathbf{P}^a(\Diamond s) \leqslant \lambda$. Of course, we can abstract over K straight away to immediately compute $\mathbf{P}^a(\Diamond s) = \mathbf{P}_K^a(as)$ and check the threshold. However, due to Corollary 1, we can now choose *any* sequence of $t \in \mathbb{N}^+$ subsets $S_1, \ldots, S_t \subseteq K$ of K as interesting as we like, and iteratively compute a sequence of (substochastic) DTMCs according to $M_0 := M$ and $M_i := M_{i-1} - S_i$ for $i = 1, 2, \ldots$ ($i \leqslant t$) until we hit a DTMC M_i that has a path from a to s with probability $> \lambda$. Then we can go back along the sequence to trace what path set (which is a subset of aK^*s) was responsible for exceeding the threshold.

Our contribution here is that $S_1, \ldots, S_t$ does not have to follow the SCC structure as in [1], but can be any finite sequence of subsets of K, thus extending the method well beyond the connectivity of the underlying digraph of the DTMC.

Example 6. Consider M_e from Fig. 1 and reachability constraint $\mathbf{P}^{s_1}(\Diamond s_7) \leqslant \frac{4}{9}$. It is not immediate from Fig. 1 that this constraint is violated by M_e; neither is it clear from the DTMCs from Figs. 5, 2, and 3, which are the intermediate abstractions in the recursive SCC-based method. Instead of following a sequence of subsets of the state space implicitly generated by any SCC-based method, we can, for

---

[9] Condition (ii) is required for the initial S_1 to ensure termination, too; see [15, App. C].
[10] Note that this holds by Theorem 3, as $S_0 \subset S_1$.

the DTMC M_e from Fig. 1, compute the abstraction $M_e - \{s_1, s_2, s_3, s_4\}$ (without having to abstract over $K = \{s_1, \ldots, s_6\}$ itself), as depicted in Fig. 6. Then, we can see that the reachability constraint $\mathbf{P}^{s_1}(\lozenge s_7) \leqslant \frac{4}{9}$ for the original DTMC M_e in Fig. 1 is violated, since the transition (s_1, s_7) in the DTMC of Fig. 6 has probability $\frac{13}{27} > \frac{4}{9}$—without having to abstract over other parts of the DTMC.

5 Numerical Methods

Let us briefly recall some numerical methodology typically used to compute reachability probabilities for DTMC model checking. Good reference material includes [3, Sec. 10.1.1] and [8]. Our goal is to design a reference implementation to compute path abstractions, more abstract than that of [1, Algorithm 3] and executable. Throughout, given $\mathbf{P} \in \mathcal{T}(S)$, let $\mathbf{P}_2 \in \mathbb{R}^{S \times S}$ denote the transition probability matrix corresponding to $\mathbf{P}$—i.e., $\mathbf{P}_2(s, t) = \mathbf{P}(st)$ for all $s, t \in S$.

Theorem 4. *(1) The matrix $\mathbf{P}_2$ is substochastic, and (2) if $S_1 \subseteq S$, $i \in \mathbb{N}^+$, and $s, r \in S_1$, then $\mathbf{P}(sS_1^{i-1}r) = \mathbf{P}_2(S_1)^i(s, r)$.*

Proof (sketch). Part 1: by Definition 2–2. Part 2: by induction on i. The full proof is included in [15, App. D].

Now, we transform the above results to reachability probabilities for DTMC.

Theorem 5. *Let $S_1 \subsetneq S$ and set*

$$\mathscr{I} := S_1 \setminus (S_1)_0^M, \qquad \mathscr{O} := \{t \in S \setminus S_1 \mid \mathbf{P}(S_1 t) > 0\},$$
$$\mathscr{U} := \{r \in S_1 \mid \mathbf{P}(rS_1^* \mathscr{O}) > 0\}, \qquad \mathscr{U}_1 := \{r \in S_1 \mid \mathbf{P}(r\mathscr{O}) > 0\}.$$

Suppose that $\mathscr{I} \cap \mathscr{U}$ and $\mathscr{O}$ are nonempty. Let $s \in \mathscr{I} \cap \mathscr{U}$ and $t \in \mathscr{O}$. Then, we have

$$\mathbf{P}^a_{S_1}(st) = \mathbf{P}(st + S_1) = \mathbf{P}(sS_1^* t) = Q(s, \mathscr{U}_1)\mathbf{P}_2(\mathscr{U}_1, t) \tag{1}$$

where $Q(\mathscr{I} \cap \mathscr{U}, \mathscr{U}_1)$ is extracted from the unique solution $Q \in \mathbb{R}^{\mathscr{U} \times \mathscr{U}_1}$ of the system of equations $(\mathbf{1} - \mathbf{P}_2(\mathscr{U})) \cdot Q = \mathbf{1}(\mathscr{U}, \mathscr{U}_1)$.

Proof (sketch). By showing that $\sum_{i=0}^{\infty} \mathbf{P}_2(\mathscr{U})^i$ converges by noting that the spectral radius of $\mathbf{P}_2(\mathscr{U})$ is smaller than 1, which follows from the definition of $\mathscr{U}$, and combining this with Definition 6. The full proof is included in [15, App. D].

Now, as the computations of the sets $\mathscr{I}$, $\mathscr{O}$ and $\mathscr{U}_1$ are obvious but the computation of $\mathscr{U}$ is not, let us present a reachability algorithm to compute $\mathscr{U}$.

Lemma 3. *Consider Theorem 5. Then, $\mathscr{U}$ can be computed as follows:*

1. *Set $\mathscr{U}_0 := \mathscr{O}$, set $\mathscr{V}_0 := \mathscr{U}_0$, and set $i := 0$.*
2. *While $\mathscr{U}_i \neq \varnothing$, set $\mathscr{U}_{i+1} := \{r \in S_1 \setminus \mathscr{V}_i \mid \mathbf{P}(r\mathscr{U}_i) > 0\}$, set $\mathscr{V}_{i+1} := \mathscr{V}_i \uplus \mathscr{U}_{i+1}$, and set $i := i + 1$.*
3. *Output $\mathscr{V}_i \setminus \mathscr{O}$, which is precisely $\mathscr{U}$.*

Proof (sketch). Straightforward. The full proof is included in [15, App. D].

The observation now is that Theorem 5 and Lemma 3 provide a full recipe for a numerical implementation of path abstraction: We compute the sets $\mathscr{I}$, $\mathscr{O}$, $\mathscr{U}_1$, and $\mathscr{U}$, and solve $(1 - \mathbf{P}_2(\mathscr{U})) \cdot Q = 1(\mathscr{U}, \mathscr{U}_1)$ for Q. Now, we build the matrix $Y \in \mathbb{R}^{S \times S}$ using matrix comprehension as follows. We iterate over $(s,t) \in S \times S$, and we set, using the "if-then-else blocks" $[\cdot, \cdot, \cdot]$,

$$Y(s,t) := [s \in S \setminus S_1 \wedge t \in (S \setminus S_1) \uplus \mathscr{I}, \mathbf{P}_2(s,t), \\ [s \in \mathscr{I} \cap \mathscr{U} \wedge t \in \mathscr{O}, Q(s, \mathscr{U}_1)\mathbf{P}_2(\mathscr{U}_1, t), 0]], \quad (2)$$

and then we have determined $(\mathbf{P}_{S_1}^a)_2 = Y$ fully. Indeed: Whenever $s, t \in S_1$ then $st + S_1 = \varnothing$; whenever $s \in S_1 \setminus \mathscr{I}$ or $t \in S_1 \setminus \mathscr{I}$ then $\mathbf{P}_{S_1}^a(st) = 0$; whenever $s, t \in S \setminus S_1$ then we keep the original probability value of the transition (s,t); whenever $s \in S \setminus S_1$ and $t \in \mathscr{I}$ then we have $\mathbf{P}_{S_1}^a(st) = \mathbf{P}(stS_1^*) = \mathbf{P}(st)$ as well as $(stS_1^*)^{\leqslant} = \{st\}$; whenever $s \in \mathscr{I} \setminus \mathscr{U}$ and $t \in \mathscr{O}$ then $\mathbf{P}_{S_1}^a(st) = 0$ because $\mathbf{P}(sS_1^*t) = 0$ in this case (by definition of $\mathscr{U}$); and whenever $s \in \mathscr{I} \cap \mathscr{U}$ and $t \in \mathscr{O}$ then we have $\mathbf{P}_{S_1}^a(st) = Q(s, \mathscr{U}_1)\mathbf{P}_2(\mathscr{U}_1, t)$ by Theorem 5.

```
pathAbstr = M -> (K -> \
    X = M[1]; \
    a = M[2]; \
    n = matsize(X)[1]; \
    II = vector(n, i, K[i] && (i == a || \
        vecsum(vector(n, j, !K[j] * X[j, i])) != 0)); \
    OO = vector(n, i, !K[i] && \
        vecsum(vector(n, j, K[j] * X[j, i])) != 0); \
    Uvec = [OO]; \
    Vvec = [OO]; \
    i = 1; \
    while (Uvec[i] != vector(n), \
        Uvec = concat(Uvec, [vector(n, j, K[j] && !Vvec[i][j] && \
            vecsum(vector(n, k, Uvec[i][k] * X[j, k])) != 0)]); \
        Vvec = concat(Vvec, \
            [vector(n, j, Vvec[i][j] || Uvec[i+1][j])]); \
        i++; \
    ); \
    U1 = Uvec[2]; \
    U = vector(n, j, Vvec[i][j] && !OO[j]); \
    XU = matrix(n, n, i, j, (U[i] && U[j]) * X[i, j]); \
    Q = matsolve(matid(n) - XU, \
        matrix(n, n, i, j, i == j && U[i] && U1[j])); \
    Y = matrix(n, n, i, j, if(!K[i] && (!K[j] || II[j]), \
        X[i, j], if(II[i] && U[i] && OO[j], Q[i,] * X[, j], 0))); \
    [Y, a]; \
);
```

Listing 1. Computing $M - K$ where $M \in \mathcal{M}(S, a)$ and $K \subseteq S$

6 Reference Implementation

Now, we are in position to apply the numerical methodology to compute path abstractions of (substochastic) DTMC. We give a high-level reference implementation in a semi-functional setting, as close as possible to the mathematical reasoning of the recipe provided at the end of Sect. 5. We use the PARI/GP computer algebra system [25], and write a GP script that can be interpreted by the GP interpreter. Here, DTMCs are represented as pairs whose first element is a square matrix carrying the probabilities and whose second element is the initial state. The state space is implicitly understood as the set $[n] = \{1, \ldots, n\}$, where n is the dimension of the matrix. Computing $M_e - S_1$ (Fig. 2) from M_e (Fig. 1), for example, can be done with

pathAbstr([X, 1])([0,1,0,0,1,1,0,0])

where X is the 8×8 matrix carrying the probabilities of M_e. The output will be [Y, 1], where Y is the matrix carrying the probabilities of the abstracted DTMC $M_e - S_1$. We refer to [15, App. E] for a full analysis of Listing 1.

7 Conclusion and Future Work

We have recast the path abstraction approach to DTMC model checking of Ábrahám et al. [1] in terms of the free monoid on the DTMC's state space. In this setting, the procedure can be expressed elegantly and concisely; in proofs of similar qualities, we have shown that path abstraction can be applied to any number of subsets of the final abstraction set in any order. This generalizes the approach of [1], and makes counterexample refinement much more flexible.

We note that path abstraction is similar in spirit, but effectively very different, from the *state elimination* method used for parametric [6,10] and exact probabilistic model checking [16]. In particular, where state elimination may incur an intermediate blowup in transitions, path abstraction never generates more transitions. We plan to properly study the complexity of path abstraction, and it may then be interesting to investigate if a combination of state elimination and path abstraction could bring together the simplicity of the former with the scalability of the latter. Just like state elimination, path abstraction should straightforwardly generalize to expect-reward properties, too.

Our reference implementation arguably comes with a relatively high level of trustworthiness given its short distance from the underlying numerical recipes, but is not practically efficient. It can however be extremely useful as a baseline for testing and comparison for future lower-level, optimized implementations e.g. in C or C# as part of the MCSTA model checker [12] of the MODEST TOOLSET [11].

To attain the highest level of trust, we shall formalize the concepts and algorithms of this paper in the interactive theorem prover Isabelle/HOL to (1) machine-check the proofs and (2) derive a verified correct-by-construction *and* fast implementation in LLVM bytecode using the Isabelle Refinement Framework [20,21], as recently done in probabilistic model checking for SCC finding [13], MEC decomposition [14], and finally the interval iteration algorithm [18].

Acknowledgments. We thank Milan Lopuhaä-Zwakenberg for inspiring us to use free monoids where suitable—specifically, in the analysis of *repeatedly* applying path abstraction—and use transition probability matrices for the numerical analysis of path abstractions as separate operations, and Benedikt Peterseim for providing detailed feedback on drafts of this paper.

Extended Version and Data Availability. An extended version of this paper, which includes an appendix with the full proofs as well as the code shown in Listing 1 in a separate file, is available on arXiv with DOI 10.48550/arXiv.2509.02393 [15].

References

1. Ábrahám, E., Jansen, N., Wimmer, R., Katoen, J.P., Becker, B.: DTMC model checking by SCC reduction. In: 7th International Conference on the Quantitative Evaluation of Systems, pp. 37–46. IEEE Computer Society (2010). https://doi.org/10.1109/QEST.2010.13
2. Andrés, M.E., D'Argenio, P., van Rossum, P.: Significant diagnostic counterexamples in probabilistic model checking. In: Chockler, H., Hu, A.J. (eds.) Hardware and Software: Verification and Testing. LNCS, vol. 5394, pp. 129–148. Springer, Heidelberg (2009). https://doi.org/10.1007/978-3-642-01702-5_15
3. Baier, C., Katoen, J.P.: Principles of model checking. MIT Press (2008)
4. Ciesinski, F., Baier, C., Größer, M., Klein, J.: Reduction techniques for model checking Markov decision processes. In: 5th International Conference on the Quantitative Evaluation of Systems (QEST 2008), pp. 45–54. IEEE Computer Society (2008). https://doi.org/10.1109/QEST.2008.45
5. Dai, P., Goldsmith, J.: Topological value iteration algorithm for Markov decision processes. In: Veloso, M.M. (ed.) 20th International Joint Conference on Artificial Intelligence (IJCAI 2007), pp. 1860–1865 (2007). http://ijcai.org/Proceedings/07/Papers/300.pdf
6. Daws, C.: Symbolic and parametric model checking of discrete-time Markov chains. In: Liu, Z., Araki, K. (eds.) ICTAC 2004. LNCS, vol. 3407, pp. 280–294. Springer, Heidelberg (2005). https://doi.org/10.1007/978-3-540-31862-0_21
7. Forejt, V., Kwiatkowska, M., Norman, G., Parker, D.: Automated verification techniques for probabilistic systems. In: Bernardo, M., Issarny, V. (eds.) SFM 2011. LNCS, vol. 6659, pp. 53–113. Springer, Heidelberg (2011). https://doi.org/10.1007/978-3-642-21455-4_3
8. Guen, H.L., Marie, R.A.: Visiting probabilities in non-irreducible Markov chains with strongly connected components. In: Amborski, K., Meuth, H. (eds.) 16th European Simulation Multiconference: Modelling and Simulation 2002, pp. 548–552. SCS Europe (2002)
9. Gui, L., Sun, J., Song, S., Liu, Y., Dong, J.S.: SCC-based improved reachability analysis for Markov decision processes. In: Merz, S., Pang, J. (eds.) ICFEM 2014. LNCS, vol. 8829, pp. 171–186. Springer, Cham (2014). https://doi.org/10.1007/978-3-319-11737-9_12

10. Hahn, E.M., Hermanns, H., Zhang, L.: Probabilistic reachability for parametric Markov models. Int. J. Softw. Tools Technol. Transf. **13**(1), 3–19 (2011). https://doi.org/10.1007/S10009-010-0146-X
11. Hartmanns, A., Hermanns, H.: The Modest Toolset: an integrated environment for quantitative modelling and verification. In: Ábrahám, E., Havelund, K. (eds.) TACAS 2014. LNCS, vol. 8413, pp. 593–598. Springer, Heidelberg (2014). https://doi.org/10.1007/978-3-642-54862-8_51
12. Hartmanns, A., Hermanns, H.: Explicit model checking of very large MDP using partitioning and secondary storage. In: Finkbeiner, B., Pu, G., Zhang, L. (eds.) ATVA 2015. LNCS, vol. 9364, pp. 131–147. Springer, Cham (2015). https://doi.org/10.1007/978-3-319-24953-7_10
13. Hartmanns, A., Kohlen, B., Lammich, P.: Fast verified SCCs for probabilistic model checking. In: André, É., Sun, J. (eds.) 21st International Symposium on Automated Technology for Verification and Analysis (ATVA 2023). LNCS, vol. 14215, pp. 181–202. Springer (2023). https://doi.org/10.1007/978-3-031-45329-8_9
14. Hartmanns, A., Kohlen, B., Lammich, P.: Efficient formally verified maximal end component decomposition for MDPs. In: Platzer, A., Rozier, K.Y., Pradella, M., Rossi, M. (eds.) 26th International Formal Methods Symposium (FM 2024). LNCS, vol. 14933, pp. 206–225. Springer (2024). https://doi.org/10.1007/978-3-031-71162-6_11
15. Hartmanns, A., Modderman, R.: DTMC model checking by path abstraction revisited (extended version). CoRR abs/2509.02393 (2025). https://doi.org/10.48550/arXiv.2509.02393
16. Hensel, C., Junges, S., Katoen, J.P., Quatmann, T., Volk, M.: The probabilistic model checker Storm. Int. J. Softw. Tools Technol. Transf. **24**(4), 589–610 (2022). https://doi.org/10.1007/S10009-021-00633-Z
17. Jansen, N., Corzilius, F., Volk, M., Wimmer, R., Ábrahám, E., Katoen, J.P., Becker, B.: Accelerating parametric probabilistic verification. In: Norman, G., Sanders, W.H. (eds.) 11th International Conference on the Quantitative Evaluation of Systems (QEST 2014). LNCS, vol. 8657, pp. 404–420. Springer (2014). https://doi.org/10.1007/978-3-319-10696-0_31
18. Kohlen, B., Schäffeler, M., Abdulaziz, M., Hartmanns, A., Lammich, P.: A formally verified IEEE 754 floating-point implementation of interval iteration for MDPs. In: Piskac, R., Rakamaric, Z. (eds.) 37th International Conference on Computer Aided Verification (CAV 2025). LNCS, vol. 15932, pp. 122–146. Springer (2025). https://doi.org/10.1007/978-3-031-98679-6_6
19. Kwiatkowska, M.Z., Parker, D., Qu, H.: Incremental quantitative verification for Markov decision processes. In: 2011 IEEE/IFIP International Conference on Dependable Systems and Networks (DSN 2011), pp. 359–370. IEEE Compute Society (2011). https://doi.org/10.1109/DSN.2011.5958249
20. Lammich, P.: Automatic data refinement. In: Blazy, S., Paulin-Mohring, C., Pichardie, D. (eds.) ITP 2013. LNCS, vol. 7998, pp. 84–99. Springer, Heidelberg (2013). https://doi.org/10.1007/978-3-642-39634-2_9
21. Lammich, P., Tuerk, T.: Applying data refinement for monadic programs to hopcroft's algorithm. In: Beringer, L., Felty, A. (eds.) ITP 2012. LNCS, vol. 7406, pp. 166–182. Springer, Heidelberg (2012). https://doi.org/10.1007/978-3-642-32347-8_12
22. Lothaire, M.: Algebraic Combinatorics on Words, Encyclopedia of Mathematics and its Applications. Cambridge University Press (2002)

23. Song, S., Gui, L., Sun, J., Liu, Y., Dong, J.S.: Improved reachability analysis in DTMC via divide and conquer. In: Johnsen, E.B., Petre, L. (eds.) IFM 2013. LNCS, vol. 7940, pp. 162–176. Springer, Heidelberg (2013). https://doi.org/10.1007/978-3-642-38613-8_12
24. Tao, T.: An Introduction to Measure Theory. American Mathematical Society (2011)
25. The PARI Group, Univ. Bordeaux: PARI/GP version 2.11.0 (2018). http://pari.math.u-bordeaux.fr/

Counterexample-Guided Abstraction Refinement for Star-Based Neural Network Verification

László Antal[(✉)], Franz Link, and Erika Ábrahám

RWTH Aachen University, Aachen, Germany
{antal,abraham}@cs.rwth-aachen.de, franz.link@rwth-aachen.de

Abstract. We consider two reachability analysis methods for feedforward neural networks, which use star sets as datatype to store network states. While the first method is complete but computationally expensive, the second method offers better scalability by using over-approximations and thus sacrificing completeness. In this paper, we propose a counterexample-guided abstraction refinement (CEGAR) framework to combine the strengths of these two procedures, starting with the second method, and iteratively refining the over-approximation based on spurious counterexamples. Our algorithm is complete and it either certifies safety or it returns a counterexample as a proof of unsafety. We suggest multiple heuristics for the refinement and evaluate them experimentally, demonstrating that our CEGAR-based approach is more efficient than a previously proposed direct abstraction refinement method, and on some benchmarks, it significantly outperforms the exact method.

Keywords: Feedforward Neural Networks · Verification · CEGAR

1 Introduction

Feedforward neural networks (FNNs) [11]. became popular tools for solving complex real-world problems, such as autonomous driving [10], object and speech recognition [6,9] and robot vision [12], to mention a few. Due to their complex, non-linear nature, reasoning about their output for a priori unknown inputs is very challenging. Therefore, their application in safety-critical or mission-critical systems is still limited. Recently developed *safety verification* techniques aim to provide formal guarantees about neural networks, for example by showing that for a restricted input set, the set of all possible outputs is disjoint from a set of unsafe states.

Related work. In this work we build upon two safety verification techniques [1–4,15,16] for FNNs with piecewise linear activation functions, which use *star sets* as a datatype to represent sets of FNN states. The first method uses *exact* computations. While it is sound and complete, and it can prove both safety and unsafety, it suffers from scalability to larger networks since they would require an exponential number of star sets to represent case distinctions in the network. In contrast, the second method uses *over-approximative (relaxed)* computations,

© The Author(s), under exclusive license to Springer Nature Switzerland AG 2026
P. Ganty and A. Mansutti (Eds.): RP 2025, LNCS 16230, pp. 202–216, 2026.
https://doi.org/10.1007/978-3-032-09524-4_14

thereby achieving efficiency but losing completeness. Thus, the relaxed method might verify safety, but is unable to prove unsafety. Leveraging abstraction refinement techniques, such as the *counterexample-guided abstraction refinement (CEGAR)*, one can regain the completeness for relaxed verification algorithms. In this work we focus on adapting CEGAR for the star-based FNN verification to dynamically refines relaxations using spurious counterexamples.

Contributions. Our contributions in this paper are as follows:

1. We employ the CEGAR framework for the star-based reachability analysis of FNNs, improving scalability while maintaining completeness.
2. Furthermore, we introduce several heuristics for the identification of the cause of spurious counterexamples to guide the refinement.
3. Lastly, we provide an open-access implementation[1] of our approach, which we used to conduct an experimental evaluation with different heuristics and to compare the results to other existing approaches [3,4].

Outline. After some preliminaries in Sect. 2, we present our CEGAR framework and multiple heuristics in Sect. 3. In Sect. 4 we report on experimental results and conclude the paper in Sect. 5.

2 Preliminaries

Let $\mathbb{N}$ and $\mathbb{R}$ be the set of all natural (incl. 0) resp. real numbers; lower indices specify subsets, e.g. $\mathbb{R}_{>0}$ stands for the positive reals. We consider elements of $\mathbb{R}^n$ (for $n \in \mathbb{N}_{\geq 2}$) to be column vectors. We recall some preliminaries from [1,2,15].

2.1 Feedforward Neural Networks

A *feedforward neural network (FNN)* [14] is an acyclic directed weighted graph. As shown in Fig. 1, an FNN has a finite set of nodes called *neurons*, which are organized into $k \in \mathbb{N}_{\geq 2}$ disjoint, non-empty, ordered sets $\ell_1, \ldots, \ell_k$ called *layers*, with ℓ_1 being the input layer, ℓ_k the output layer, and $\ell_2, \ldots, \ell_{k-1}$ the hidden layers. Each neuron of a non-input layer has a connection from each neuron in the preceding layer. The weight matrices $\mathbf{W}^{(i)} \in \mathbb{R}^{|\ell_i| \times |\ell_{i-1}|}$ with entries $w_{rc}^{(i)}$ store the weight of the connection from neuron c in layer

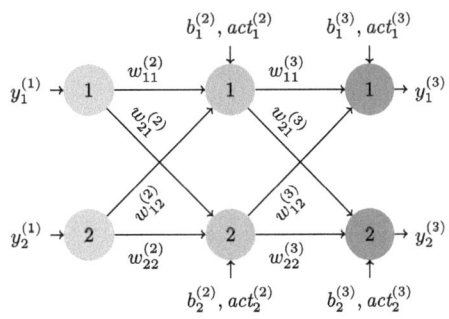

Fig. 1. Example FNN with an input layer (green/left), one hidden layer (blue/middle), and an output layer (red/right).

---
[1] See https://github.com/hypro/hypro for the code, and README.md under Case Studies/Neural Network Verification on HyPro's GitHub page for the experiments.

ℓ_{i-1} to neuron r in layer ℓ_i. Additionally, in each non-input layer, each neuron is annotated with a bias value from $\mathbb{R}$ and an activation function of type $\mathbb{R} \to \mathbb{R}$.

We jointly represent the bias values in layer ℓ_i by the bias vector $\mathbf{b}^{(i)} \in \mathbb{R}^{|\ell_i|}$, and the activation functions as $\mathbf{act}^{(i)} : \mathbb{R}^{|\ell_i|} \to \mathbb{R}^{|\ell_i|}$ with $\mathbf{act}^{(i)}(\mathbf{x}) = \left(\text{act}_1^{(i)}(x_1), \ldots, \text{act}_{|\ell_i|}^{(i)}(x_{|\ell_i|})\right)^T$. A representative example for an activation function is the rectified linear unit, defined as $ReLU(x) = \max(0, x)$. In our formulations, we assume ReLU activations, but note that our approach can be extended to any other piecewise linear function [1,2,16].

For an *input* $\mathbf{y}^{(1)} \in \mathbb{R}^{|\ell_1|}$, we define the *state* of each non-input layer ℓ_i recursively as $\mathbf{y}^{(i)} = \mathbf{act}^{(i)}\left(\mathbf{b}^{(i)} + \mathbf{W}^{(i)}\mathbf{y}^{(i-1)}\right)$. Thus, one can see the FNN as a function $f : \mathbb{R}^{|\ell_1|} \to \mathbb{R}^{|\ell_k|}$, mapping each input to the output layer's state, which we call the *output*. For an FNN and an input set $\mathcal{R}_1 \subseteq \mathbb{R}^{|\ell_1|}$, the *(FNN) reachability problem* is the task to compute all possible states of all layers $1 < i \leq k$:

$$\mathcal{R}_i = \left\{\mathbf{act}^{(i)}\left(\mathbf{b}^{(i)} + \mathbf{W}^{(i)}\mathbf{y}^{(i-1)}\right) \,\middle|\, \mathbf{y}^{(i-1)} \in \mathcal{R}_{i-1}\right\}. \tag{1}$$

In this work, we assume input sets to be convex polyhedra.

2.2 Star Sets

We use *star sets* to represent FNN state sets [2].

Definition 1. *For $n, m \in \mathbb{N}_{\geq 1}$ an (n, m)-dimensional star set (shortly star) is a tuple $\Theta = \langle \mathbf{c}, \mathbf{G}, \mathcal{P} \rangle$ of (i) a center $\mathbf{c} \in \mathbb{R}^n$, (ii) a generator matrix $\mathbf{G} \in \mathbb{R}^{n \times m}$ whose columns $\mathbf{g}^{(1)}, \ldots, \mathbf{g}^{(m)} \in \mathbb{R}^n$ are called generators, and (iii) a predicate $\mathcal{P} \subseteq \mathbb{R}^m$. The star Θ represents the set $\{\mathbf{c} + \mathbf{G}\boldsymbol{\alpha} \mid \boldsymbol{\alpha} \in \mathcal{P}\}$.*

We overload notation and write Θ both for the star and the set it represents. Moreover, as in [2,15], we restrict the predicate to be a convex polyhedron $\mathcal{P} = \{\boldsymbol{\alpha} \in \mathbb{R}^m \mid \mathbf{A}\boldsymbol{\alpha} \leq \mathbf{d}\}$ for some $p \in \mathbb{N}_{\geq 1}$, $\mathbf{A} \in \mathbb{R}^{p \times m}$ and $\mathbf{d} \in \mathbb{R}^p$. Consequently, the stars considered in this paper represent convex polyhedra in $\mathbb{R}^n$.

For some operations on stars (e.g., affine mapping, set operations, bounding box computation) and some relevant properties we refer to [2,4,15] (but their consultation is not mandatory to understand the following contributions).

2.3 Reachability Analysis

From now on, let f be an FNN with ReLUs, input set $\mathcal{R}_1$, and unsafe set $\mathcal{U}$.

We recall an *exact* and a *relaxed (over-approximative)* algorithm to solve the reachability problem. Both algorithms compute for each layer all reachable states. While the exact method provides this as a union of stars, the relaxed method over-approximates reachability by a single star.

Considering a star input $\Theta = \langle \mathbf{c}, \mathbf{G}, \mathcal{P} \rangle$ to layer ℓ_i, both algorithms compute the ReLU input through the affine mapping $\Theta' := \mathit{Affine}^{(i)}(\Theta) = \langle \mathbf{c}', \mathbf{G}', \mathcal{P} \rangle$ with $\mathbf{c}' = \mathbf{b}^{(i)} + \mathbf{W}^{(i)}\mathbf{c}$ and $\mathbf{G}' = \mathbf{W}^{(i)}\mathbf{G}$. Then $\mathcal{R}_i = ReLU_{|\ell_i|}(\cdots(ReLU_1(\Theta'))\cdots)$

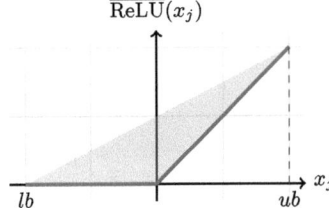 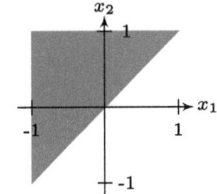 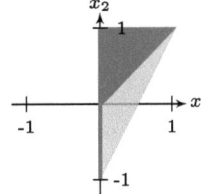

Fig. 2. Left: Relaxation for ReLU [2,5]. Middle: Example input set. Right: ReLU over-approximation for the input set at dimension x_1 (the dark part being the exact result).

is the reachable set of layer ℓ_i, where $ReLU_j$ is the ReLU operation applied at dimension j. Assuming Θ' is bounded in the jth dimension to values from $[lb, ub]$, we distinguish the cases (i) $lb \geq 0$, (ii) $ub \leq 0$ and (iii) $lb < 0 < ub$.

In cases (i) and (ii), the domain $[lb, ub]$ is either non-positive or non-negative, hence the result of the ReLU can be exactly represented using a single convex star: in case (i) we have $ReLU_j(\Theta') = \Theta'$, while in case (ii) $ReLU_j(\Theta')$ is computed from Θ' by updating **c** and **G**, resetting all values in dimension j to 0.

In case (iii), however, $ReLU_j(\Theta')$ might be non-convex and thus not representable by a single star. The *exact* method splits Θ' into $\Theta'_{j-} = \{(x_1,\ldots,x_n)^T \in \Theta' \mid x_j < 0\}$ handled as in (i), and $\Theta'_{j+} = \{(x_1,\ldots,x_n)^T \in \Theta' \mid x_j \geq 0\}$ handled as in (ii). The exact result of $ReLU_j(\Theta')$ is the union of the two stars.

Instead of case splitting, the *relaxed* algorithm constructs a single star that conservatively over-approximates both cases. To do so, it extends the predicate with a new variable α_{m+1} and three new constraints on α_{m+1} defining the convex triangular over-approximation $\overline{ReLU_j}(\Theta')$ of $ReLU_j(\Theta')$ (see Fig. 2).

3 Counterexample-Guided Abstraction Refinement

3.1 Reach Trees and Counterexamples

Applying ReLU on a star might yield two stars, which need to be processed separately. Therefore, as illustrated in Fig. 3, the exact method computes for f and $\mathcal{R}_1$ a binary *reach tree*, whose nodes store the computed stars. In contrast, the relaxed method always yields a single-path reach tree. We get the exact resp. over-approximated reachable output set by uniting the leaf stars.

We are going to introduce a method which combines exact and relaxed computation steps, yielding more general reach trees as formalized next. For simplicity, we assume an identification mechanism and refer to reach tree nodes by the sets they store. We use standard tree notions, node depth being 0 for the root.

Definition 2 (Reach tree). *A reach tree $\mathcal{T}$ (for f and $\mathcal{R}_1$) is an ordered binary tree, each node storing a star, the root node storing $\mathcal{R}_1$, with the following properties. For $i \in \{2,\ldots,k\}$ and $j \in \{0,\ldots,|\ell_i|\}$ let $d(i,j) = j + \sum_{h=2}^{i-1}(1+|\ell_h|)$, and let $D = 1 + d(k, |\ell_k|)$ be the height of $\mathcal{T}$.*

1. *Nodes Θ at depth $d(i,0)$ for some $i \in \{2,\ldots,k\}$ are called* base *and have only a left child $Affine^{(i)}(\Theta)$.*

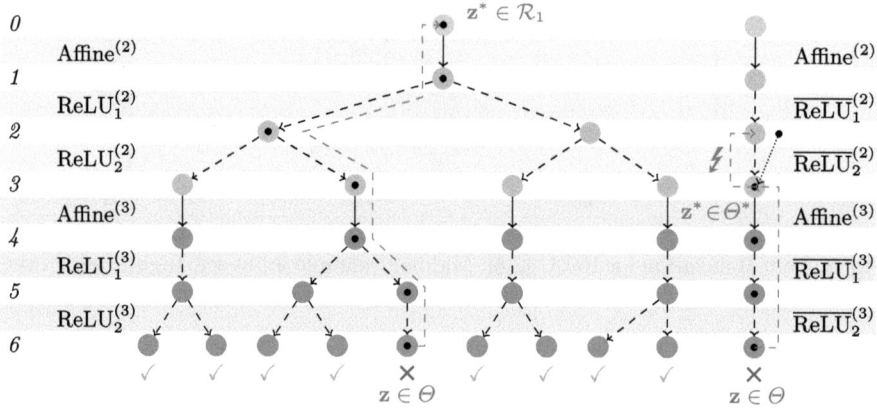

Fig. 3. Reach trees for the FNN in Fig. 1, computed for input $\mathcal{R}_1$ with the exact (left) resp. relaxed (right) method. Levels 1–3 and 4–6 cover the 1st resp. 2nd layer. Safe and unsafe leaves are marked with ✓ resp. ×. A counterexample z is back-traced either to $\mathcal{R}_1$ proving unsafety (left) or to a relaxed star Θ^*, triggering refinement (right).

2. Nodes Θ at depth $d(i,j)$ for some $i \in \{2,\ldots,k\}$ and $j \in \{1,\ldots,|\ell_i|\}$ are (i) either exact, having a left child $ReLU_j(\Theta_{j-})$ and a right child $ReLU_j(\Theta_{j+})$, (ii) or relaxed, having only a left child $\overline{ReLU_j}(\Theta)$, in which case we require $\Theta \neq \Theta_{j-}$ and $\Theta \neq \Theta_{j+}$.
3. All nodes at depth D are leaves.

A path in $\mathcal{T}$ is a sequence $\Theta_s, \Theta_{s+1}, \ldots, \Theta_t$ of nodes from $\mathcal{T}$ such that Θ_i is the parent of Θ_{i+1} for all $i \in \{s,\ldots,t-1\}$.

A counterexample in $\mathcal{T}$ is an unsafe state z in a leaf Θ of $\mathcal{T}$. A counterexample is real if $f(\mathbf{z}^*) = \mathbf{z}$ for some $\mathbf{z}^* \in \mathcal{R}_1$; otherwise it is spurious.

At each depth $d(i,0)$ with base nodes, the union of all stars constitute (an over-approximation of) the output of layer ℓ_{i-1} (or the input set for $i=2$). For the non-base nodes, the union of all stars at depth $d(i,j)$ stores the inputs to the jth ReLU activation of layer ℓ_i.

To simplify the formalisms, we allow reach trees to contain empty nodes, potentially appearing under exact ReLU computations in purely negative or purely positive cases. However, such nodes are omitted in our implementation.

3.2 Our CEGAR Algorithm

On the one hand, the exact method is complete but it suffers from the potentially exponential growth of the number of stars computed during the analysis. On the other hand, the relaxed method mitigates the scalability issue by over-approximating the ReLU function applications, however this approach sacrifices completeness. Therefore, we propose to combine these methods in a *counterexample-guided abstraction refinement* (CEGAR) framework, starting

Algorithm 1: CEGAR-based Verification of Neural Networks

Input	: Neural network f, input set $\mathcal{R}_1$, unsafe set $\mathcal{U}$
Parameters	: Back-tracing method BT, Mode heuristic MODE, Safe histories SH
Output	: SAFE or UNSAFE with real counterexample

1: $\mathcal{T} \leftarrow$ BuildRelaxedReachTree(f, $\mathcal{R}_1$)   // Relaxed method
2: **while** $\mathcal{T}$ *has a leaf* Θ *with an unsafe state* $\mathbf{z} \in \Theta \cap \mathcal{U}$ **do**
3:      $(\Theta^*, \mathbf{z}^*) \leftarrow$ FindOrigin($\mathcal{T}$, Θ, $\mathbf{z}$, BT)
4:      **if** Θ^* *is the root of* $\mathcal{T}$ **then**
5:          $\lfloor$ **return** UNSAFE, $\mathbf{z}^*$ // $\mathbf{z}^*$ is an unsafe input
6:      Set computation method at $parent(\Theta^*)$ to EXACT   // Refine Θ^*-predec.
7:      RecomputeSubTreeRootedAt($\mathcal{T}$, $parent(\Theta^*)$, MODE, SH)
8: **return** SAFE

---

with the efficient over-approximation and utilizing spurious counterexamples for gradual refinement, thereby regaining completeness.

A pseudocode-description of the method is shown in Algorithm 1. It incorporates heuristic parameters (BT, MODE, SH), which configure the refinement strategy (cf. Sections 3.3, 3.4 and 3.5) and access shared global data. First, the fully relaxed reach tree $\mathcal{T}$ is constructed in Line 1. If $\mathcal{T}$ is safe, the algorithm terminates; otherwise, the refinement loop (Lines 2 and 7) is entered. There, the procedure iteratively refines $\mathcal{T}$ until we can prove either safety or unsafety. Each detected counterexample $\mathbf{z}$ is analyzed via the FindOrigin method, which returns a node Θ^* in the reach tree. If the returned node is the root, then the counterexample is real and the algorithm reports UNSAFE. Otherwise, the abstraction is refined by replacing the over-approximating ReLU at the parent node of Θ^* with its exact counterpart, the corresponding subtree is recomputed, and safety is re-checked.

For each refinement step, there are two main heuristic choices to be made: (1) which relaxed node to refine to be exact, and (2) in the subtree re-computation, which nodes to handle exactly and which relaxed. The former is fixed by a *back-tracing method*, and the latter by a *mode heuristic*.

3.3 Back-Tracing Methods

Our back-tracing idea takes a counterexample $\mathbf{z}$ in a leaf Θ of $\mathcal{T}$ and tries to find on the path from the root to Θ the node Θ^* at the smallest depth d from which $\mathbf{z}$ is reachable using exact computations, which is formalized as follows.

Definition 3 (Sequence of operations). *For $i \in \{2, \ldots, k\}$, $j \in \{0, \ldots, |\ell_i|\}$ and $d = d(i,j)$, we define the function $op_d = \text{Affine}^{(i)}$ if $j = 0$ and $op_d = ReLU_j$ otherwise. Moreover, for $d' \in \{d, \ldots, D-1\}$ let $op_{d,d'} = op_{d'}(op_{d'-1}(\ldots(op_d))\ldots)$.*

For a node Θ at level $d = d(i,j)$ in $\mathcal{T}$ we set op_Θ to be $\overline{ReLU_j}$ if Θ is relaxed and op_d otherwise. For paths $\Theta_s, \Theta_{s+1}, \ldots, \Theta_t$ in $\mathcal{T}$ we set $op_{\Theta_s, \Theta_t} = op_{\Theta_t}(\ldots(op_{\Theta_{s+1}}(op_{\Theta_s}))\ldots))$.

Definition 4 (Source of a counterexample). *Assume a path $\Theta_s \ldots, \Theta_t$ in $\mathcal{T}$ and a point $\mathbf{z} \in \Theta_t$. Let d_s and d_t be the depths of Θ_s resp. Θ_t. A point $\mathbf{z}^* \in \Theta_s$ with $\mathrm{op}_{d_s, d_t}(\mathbf{z}^*) = \mathbf{z}$ is called a source of $\mathbf{z}$. The origin of $\mathbf{z}$ is the source of $\mathbf{z}$ at the smallest depth.*

Points in stars resulting from affine and exact ReLU operations always have a source in their parent nodes. However, this does not hold for $\overline{ReLU_j}$ operations. An example is depicted in (the right of) Fig. 3, the black dots illustrating sources, and the dotted arrow rooted in a point which would be the source, since applying the exact ReLU step would give $\mathbf{z}^*$, but it does not lie inside the parent node.

Moreover, for a leaf Θ of $\mathcal{T}$, if a counterexample $\mathbf{z} \in \Theta$ has a source $\mathbf{z}^* \in \mathcal{R}_1$, then $\mathbf{z}$ is a real counterexample. Otherwise, if $\mathbf{z}^* \in \Theta^* \neq \mathcal{R}_1$, then the counterexample is spurious, and the spuriosity was injected by a relaxed ReLU applied to the origin's parent node Θ'. It is easy to see that replacing that relaxed ReLU by its exact counterpart assures that $\mathbf{z}^*$ will not be included in the new children of Θ'.

If we cannot identify $\mathbf{z}$ as real, then we try to identify its origin by *back-tracing*. As a naive approach, we could back-trace level-by-level, finding sources $\mathbf{z}^*$ in the corresponding ancestor nodes and iteratively reusing each source for finding a previous source at smaller depth. But this would involve in the worst-case D feasibility checks, which could be computationally expensive.

Instead, we consider a path $(\Theta_s, \ldots, \Theta_t)$. If the earliest source of $\mathbf{z} \in \Theta_t$ in this path is $\mathbf{z}^* \in \Theta_s$, then we consider a new path ending in Θ_s and search for the earliest source of $\mathbf{z}^*$ in it. We repeat this until we find the origin of $\mathbf{z}$.

Based on this idea, we propose two back-tracing approaches BT1 and BT2.

BT1: UNSAT-Core-Based Back-Tracing. Given a path $(\Theta_s, \Theta_{s+1}, \ldots, \Theta_t)$ in $\mathcal{T}$ and some $\mathbf{z} \in \Theta_t$, the BT1 method encodes the existence of a source $\mathbf{z}^* \in \Theta_s$ logically by a set $\Phi_{s,t}$ of constraints. We label each constraint, such that we can identify the node whose operation it describes.

If these constraints are together satisfiable, then either Θ_s is the root and the counterexample is real, or we proceed with another path to Θ_s. Otherwise, if the constraint set is infeasible, then we make use of an UNSAT core, which can be generated by e.g. the SMT solver z3.

Definition 5 (UNSAT Core). *An UNSAT core of an unsatisfiable set $\Phi = \{\varphi_1, \varphi_2, \ldots, \varphi_n\}$ of constraints is an unsatisfiable subset $\Phi' \subseteq \Phi$.*

Note that UNSAT cores are not required to be minimal. From an UNSAT core we identify the first star Θ_p (with the smallest index) on the considered path $(\Theta_s, \Theta_{s+1}, \ldots, \Theta_t)$, whose operation is encoded by one of the constraints in the UNSAT core. Then, we repeat the procedure on $(\Theta_{p+1}, \ldots, \Theta_t)$ iteratively until we find an origin. Once the origin has been found, BT1 returns the origin $\mathbf{z}^*$ and the node Θ^* containing the origin.

The correctness of BT1 relies on the fact, that if there is no source of $\mathbf{z}$ in Θ_s and op_{Θ_s} is an exact ReLU or affine operation, then there is also no source of $\mathbf{z}$ in Θ_{s+1}. Iterating this argument yields the following proposition.

Proposition 1. *Given a counterexample $\mathbf{z}$ in a reach tree $\mathcal{T}$ as input, the BT1 method always terminates and returns the origin of $\mathbf{z}$ in $\mathcal{T}$.* ⌟

BT2: Predicate-Based Back-Tracing. Our second heuristic BT2 exploits how the stars are processed during the reachability analysis. For each node Θ whose parent Θ' has an exact ReLU or affine operation $\mathrm{op}_{\Theta'}$, the parent Θ' has the same predicate as the child Θ, and for any point in the child node we can generate a source in the parent using the same predicate assignment. For relaxed ReLU computations, this does not hold in general, but we can still construct for each point in the child a point in the parent by projecting out the last dimension of the predicate value. However, in this case, the constructed point in the parent might be but not guaranteed to be a source anymore.

Proposition 2 (Re-using Predicate Assignments). *Assume a non-root node $\Theta_t = \langle \mathbf{c}_t, \mathbf{G}_t, \mathcal{P}_t \rangle$ in $\mathcal{T}$, and a point $\mathbf{z}_t = \mathbf{c}_t + \mathbf{G}_t \boldsymbol{\alpha}_t \in \Theta_t$ for some $\boldsymbol{\alpha}_t = (\alpha_1,\ldots,\alpha_m,\alpha_{m+1})^T \in \mathcal{P}_t$. Let $\Theta_s = \langle \mathbf{c}_s, \mathbf{G}_s, \mathcal{P}_s \rangle$ be the parent node of Θ_t in $\mathcal{T}$.*

- *If op_{Θ_s} is an exact ReLU or an affine operation, then $\boldsymbol{\alpha}_t \in \mathcal{P}_s$ and $\mathbf{z}_s = \mathbf{c}_s + \mathbf{G}_s \boldsymbol{\alpha}_t \in \Theta_s$ is a source of $\mathbf{z}_t$.*
- *If op_{Θ_s} is a relaxed ReLU operation, then $\boldsymbol{\alpha}_s = (\alpha_1,\ldots,\alpha_m)^T \in \mathcal{P}_s$ and thus $\mathbf{z}_s = \mathbf{c}_s + \mathbf{G}_s \boldsymbol{\alpha}_s \in \Theta_s$.* ⌟

For any point $\mathbf{z}_t \in \Theta_t$, using Proposition 2 we can always find an element $\mathbf{z}_s \in \Theta_s$ in the parent Θ_s of Θ_t, which *might* be a source. It is guaranteed to be a source if the operation op_{Θ_s} of the parent is exact, or if it is inexact but $\mathbf{z}_t = \mathrm{op}_{\Theta_s}(\mathbf{z}_s)$ still holds.

We could iterate this procedure on the respective parent nodes, as long as we can still generate a source one level higher. However, we design a more efficient procedure by extending the check to paths $(\Theta_s,\ldots,\Theta_t)$ that span from some depth d_s to d_t. Given a predicate assignment $\boldsymbol{\alpha}_t = (\alpha_1, \alpha_2, \ldots, \alpha_m)^T$, our algorithm takes the predicate assignment $\boldsymbol{\alpha}_s = (\alpha_1, \alpha_2, \ldots, \alpha_{m-r})^T$, where r denotes the number of relaxed ReLUs on the path, i.e., in $\mathrm{op}_{\Theta_s,\Theta_t}$. Now, we know that $\mathbf{z}_s = \mathbf{c}_s + \mathbf{G}_s \boldsymbol{\alpha}_s \in \Theta_s$, and if the condition $\mathbf{z}_t = \mathrm{op}_{d_s,d_t}(\mathbf{z}_s)$ holds, then $\mathbf{z}_s$ is a source of $\mathbf{z}_t$.

Otherwise, for any dimension in which $\mathbf{z}_t = \mathrm{op}_{d_s,d_t}(\mathbf{z}_s)$ does not hold, we could choose the corresponding relaxed ReLU operation(s) for refinement. In our experiments we tested three options: (i) taking the *first* dimension (closest to the root), (ii) *last* dimension (closest to the leaves) and (iii) *all* dimensions where the equality does not hold for refining the reach tree.

Note that this method does not guarantee to refine at the origin of the counterexample: even if we could not (heuristically) generate a source, it is not assured that there is no source in the parent node. On the one hand, this is an efficient way of back-tracing counterexamples since we do not have to compute time-consuming feasibility checks, in contrast to the UNSAT-core-based tracing. On the other hand, for this refinement we cannot guarantee any more to exclude the given counterexample on the considered path. Still, due to the monotonicity, the main algorithm remains complete also with this heuristics, refining the reach tree at the origin at a later stage if required.

3.4 Mode Heuristic

The back-tracing methods BT1 and BT2 return a node (or for option (iii) of BT2 potentially a set of nodes) for refinement. This means, we replace a relaxed ReLU by the corresponding exact operation. This introduces a split in the reach tree, replacing a single child by two children. Subsequently, the subtrees rooted at those two children need to be computed. During these computations, we are free to choose relaxed or exact computations for each node individually, as the method is sound and complete with any heuristics.

A naive approach would be to *relax* all computations, in the hope of fewer nodes to be computed in the new subtree, but risking potentially more refinements needed later to achieve a conclusive answer. However, in our experiments, this did not yield good results.

More successful was to *inherit* the computational modes individually for each node from the previous subtree subject to refinement, such that both subtrees rooted at the new children will have the same structure with the same node types as the subtree rooted at the previous unrefined single child. This we call the *tree-based inheritance* in Sect. 4. Alternatively, we also tried to apply exact computations at all tree depths, at which there was at least one exact computation in the previous tree. This we refer to as *depth-based inheritance*.

3.5 Safe Histories

We further elaborate on the idea to remember parts of the search space that have already been detected safe, and avoid re-computations in those areas.

Definition 6 (Safe history). *A node $\Theta \in \mathcal{T}$ is safe if $\Theta' \cap \mathcal{U} = \emptyset$ holds for all leafs Θ' of the subtree rooted in Θ.*

Let $(\Theta_1, \ldots, \Theta_{n+1})$ be a path from the root $\mathcal{R}_1 = \Theta_1$ to the node $\Theta = \Theta_{n+1}$ in $\mathcal{T}$. The branching history of Θ in $\mathcal{T}$ is the sequence $b_\Theta = (b_1, \ldots, b_n) \in \{L, R\}^n$ such that for all $i \in \{1, \ldots, n\}$, $b_i = L$ if Θ_{i+1} is the left child of Θ_i, and $b_i = R$ otherwise. A safe history is a pair $(op_{\mathcal{R}_1, parent(\Theta)}, b_\Theta)$ for a safe node Θ.

Proposition 3 (Safe history based pruning). *Assume two reach trees $\mathcal{T}$ and $\mathcal{T}'$ for f and $\mathcal{R}_1$. Assume a node $\Theta \in \mathcal{T}$ with $op_{\mathcal{R}_1, parent(\Theta)} = op_n(\ldots(op_1)\ldots)$ and branching history $(b_1, \ldots, b_n)$, and let $(op'_n(\ldots(op'_1)\ldots), (b_1, \ldots, b_n))$ be a safe history. Assume furthermore, that for any $i \in \{1, \ldots, n\}$, if op'_i is an exact ReLU operation then also op_i is an exact ReLU. Then Θ is safe.*

3.6 Soundness and Completeness

Theorem 1 *Algorithm 1 is sound and complete, i.e., it always terminates with the correct safety answer.*

Based on the soundness of the relaxed and the exact methods, soundness is easy to prove.

Regarding completeness, we show that there is an order over reach trees, according to which the reach trees generated by Algorithm 1 are strict monotonically decreasing. Since there are finitely many reach trees (for f and $\mathcal{R}_1$), the algorithm must always terminate.

To define such a reach tree order, we first set the *profile* of $\mathcal{T}$ to be the vector $profile(\mathcal{T}) = (\#rel_1, \ldots, \#rel_D) \in \mathbb{N}^D$ where $\#rel_i$ is the number of relaxed nodes at depth i in $\mathcal{T}$. We use this notion to define the order $\mathcal{T} \prec \mathcal{T}'$ iff $profile(\mathcal{T}) < profile(\mathcal{T}')$ (according to lexicographic order, i.e., if there exists $1 \leq i \leq D$ such that $profile(\mathcal{T})_i < profile(\mathcal{T}')_i$ and $profile(\mathcal{T})_j = profile(\mathcal{T}')_j$ for all $1 \leq j < i$).

In each iteration, Algorithm 1 identifies a relaxed node Θ in the current reach tree $\mathcal{T}$, changes its type to exact, and changes the subtree below it. I.e., we re-compute reachability from Θ onwards, using exact computation at Θ but allowing any heuristic - whether to choose exact or relaxed computations - for later steps. The type switch at Θ and allowing changes only at larger depth is sufficient to assure strictly decreasing reach tree order with respect to $\prec$, and thus completeness.

4 Experimental Results

We performed an experimental evaluation that consists of three parts: (i) comparison of our CEGAR method to the *direct abstraction refinement* (DAR) [3] approach, (ii) comparison of CEGAR to the state-of-the-art tool nnenum [3], and (iii) comparisons of different heuristic configurations of CEGAR. For the experiments we used the publicly available *drones benchmark* suite [2,7,8], which includes eight FNNs of varying complexity with ReLU activations. For each network, we verify one safety property ACX_2, where $X \in \{1, 2, \ldots, 8\}$ [2].

All experiments were conducted on a machine equipped with an Intel Core i7-9700K CPU (3.6GHz with 8 cores) and 32GB of RAM, with a timeout of 1 h. No parallelization was used for any of the experiments.

4.1 Comparison of CEGAR and Direct Abstraction Refinement

We compared CEGAR against the *direct abstraction refinement* (DAR) technique [3], which also starts with a fully relaxed analysis. If this fails, the approach refines a relaxed node at the smallest possible depth and recursively recomputes the two corresponding subtrees fully relaxed. If safety still cannot be verified, the process is repeated until a conclusive result is obtained.

We implemented the DAR approach in our C++ tool HyPro [13] and compared it against our implementations of the exact and CEGAR approaches (see Table 1). For CEGAR we used BT2 with refining closest to leaves where the back-tracing failed, depth-based inheritance, and pruning via safe histories.

In all of our experiments, CEGAR reached a conclusive answer in a shorter time than the direct abstraction refinement approach. Nonetheless, for some benchmarks, the exact method verified the networks faster than CEGAR.

Table 1. Comparison of different verification approaches and different tools (times in sec, TO=timeout, boldface=fastest in HyPro). The first 4 networks have two, last 4 have three hidden layers. **Hidden size** shows the number of neurons per layer.

Net.	Hidden size	DAR in nnenum	DAR in HyPro	CEGAR in HyPro	Exact in HyPro
$AC1_2$	32×16	0.17	4.09	**3.52**	6.31
$AC2_2$	64×32	0.13	2.35	1.94	**1.34**
$AC3_2$	128×64	0.13	43.17	**33.27**	91.18
$AC4_2$	256×128	0.77	2809.87	**196.88**	TO
$AC5_2$	$32 \times 16 \times 8$	0.13	1.58	1.07	**0.45**
$AC6_2$	$64 \times 32 \times 16$	0.12	2.13	**1.42**	1.94
$AC7_2$	$128 \times 64 \times 32$	0.20	205.50	115.03	**76.44**
$AC8_2$	$256 \times 128 \times 64$	0.16	177.02	104.20	**48.89**

4.2 Comparison with the State-of-the-Art Tool nnenum

Table 1 shows results for our implementation of DAR in HyPro and for the nnenum tool [3], which also uses direct abstraction refinement. For a fair comparison, we disabled additional heuristics and optimizations in both tools that are not shared between them, such as prefiltering, eager bound computation, and parallelization.

The most notable technical difference between the two tools is the numerical representation. While nnenum uses floating-point arithmetic, HyPro performs computations with exact arithmetic, which comes with high computational costs. This makes HyPro significantly slower than nnenum. Additionally, other components of our implementation, such as the state set representation and internal optimizations, are still under development and does not yet match the efficiency of mature tools like nnenum.

In all instances of the benchmark, nnenum using DAR produced faster results than DAR in HyPro. However, this experiment indicates that incorporating our CEGAR-based verification into state-of-the-art tools such as nnenum could provide a promising opportunity for running-time reductions, as demonstrated by the performance of CEGAR in our other experiments (Sects. 4.1 and 4.3).

4.3 CEGAR with Different Heuristic Configurations

Figure 4 shows results for the exact method and different CEGAR heuristic configurations on the drones benchmark set. We experimented with different heuristic configurations within CEGAR, from which we report here on the more effective ones. For BT1, we always selected for refinement the *highest* relaxed star whose encoding is involved in the UNSAT core. Conversely, for BT2 we consistently refined at the *deepest* star where the predicate back-tracing condition failed, as this strategy led to slightly better results.

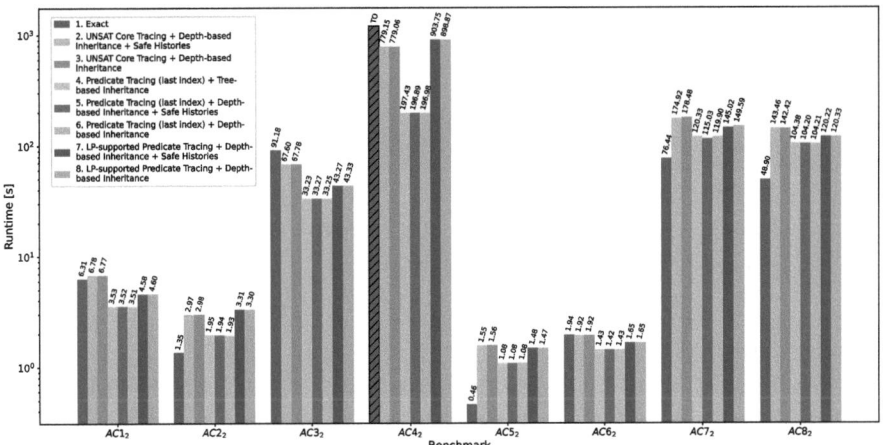

Fig. 4. Running times (sec on log-scale) for $AC1_2$-$AC8_2$. Timeouts (TO) are hatched.

Regarding the mode heuristic, we observe no significant difference between *tree-based* and *depth-based* inheritance for the benchmarks considered. However, for larger neural networks, tree-based inheritance might provide stronger improvements.

Unfortunately, the safe histories heuristic was triggered only in $AC7_2$ twice, yielding a small improvement that is visible in Fig. 4 for $AC7_2$ comparing the dark and light green bars. The reason might be our depth-first implementation, where the detection of a spurious counterexample directly triggers a refinement, postponing the computations of the remaining tree parts, such that the whole reach tree is computed only in the very last iteration. Nonetheless, also this heuristic could reduce the running time more significantly for larger benchmarks.

We also evaluate a BT2 variant where, instead of deterministically choosing the last failed index for refinement, a feasibility check is used to precisely identify the operation that is an origin of the counterexample. While this approach aims to improve the refinement location, the results indicate that the additional overhead introduced by solving the feasibility check outweighs the potential benefits.

Table 2 displays additional observations on the computations with different methods. Under "Dim.", an entry $p \times m$ means that the last feasability problem checked had p constraints in m variables. These results confirm our earlier observations. First, this benchmark set is rather unfavorable for fully relaxed analysis, which fails to provide definitive answers due to excessive over-approximation error in all instances. Second, while exact analysis can prove both safety and unsafety for all but one network, it requires a much larger number of leaf nodes in case of safe networks. Both the CEGAR and direct abstraction-refinement approaches alleviate this by reducing the number of leaves, though at the cost of potentially increasing the complexity (Dim.) of the feasibility checks within each leaf. Finally, the results show that CEGAR either performs fewer refinement

Table 2. Statistics for $AC1_2$ to $AC8_2$ (Result: ✓ means safe, × unsafe, ? inconclusive; #Leaves: num. leaf nodes computed; Dim.: size of the last feasability problem checked; #Refs: num. refinements).

	Metric	$AC1_2$	$AC2_2$	$AC3_2$	$AC4_2$	$AC5_2$	$AC6_2$	$AC7_2$	$AC8_2$
Exact	Result	✓	✓	✓	?	×	×	✓	×
	Time [s]	6.31	**1.35**	91.18	TO	**0.46**	1.94	**76.44**	48.90
	#Leaves	25	5	81	397	3	4	88	3
	Dim.	29×12	25×12	27×12	33×12	27×12	26×12	30×12	30×12
Approx.	Result	?	?	?	?	?	?	?	?
	Time [s]	1.10	0.75	14.79	58.01	0.61	1.02	32.86	40.43
	Dim.	42×18	33×15	48×20	57×23	33×15	30×14	48×20	54×22
CEGAR	Result	✓	✓	✓	✓	×	×	✓	×
	Time [s]	**3.52**	1.94	**33.27**	196.89	1.08	**1.42**	115.03	104.21
	#Leaves	4	4	2	4	2	2	13	4
	Dim.	35×15	29×13	46×19	53×21	31×14	28×13	40×16	45×18
	#Refs.	2	2	1	2	1	1	4	4
DAR	Result	✓	✓	✓	✓	×	×	✓	×
	Time [s]	4.10	2.36	43.18	2809.87	1.59	2.14	205.51	177.02
	#Leaves	2	4	2	40	5	3	6	10
	Dim.	37×16	29×13	46×19	39×16	27×12	26×12	39×16	31×12
	#Refs.	1	3	1	39	3	2	5	7

steps (e.g., $AC4_2$) or lowers complexity (e.g., $AC1_2$). Both factors contribute to CEGAR achieving faster running times than the direct approach.

5 Conclusion

In this work, we proposed a CEGAR framework for neural network verification. Our current implementation is still prototypical and cannot compete with state-of-the-art tools. Nevertheless, our experiments demonstrate that CEGAR holds significant potential, and we expect that other state-of-the-art tools would benefit from its incorporation. As a next step, we plan to integrate our approach into other verification tools (such as **nnenum**) to validate this assumption, and experiment with larger benhcmarks.

Acknowledgments. This project has received funding from Horizon 2020 project REMARO (956200) and the DFG project RealySt (471367371).

Disclosure of Interests. The authors have no competing interests to declare that are relevant to the content of this article.

References

1. Antal, L., Masara, H., Ábrahám, E.: Extending neural network verification to a larger family of piece-wise linear activation functions. In: Proceedings of the 5th International Workshop on Formal Methods for Autonomous Systems (FMAS@iFM 2023). EPTCS, vol. 395, pp. 30–68 (2023). https://doi.org/10.4204/EPTCS.395.4
2. Antal, L., Ábrahám, E., Masara, H.: Generalizing neural network verification to the family of piece-wise linear activation functions. Sci. Comput. Program. **243**, 103269 (2025)
3. Bak, S.: nnenum: Verification of ReLU neural networks with optimized abstraction refinement. In: Proceedings of the 13th International Symposium on NASA Formal Methods (NFM 2021). LNCS, vol. 12673, pp. 19–36. Springer (2021). https://doi.org/10.1007/978-3-030-76384-8_2
4. Bak, S., Tran, H.-D., Hobbs, K., Johnson, T.T.: Improved geometric path enumeration for verifying ReLU neural networks. In: Lahiri, S.K., Wang, C. (eds.) CAV 2020. LNCS, vol. 12224, pp. 66–96. Springer, Cham (2020). https://doi.org/10.1007/978-3-030-53288-8_4
5. Ehlers, R.: Formal verification of piece-wise linear feed-forward neural networks. In: D'Souza, D., Narayan Kumar, K. (eds.) ATVA 2017. LNCS, vol. 10482, pp. 269–286. Springer, Cham (2017). https://doi.org/10.1007/978-3-319-68167-2_19
6. Erhan, D., Szegedy, C., Toshev, A., Anguelov, D.: Scalable object detection using deep neural networks. In: Proceedings of the 2014 IEEE Conference on Computer Vision and Pattern Recognition (CVPR 2014), pp. 2155–2162. IEEE (2014). https://doi.org/10.1109/CVPR.2014.276
7. Guidotti, D.: Verification of neural networks for safety and security-critical domains. In: Proceedings of the 10th Italian Workshop on Planning and Scheduling (IPS 2022), RCRA Incontri E Confronti (RiCeRcA 2022), and the Workshop on Strategies, Prediction, Interaction, and Reasoning in Italy (SPIRIT 2022). CEUR Workshop Proceedings, vol. 3345, pp. 1–10. CEUR-WS.org (2022). https://ceur-ws.org/Vol-3345/paper10_RiCeRCa3.pdf
8. Guidotti, D., Demarchi, S., Pulina, L., Tacchella, A.: Evaluating reachability algorithms for neural networks on NeVer2 (2022). https://www.researchgate.net/publication/363845518_Evaluating_Reachability_Algorithms_for_Neural_Networks_on_NeVer2
9. Hinton, G., et al.: Deep neural networks for acoustic modeling in speech recognition: The shared views of four research groups. IEEE Signal Process. Mag. **29**(6), 82–97 (2012). https://doi.org/10.1109/MSP.2012.2205597
10. Kuutti, S., Bowden, R., Jin, Y., Barber, P., Fallah, S.: A survey of deep learning applications to autonomous vehicle control. IEEE Trans. Intell. Transp. Syst. **22**(2), 712–733 (2021). https://doi.org/10.1109/TITS.2019.2962338
11. LeCun, Y., Bengio, Y., Hinton, G.: Deep learning. Nature **521**(7553), 436–444 (2015). https://doi.org/10.1038/nature14539
12. Lee, A.: Comparing deep neural networks and traditional vision algorithms in mobile robotics. Swarthmore University (2015). https://api.semanticscholar.org/CorpusID:10011895
13. Schupp, S., Ábrahám, E., Makhlouf, I., Kowalewski, S.: HyPro: a C++ library of state set representations for hybrid systems reachability analysis. In: Proceedings of the 9th International Symposium NASA Formal Methods (NFM 2017). LNCS, vol. 10227, pp. 288–294. Springer (2017). https://doi.org/10.1007/978-3-319-57288-8_20

14. Svozil, D., Kvasnicka, V., Pospichal, J.: Introduction to multi-layer feed-forward neural networks. Chem. Intell. Lab. Syst. **39**(1), 43–62 (1997). https://doi.org/10.1016/S0169-7439(97)00061-0
15. Tran, H.D., et al.: Star-based reachability analysis of deep neural networks. In: Formal Methods – The Next 30 Years, pp. 670–686. Springer (2019). https://doi.org/10.1007/978-3-030-30942-8-39
16. Tran, H.D., et al.: Verification of piecewise deep neural networks: a star set approach with zonotope pre-filter. Formal Aspects Comput. **33**(4), 519–545 (2021). https://doi.org/10.1007/s00165-021-00553-4

Maximum Path Sets in Trees

A. Subramani[1], K. Subramani[2](✉), and Jacob Restanio[2]

[1] Candor Care, Morgantown, WV, USA
anand.subramani@candorcare.org, jacob.restanio@agile5technologies.com
[2] LDCSEE, West Virginia University, Morgantown, WV, USA
k.subramani@mail.wvu.edu

Abstract. This paper proposes linear time algorithms for the Maximum Path Set (MPS) problem in undirected trees and arborescences. In the MPS problem, we are given a graph $G = (V, E)$ and asked to find a maximum cardinality set of edges $E' \subseteq E$, such that $G' = (V, E')$ is a collection of vertex-disjoint paths. The MPS problem finds applications in a number of logistics domains. In [3], it was shown that this problem is **NP-complete** in general graphs. This paper demonstrates that the MPS problem is solvable in linear time in undirected trees. Additionally, we reduce the MPS problem to the b-matching problem, which in turn can be reduced to the Maximum Flow problem. From a polyhedral perspective, we design an integer program for the MPS problem in trees and prove that the constraint matrix of this formulation is totally unimodular. In other words, solving the linear programming relaxation provides an integral solution. We also design a linear time algorithm to solve the MPS problem in arborescences, which form a class of directed trees. Finally, we empirically analyze the various algorithms discussed in this paper.

Keywords: Maximum Path Set · Linear time algorithm · Linear Program · Empirical Analysis

1 Introduction

In this paper, we investigate the Maximum Path Set (MPS) problem in undirected trees and arborescences. In the Maximum Path Set problem, we are given a graph G, and we want to find a maximum cardinality path set, i.e., a subset of edges that forms a collection of non-intersecting paths. MPS is known to be **NP-complete** in general graphs [3]. We show that this problem is polynomial-time solvable in weighted trees and linear-time solvable in undirected trees and arborescences.

As part of our investigation, we analyze a natural integer programming formulation for the MPS problem in trees. We show that the constraint matrix of our integer programming formulation is totally unimodular. We can then solve the MPS problem in trees by solving the linear programming relaxation of the integer program. This approach works even when the edges of the graph are weighted.

We also investigate the algorithms discussed in this paper from an empirical perspective. The empirical analysis reinforces the efficacy of our linear time algorithm.

The MPS problem (henceforth, MPS) finds applications in Air Force logistics [1,3,6]. Suppose we wish to strike down some targets with fighter jets. After eliminating a target, a fighter jet must immediately move on to the next target. A fighter jet must never return to a target once struck. Returning to the location may lead to several fighter jets shooting at each other requiring increased defenses and alertness. Consider the graph where the targets are vertices, and two targets have an edge if a pilot can move from the first target to the second. If we assign every target a pilot, this could create confusion about where to go next. Furthermore, this will be extremely costly, since pilots are expensive. We could give each pilot a path if we find a small collection of paths that covers the graph. Note that this problem is equivalent to MPS, as a path set with a large number of edges has a small number of paths. Once a target has been struck down, it is clear which target the pilot should attack next.

The concept of reachability is critical in network verification and communication network robustness [2,5,7]. Identifying feasible paths between network components can determine which destination is reachable from a given source. Our research applies to connectivity under edge removal. We capture this phenomenon through the MPS problem. The MPS problem can be considered as a minimum edge removal problem, and the solutions can be examined as edge failure cases to see how networks could become disconnected. The network may be less robust if total disconnection is achieved by eliminating only a few edges. Such a network would require additional redundancies to increase fault tolerance and make the network less susceptible to attack [3].

The principal contributions of this paper are as follows:

1. A (practical) linear time algorithm for the MPS problem in undirected trees.
2. An integer program for the MPS problem in undirected trees.
3. A proof that all optimal solutions to the linear programming relaxation of our integer program are integral.
4. A linear time algorithm for the MPS problem in arborescences.
5. An empirical analysis of the different techniques discussed to solve the MPS problem in undirected trees.

The rest of this paper is organized as follows: In Section 2, we discuss the problems described in this paper and define the required terms. In Sect. 3, we describe and analyze an algorithm for the MPS problem in undirected trees. Section 4 outlines a reduction from MPS to maximum flow, which is then used as an alternative algorithm for comparison and analysis. Section 5 discusses a polyhedral analysis for the MPS problem in trees. In Sect. 6, we describe algorithms to solve the MPS problem in two types of directed trees. Section 7 describes implementation details and presents results comparing the performance metrics of the different algorithms discussed in this paper. We conclude in Sect. 8 by summarizing our contributions and identify avenues for future research.

2 Statement of Problems

This section formally describes the problems that we study in this paper.

A path set in a graph $G = (V, E)$ is a set of edges $P \subseteq E$ such that every connected component in $G' = (V, P)$ is a simple path. Note that a single isolated vertex counts as a simple path.

Definition 1. Maximum Path Set: *Given a finite graph $G = (V, E)$, find a maximum cardinality path set, where the cardinality of a path set is the number of edges in it.*

A graph $G = (V, E)$ is a **tree**, if it is acyclic and connected. We focus on the Maximum Path Set (MPS) problem in trees. Without loss of generality, we assume that the input tree is rooted at a root vertex r. Rooting an undirected tree does not change its MPSs because the rooted tree is isomorphic to the original tree.

For example, consider the tree in Fig. 1.

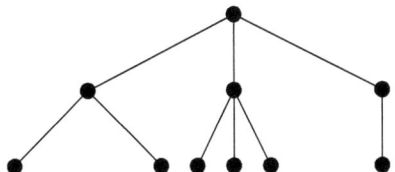

Fig. 1. An example tree.

Two MPSs are shown in Fig. 2.

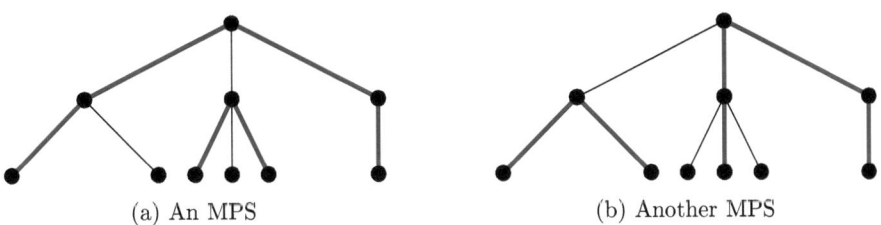

(a) An MPS (b) Another MPS

Fig. 2. Example MPSs for the tree in Fig. 1.

Throughout the paper, we assume $d(v)$ denotes the degree of vertex v. Also, we let $\Delta(G)$ denote the degree of the vertex with the maximum degree in the graph.

We define the Maximum Flow problem as follows:

Definition 2 (Maximum Flow problem). *Let $N = (V, E, \mathbf{c})$ be a network (that is, a directed graph and a capacity function $\mathbf{c}\colon E \mapsto \mathbb{R}^+$) with $s, t \in V$ being the source and sink of N, respectively. A flow is a mapping $f\colon E \mapsto \mathbb{R}$ that satisfies:*

1. **The capacity constraint:** *$f(v_i, v_j) \le c(v_i, v_j)$ for all $(v_i, v_j) \in E$. In other words, the flow along an edge cannot exceed its capacity.*
2. **The flow conservation constraint:**

$$\sum_{v_j \in V, (v_i, v_j) \in E} f(v_i, v_j) - \sum_{v_j \in V, (v_j, v_i) \in E} f(v_j, v_i) = 0,$$

for all $v_i \in V \setminus \{s, t\}$. In other words, the sum of flows entering a node must be equal to the sum of flows exiting that node for all nodes except for the source and the sink.

The problem is to find the flow that maximizes $\sum_{v_i \in V, (s, v_i) \in E} f(s, v_i)$. In other words, the flow that maximizes the flow out of the source.

Let $A = (V, E)$ denote a directed graph. A is considered a directed tree if the underlying undirected graph is a tree. Note that this is not equivalent to considering a graph without directed cycles. Without loss of generality, we consider directed trees with a root r. An arborescence, or out-tree, is a directed tree where every vertex v has a unique, directed path that begins at r and ends at v.

3 An Iterative Improvement Algorithm for the MPS Problem in Trees

This section discusses a linear time algorithm for the MPS problem in trees. We first explain our algorithmic insights through a series of reduction rules. These will then be put together to form the algorithm.

Let $T = (V, E)$ denote a rooted tree with root r. Let L be the vertices of degree one, i.e., the leaves.

Reduction Rule 1 (RR1): If the graph has fewer than two edges, return all edges in the graph.

Soundness: If the graph has zero edges, there are no edges to put in any path set, so the MPS is $\emptyset$. If the graph has one edge, this is the only edge that can be put in the path set, and it forms a path set, so that edge is an MPS.

Reduction Rule 2 (RR 2): Let $l_i \in L$ be a leaf of maximum depth. If the parent p of l_i has degree two, then (p, l_i) is part of every MPS. Furthermore, any MPS in the tree $G = (V, E)$ consists of an MPS in the tree $G' = (V \setminus \{l_i\}, E \setminus \{(p, l_i)\})$ and the edge (p, l_i).

Soundness: Observe that (p, l_i) can always be added to a path set to make it larger. This is because, before (p, l_i) is added, p could have a maximum degree of one in the maximum path set. This means that upon its addition to the path set, p would have a maximum degree of two. Similarly, l_i would have degree exactly one upon adding (p, l_i).

Reduction Rule 3 (RR 3): Let $l_i \in L$ be a leaf of maximum depth. Let the parent of l_i, say p, have degree three or more. This means another leaf $l_j \in L$ is a child of p. There exists an MPS with (p, l_i) and (p, l_j). Furthermore, any MPS with (p, l_i) and (p, l_j) consists of an MPS in the graph without p and all of its children combined with (p, l_i) and (p, l_j).

Soundness: Consider a maximum path set without (p, l_i) or (p, l_j). First, we remove all edges adjacent to p in the maximum path set. It is important to note that there are at most two of these edges. Then, we can simply add (p, l_i) and (p, l_j). A similar argument can be used if only one of these edges is missing; we remove one edge incident on p in the maximum path set. A cut and replace argument can be used to prove the second part of the reduction rule. If there was a larger path set in the graph without p and all of its children we could improve the current solution, contradicting the hypothesis that the path set with (p, l_i) and (p, l_j) is maximum.

Algorithm 1. An algorithm for the MPS problem in a tree $T = (V, E)$

1: **Input:** A rooted tree $T = (V, E)$ with n vertices and a root r and the current path set P (P is initialized to $\emptyset$)
2:
3: **Output:** A maximum path set in T.
4: Apply **RR 1** if the graph is a single vertex.
5:
6: Perform breadth-first search on the vertices and store all of their distances from the root. ▷ Note that this is only done on the first recursion
7:
8: Pick a leaf l_i of maximum depth.
9:
10: Let the parent of l_i be p.
11:
12: **if** p has degree two **then**
13:     Apply **RR 2**, remove l_i, and recurse on the remaining tree.
14:
15: **else**
16:     Apply **RR 3** to add (p, l_i) and (p, l_j) for two leaves l_i and l_j adjacent to p, remove p and all of its children, and recurse on the remaining tree.
17:
18: **end if**

Algorithm 1 summarizes these discussions.

3.1 Proof of Correctness

In this subsection, we prove that our algorithm is correct.

Theorem 1. *Algorithm 1 is correct.*

Proof. We prove that the algorithm is correct via the second principle of induction on the number of vertices in the tree T. The soundness proof of **RR 1** proves that the algorithm is correct on trees with one vertex. Let it be the case that the algorithm returns the correct answer for all trees with at most k vertices. Consider a tree with $(k+1)$ vertices. Consider a leaf l_i of maximum depth in this tree with parent p. Note that we pick the vertex of maximum depth to ensure all of p's children are leaves so the reduction rules can be applied correctly. Furthermore, if p is of degree one because the graph is a single edge, the maximum path set is returned via **RR 1**.

If p is of degree two, we remove l_i using **RR 2**. As argued before, this removal is sound. We know that the algorithm will find an MPS in the remaining tree according to the inductive hypothesis. Thus, if the parent of the leaf node of maximum depth (i.e., p) has degree two, the algorithm is correct by induction.

Let p have degree three or more. Using **RR 3**, we can remove p and all of its descendants, while adding (p, l_i) and (p, l_j) to our path set, for some children l_i and l_j of p. As argued before, this transformation is sound. Again, we removed at least one vertex, and we know that the algorithm will find an MPS in the residual tree without p and its descendants, via induction. Thus, Algorithm 1 is correct. □

3.2 Analysis

In this subsection, we analyze our algorithm and show that it runs in linear time.

Finding and then storing the distances of all vertices from the root can be done in linear time. Even when a vertex is removed, the parent of a vertex of maximum depth can only have leaf children. Therefore, Theorem 1 holds even when a breadth-first search is not used for every recursion.

Using the previously mentioned breadth first search, we store the parent of every vertex in the tree in a table. Identifying a leaf and its parent can be then done in constant time if the set of leaves is being updated in the algorithm. A simple table lookup will suffice to find the parent. If the parent has degree two, we only need to remove the leaf. Removing a vertex with degree one (the leaf), given that we know the vertex is adjacent (p), can be done in constant time if the input is given in an adjacency list using linked lists. Note that this is an application of **RR 2**.

In the case that the degree of the parent is more than two, we use $O(d(p))$ time to go through and remove each of p's children, applying **RR 3**. Even if the degree of every parent (vertex) that we touched in the recursion was more than two, we would use $O(\sum_{i=1}^{n} d(v_i))$ time. However, this is $O(m)$ time, where m is the number of edges. Thus, our algorithm runs in linear time.

4 Maximum Flow Based Algorithm

This section describes an alternate way of solving the MPS problem in trees. We design a reduction to the b-matching problem and then solve the b-matching instance using maximum flow. The b-matching problem is defined as follows:

Definition 3 (b-matching). *Given a graph G, and capacity function $\mathbf{b}: V \to \mathbb{N}$, find an assignment $\mathbf{x} : E \to \{0,1\}^m$ that maximizes $\sum_{(v_i,v_j) \in E} x_{i,j}$ such that $\sum_{(v_i,v_j) \in E} x_{i,j} \leq b(v_i)$ for all v_i.*

The b-matching problem is a generalization of MPS in trees. In particular, it becomes MPS in trees when $b(v_i) = 2$ for all vertices v_i. Normally, a b-matching in this instance can contain cycles, as it is simply a maximum degree two subgraph. However, since trees do not have cycles, a b-matching **must** be a path set. Let T be our input to the MPS problem. More formally, we create our capacity function $\mathbf{b}$ by taking each vertex $v \in V$ and setting $b(v) = 2$. Next, find a maximum b-matching assignment $\mathbf{x}$ on T, and $\mathbf{b}$. The edges $e \in E$ where $x_e = 1$ are the edges selected to be in our MPS solution.

Next, we reduce b-matching to maximum flow using a well-known reduction [4]. The following reduction is only valid for b-matching on bipartite graphs. We create two sets $R = R_1, R_2, \ldots, R_k$ and $B = B_1, B_2, \ldots, B_l$ which are alternating layers of the trees starting from the leaves. This can be thought of as painting the vertices of the tree red and blue. This is known to form a bipartition of the tree [4], as every edge is between a vertex in one layer and a vertex in the next, which will be assigned the opposite color. Next, we create the network N. The source vertex s has an edge leaving it and entering each vertex $R_i \in R$ with capacity $b(R_i)$. In our special instance of b-matching, each of these edges will have capacity two. For each edge $(R_i, B_j) \in E$, a corresponding edge is added from R_i to B_j with capacity 1. The sink vertex t is joined to each vertex $B_j \in B$ with capacity $b(B_j)$ (which is again two in our instance of b-matching). A maximum flow is found from s to t. Every edge crossing the cut (R, B) corresponds to an edge of T and vice versa. So, any flow in N can be made into a subset of edges in T by taking the edges assigned a flow of one. The number of edges in this set will equal the value of the flow. Similarly, any b-matching can be transformed in to a flow with value $\sum_{(v_i,v_j) \in E} x_{i,j}$.

We first reduce MPS to b-matching and then to maximum flow. An example of transforming an MPS instance to a maximum flow problem and how their solutions are related can be found in Fig. 3. Solving maximum flow allows the use of an algorithm with better running time than other network flow formulations.

5 A Polyhedral Analysis of the MPS Problem in Trees

This section proposes an integer program for the MPS problem in trees. We also show that all optimal extreme points of the linear programming relaxation of

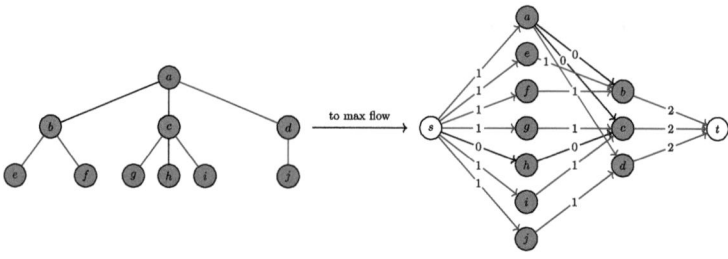

Fig. 3. An instance of MPS and the corresponding maximum flow problem. The red edges denote the solution to the MPS and the red arrows denote the corresponding solution in the maximum flow problem. (Color figure online)

the integer program contain only integral components. We choose the following decision variables:

$$x_e = \begin{cases} 1, & \text{if the edge } e \text{ is in the MPS} \\ 0, & \text{if the edge } e \text{ is not in the MPS} \end{cases}$$

Our integer program (IP) is as follows:

$$\max \sum\nolimits_{e \in E} x_e$$

subject to

$$\sum_{v_j \in V, (v_i, v_j) \in E} x_{i,j} \leq 2, \text{ for all vertices } v_i \in V \text{ that are not the leaves}$$

$$\mathbf{x} \in \{0, 1\}^m$$

The correctness of our reduction follows from Theorem 2.

Theorem 2. *Every path set of size k in the input tree T can be converted into a feasible solution to the integer program I with objective value k,* **and** *every feasible solution to I with objective value z can be converted into a path set of size z in T.*

Proof. Consider a path set. Let k denote the size of the path set. Recall that, as demonstrated in the reduction to b-matching, a path set is simply a maximum degree two subgraph (in a tree). Therefore, a path set corresponds to a feasible solution. Furthermore, this solution has the same value as the size of the MPS, as every edge corresponds to a variable being assigned one, which adds one to the objective value. In other words, a path set of size k corresponds to a solution of value k in the integer program.

Conversely, consider a feasible solution of objective value z. As mentioned, this solution corresponds to a subset of edges whose induced subgraph has a maximum degree of two (i.e., a path set). Therefore, a feasible solution with objective value z, corresponds to a path set of size z. □

Definition 4. *A totally unimodular matrix is one where every square submatrix has determinant* 0, 1, *or* −1.

If a matrix **A** is totally unimodular, the polyhedron $\{\mathbf{A} \cdot \mathbf{x} \geq \mathbf{b}\}$ has integral vertices for any right hand side vector **b** [8].

Now, we show that the constraint matrix of our IP is totally unimodular. Observe that it is the node edge incidence matrix of the original tree because every column (edge) has two 1s (corresponding to the two constraints for each decision variable). Since a tree is a bipartite graph, the constraint matrix of our IP is the node edge incidence matrix of a bipartite graph. It is known that the node edge incidence matrix of a bipartite graph is totally unimodular [8]. Note that the constraints corresponding to the edges around the leaves are omitted. It is known that removing rows that have a singular one and all other entries being zero preserves total unimodularity. Thus, the constraint matrix of our IP for the MPS problem in trees is totally unimodular.

Corollary 1. *The MPS problem is polynomial time solvable in weighted undirected trees.*

Proof. Observe that in the case of a weighted graph, only the objective function changes, as per the above formulation. In other words, the constraint matrix is totally unimodular. We can thus solve the problem by resorting to linear programming, which is in **P**. □

6 Directed Trees

In this section, we design and analyze algorithms to solve the Maximum Path Set problem in arborescences. Recall that arborescences are out-trees, with all edges pointing away from the root.

6.1 Algorithm and Proof of Correctness

Let $A = (V, E)$ denote an arborescence. First, we prove Theorem 3.

Theorem 3. *Consider any edge e in A. There exists an MPS containing e.*

Proof. Consider an MPS P without $e = (v_i, v_j)$. Consider adding e to P. Since P is an MPS, performing this operation must have created a cycle, a vertex with out-degree two, or a vertex with in-degree two. However, the input graph is an arborescence, so adding P could only have created a vertex with out-degree two. This is because, in the original arborescence, every vertex only has one edge entering, which is that of its parent. Furthermore, this vertex is v_i. Let the edge exiting v_i in P be v_k. Consider the path set P' made by removing (v_i, v_k) from P and replacing it with e. This is shown in Fig. 4. Observe that this path set has the same size at P. Thus, we can always construct an MPS containing any edge. □

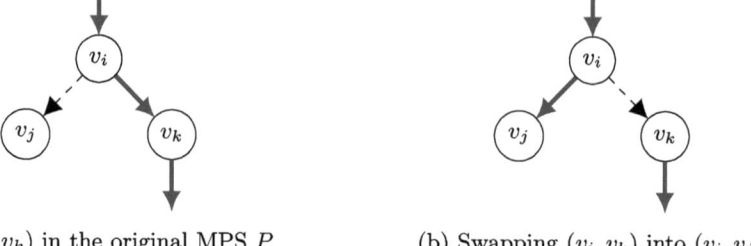

(a) (v_i, v_k) in the original MPS P (b) Swapping (v_i, v_k) into (v_i, v_j)

Fig. 4. How to create an MPS with e.

Algorithm 2. An algorithm for the MPS problem in a tree $T = (V, E)$

1: **Input:** A rooted arborescence $A = (V, E)$ with n vertices and a root r with λ children.

2: **Output:** A maximum path set in A.

3: Set $P = \emptyset$.

4: **if** (There is one vertex in the tree) **then**
5:    Return P.
6: **end if**

7: Take an arbitrary edge $e = (r, v_i)$ exiting the root.

8: Recurse on the subtrees rooted at the children of r to obtain path sets P_1 through P_λ.

9: Set $P = P \cup \{e\} \cup \bigcup_{i=1}^{\lambda} P_i$.

10: Return P.

Algorithm 2 summarizes the observations made in Theorem 3.

The correctness of this greedy algorithm directly follows from Theorem 3. This is because, as demonstrated in the proof of the theorem, for any edge e there exists an MPS that consists of that edge and the MPS in the remaining graph. We thus recurse on the connected components in the graph without r, since r cannot have multiple edges leaving it in the MPS.

It is important to note that this algorithm will not work for undirected trees. In this case, Theorem 3 does not apply.

6.2 Space and Time Analysis

The algorithm uses only the space used to store the maximum path set. Let P denote a maximum path set. Thus, our algorithm uses $O(|P|)$ space.

For the time analysis, note that, every vertex is touched twice (once when descending down the recursion tree and once when the path sets are combined) and each edge is touched once. Thus, the running time of the algorithm is $O(m+$

n) which is $O(n)$. Thus, the algorithm runs in $O(n)$ time and uses $O(|P|)$ space for an MPS P.

7 Implementation

In this section, we describe the hardware used to run the experiments, implementation details, and present our results. The code can be viewed and downloaded at https://github.com/JacobRestanio/mps.

7.1 Experimental Setup

The experiments were run on an 11th Gen Intel(R) Core(TM) i7-1165G7 processor with a clock speed of 2.80GHz. A single core was used for each algorithm. The implementation was written in Python 3.11.3. The NetworkX 3.2.1 library was used to create and manipulate the graph structures. The PuLP 2.7.0 library was used to model the linear program from the graph. CPLEX 22.1.1 from IBM ILOG CPLEX Optimization Studio with default settings was used as the solver for the linear program. All other libraries used were standard libraries from Python, which were used mainly for measuring and logging. Each algorithm was run on the same randomly generated tree. A random tree was created using the NetworkX random_tree() tree generation function. The random_tree() function generates a uniformly random Prüfer sequence that is converted into a tree. The algorithms were tested on trees of size 100, 1000, 10000, 100000. All algorithms were tested on the same set of trees to confirm their results and reinforce correctness. All algorithms produced the same cardinality maximum path set.

7.2 Implementation Details

Algorithm 1 was implemented in two ways: recursively and iteratively. The recursive implementation (referred to as recursive) used a BFS to assign each node its distance from the root. A dictionary of the nodes by distance was constructed during the BFS. The keys (node distances) were assigned a list of all nodes having that distance from the root. A node at maximum current depth from the root was selected and removed from the node distance dictionary. The reduction rules were applied to that node. Any other leaves that were removed due to **RR 3** were removed from the node distance dictionary and the tree.

The iterative implementation of the iterative improvement (referred to as iterative) algorithm was done by using a BFS to create a node list in reverse order of visitation from the root. The list was then iterated through with a check to see if the node was still in the tree before applying a reduction rule. The reduction rules were applied without recursion; any removed node was removed from the tree.

The maximum flow algorithm took advantage of NetworkX libraries. The tree was divided into bipartite sets R and S using the NetworkX bipartite.sets() function. A start node s was added with edges to each node in R with capacity 2, and an end node t was added to each node in S with capacity 2. Edges existing in the tree between R and S were added with capacity 1. Then, the maximum flow was found using the NetworkX `maximum_flow()` function. The maximum flow algorithm used was the preflow-push algorithm (alternatively, push-relabel) with a given running time of $O(n^2 \cdot \sqrt{m})$ for n nodes and m edges.

For the linear program, a model was created using PuLP. A decision variable with lower bound 0 and upper bound 1 was created for each edge in the tree. For each node in the tree, the sum of the decision variables representing the edges adjacent to that node being less than or equal to 2 was added as a constraint. Then, the model was solved using CPLEX.

7.3 Results

Results indicate that the iterative algorithm dominates the other algorithms in terms of execution time and memory usage. Execution time refers to the wall clock time for the algorithm to complete on a graph already loaded into memory. None of the algorithms required preprocessing. Execution time was measured at the function call for each algorithm when the results were returned. Table 1 shows the execution times of each algorithm on different sizes of trees. The superiority of the iterative algorithm is evident as it dominates the other algorithms in running time. Time was not recorded for algorithms that took more than 5 min to execute on a given size. Since the recursion is tied to the number of vertices in the graph, the recursion stack becomes very deep and significantly impacts the running time.

Likewise, the use of auxiliary data structures in each recursive step impacted the memory. Table 2 shows the average peak memory usage of each algorithm on different sizes of trees. Stack size was not included in memory tracing; only the memory that was allocated from the heap. Again, the superiority of the iterative algorithm is evident as it dominates the other algorithms in terms of memory usage.

Table 1. Average execution time in seconds for each algorithm on selected size categories. Time greater than 5 minutes is represented with a (-).

Size	Recursive	Iterative	Max Flow	CPLEX
100	0.0017	**0.0008**	0.0080	0.1128
1000	0.039	**0.010**	0.097	7.076
10000	3.10	**0.12**	1.46	-
100000	-	**1.22**	36.14	-

Table 2. Memory usage in MB for each algorithm on selected size categories. Memory not recorded for execution time greater than 5 minutes is represented with a (-).

Size	Recursive	Iterative	Max Flow	CPLEX
100	0.028	**0.013**	0.436	0.365
1000	0.244	**0.050**	4.50	2.82
10000	2.60	**0.91**	44.5	-
100000	-	**6.97**	459	-

8 Conclusion

This paper proposed a linear time algorithm for the Maximum Path Set problem in undirected trees and arborescences. The Maximum Path Set problem finds applications in Air Force logistics among other domains. Additionally, we designed an integer program for the MPS problem in trees. Let **A** denote the constraint matrix of our integer program. We proved that **A** is totally unimodular. Furthermore, we implemented the different approaches discussed in this paper.

From our perspective, the following avenues of research are worth pursuing:

1. Are there classes of undirected graphs other than trees for which the MPS problem can be solved in polynomial time? - This question is interesting, because the MPS problem is closely related to the Hamilton Path problem and the latter is **NP-hard** on most well-known graph classes.
2. Are there **fixed-parameter tractable** algorithms for the MPS problem in general graphs? - We are currently investigating the possibility of **FPT** algorithm for MPS, parameterized by the maximum path set cardinality.
3. Is a more substantial empirical analysis warranted? - In the current work, we implemented the naive versions of our algorithms. A more substantial study could incorporate heuristics to improve performance.

References

1. Amouzegar, M.A., Tripp, R.S., Galway, L.A.: Integrated logistics planning for the air and space expeditionary force. J. Oper. Res. Soc. **55**(4), 422–430 (2004)
2. Bayless, S., et al.: Debugging network reachability with blocked paths. In: Silva, A., Rustan, K., Leino, M. (eds.) Computer Aided Verification. Lecture Notes in Computer Science, pp. 851–862. Springer International Publishing, Cham (2021)
3. Cheng, Y., Cai, Z., Goebel, R., Lin, G., Zhu, B.: The radiation hybrid map construction problem: recognition, hardness, and approximation algorithms. Unpublished Manuscript (2008)
4. Cormen, T.H., Leiserson, C.E., Rivest, R.L., Stein, C.: Introduction to Algorithms, 3rd edn. The MIT Press, Cambridge (2009)
5. Parlangeli, G., Notarstefano, G.: On the reachability and observability of path and cycle graphs. IEEE Trans. Auotmatic Control 743–748 (2012)

6. Rodrigues, M.B., Karpowicz, M., Kang, K.: A readiness analysis for the argentine air force and the Brazilian navy A-4 fleet via consolidated logistics support. In: Proceedings of the 32nd Conference on Winter simulation, WSC 2000, Wyndham Palace Resort & Spa, Orlando, FL, USA, December 10–13, 2000, pp. 1068–1074. WSC (2000)
7. Sun, P., Kooij, R.E., Van Mieghem, P.: Reachability-based robustness of controllability in sparse communication networks. IEEE Trans. Netw. Serv. Manage. **18**(3), 2764–2775 (2021)
8. Wolsey, L.A.: Integer Programming. John Wiley & Sons, New York (1998)

Presentation-Only Contributions

Stationary Regimes of Piecewise Linear Dynamical Systems with Priorities

Xavier Allamigeon, Pascal Capetillo$^{(\boxtimes)}$, and Stéphane Gaubert

INRIA, CMAP École Polytechnique IP Paris, Paris, France
pascal.capetillo@inria.fr

Abstract. Dynamical systems governed by priority rules appear in the modeling of emergency organizations and road traffic. These systems can be modeled by piecewise linear time-delay dynamics, specifically using Petri nets with priority rules. A central question is to show the existence of stationary regimes (i.e., steady state solutions)—taking the form of invariant half-lines—from which essential performance indicators like the throughput and congestion phases can be derived. Our primary result proves the existence of stationary solutions under structural conditions involving the spectrum of the linear parts within the piecewise linear dynamics. This extends to a broader class of systems a fundamental theorem of Kohlberg (1980) dealing with nonexpansive dynamics. The proof of our result relies on topological degree theory and the notion of "Blackwell optimality" from the theory of Markov decision processes. Finally, we validate our findings by demonstrating that these structural conditions hold for a wide range of dynamics, especially those stemming from Petri nets with priority rules. This is illustrated on real-world examples from road traffic management and emergency call center operations.

This presentation is based on a HSCC 2025 paper [1].

Reference

1. Allamigeon, X., Capetillo, P., Gaubert, S.: Stationary regimes of piecewise linear dynamical systems with priorities. In: HSCC 2025 (2025). https://doi.org/10.1145/3716863.3718053

Membership and Conjugacy in Inverse Semigroups

Lukas Fleischer[1], Florian Stober[1(✉)], Alexander Thumm[2], and Armin Weiß[1]

[1] University of Stuttgart, Stuttgart, Germany
florian.stober@fmi.uni-stuttgart.de
[2] University of Siegen, Siegen, Germany

Abstract. The membership problem for an algebraic structure asks whether a given element is contained in some substructure, which is usually given by generators. For finite semigroups this problem was shown to be PSPACE-complete in the transformation model by Kozen [5] and NL-complete in the Cayley table model by Jones, Lien, and Laaser [4].

In this work we study the membership problem as well as the conjugacy problem for finite inverse semigroups. In the partial bijection model, these were shown to be PSPACE-complete by Birget and Margolis [1] and by Jack [3]. Here we present a more detailed analysis of the complexity of the membership and conjugacy problems parametrized by varieties of finite inverse semigroups. We establish dichotomy theorems for the partial bijection model and for the Cayley table model. In the former model these problems are in NC (resp. NP for conjugacy) for strict inverse semigroups and PSPACE-complete otherwise. In the latter model we obtain general L-algorithms as well as NPOLYLOGTIME upper bounds for Clifford semigroups and L-completeness otherwise. By applying our findings, we also show the following: the intersection non-emptiness problem for inverse automata is PSPACE-complete even for automata with only two states, the subpower membership problem is in NC for every strict inverse semigroup and PSPACE-complete otherwise, and the minimum generating set and the equation satisfiability problems are in NP for varieties of finite strict inverse semigroups and PSPACE-complete otherwise.

This presentation is based on an ICALP 2025 paper [2].

References

1. Birget, J.C., Margolis, S.W.: Two-letter group codes that preserve aperiodicity of inverse finite automata. Semigroup Forum **76**(1), 159–168 (2008)
2. Fleischer, L., Stober, F., Thumm, A., Weiß, A.: Membership and conjugacy in inverse semigroups. In: ICALP (2025). https://doi.org/10.4230/LIPIcs.ICALP.2025.156
3. Jack, T.: On the complexity of inverse semigroup conjugacy. Semigroup Forum **106**(3), 618–632 (2023)

4. Jones, N.D., Lien, Y.E., Laaser, W.T.: New problems complete for nondeterministic log space. Math. Syst. Theory **10**(1), 1–17 (1976)
5. Kozen, D.: Lower bounds for natural proof systems. In: SFCS (1977). https://doi.org/10.1109/SFCS.1977.16

The Ultimate Signs of Second-Order Holonomic Sequences

Fugen Hagihara[1](✉) and Akitoshi Kawamura[2]

[1] Graduate School of Science, Kyoto University, Kyoto, Japan
[2] Research Institute for Mathematical Sciences, Kyoto University, Kyoto, Japan

Abstract. A real-valued sequence $f = \{f(n)\}_{n \in \mathbb{N}}$ is said to be second-order holonomic if it satisfies a linear recurrence $f(n+2) = P(n)f(n+1) + Q(n)f(n)$ for all sufficiently large n, where $P, Q \in \mathbb{R}(x)$ are rational functions. We study the ultimate sign of such a sequence, i.e., the repeated pattern that the signs of $f(n)$ follow for sufficiently large n. In our paper, for each P, Q we determine all the ultimate signs that f can have, and show how they partition the space of initial values of f. This completes the prior work [3], which settled some restricted cases. As a corollary, it follows that when P and Q have rational coefficients, f either has an ultimate sign of length 1, 2, 3, 4, 6, 8 or 12, or never falls into a repeated sign pattern. We also provide a partial algorithm that finds the ultimate sign of f (or tells that there is none) in almost all cases. In other words, we prove that the problem of finding the ultimate sign of a given second-order holonomic sequence f Turing-reduces to that of deciding minimality of f, i.e., whether $\lim_{n \to \infty} f(n)/g(n) = 0$ for any linearly independent sequence g satisfying the same recurrence. This result is an extension of [2].

This presentation is based on an ICALP 2025 paper [1].

References

1. Hagihara, F., Kawamura, A.: The ultimate signs of second-order holonomic sequences. In: ICALP (2025). https://doi.org/10.4230/LIPIcs.ICALP.2025.159
2. Kenison, G., et al.: On positivity and minimality for second-order holonomic sequences. In: MFCS (2021). https://doi.org/10.4230/LIPIcs.MFCS.2021.67
3. Neumann, E., Ouaknine, J., Worrell, J.: Decision problems for second-order holonomic recurrences. In: ICALP (2021). https://doi.org/10.4230/LIPIcs.ICALP.2021.99

Quantitative Language Automata

Thomas A. Henzinger[1], Pavol Kebis[1], Nicolas Mazzocchi[2], and N. Ege Saraç[1(✉)]

[1] Institute of Science andTechnology (ISTA), Klosterneuburg, Austria
[2] Slovak University of Technology in Bratislava, Bratislava, Slovak Republic
negesarac@gmail.com

Abstract. A *quantitative word automaton* (QWA) defines a function from infinite words to values. For example, every infinite run of a limit-average QWA $\mathcal{A}$ obtains a mean payoff, and every word $w \in \Sigma^\omega$ is assigned the maximal mean payoff obtained by nondeterministic runs of $\mathcal{A}$ over w. We introduce *quantitative language automata* (QLAs) that define functions from language generators (i.e., implementations) to values, where a language generator can be nonprobabilistic, defining a set of infinite words, or probabilistic, defining a probability measure over infinite words. A QLA consists of a QWA and an aggregator function. For example, given a QWA $\mathcal{A}$, the infimum aggregator maps each language $L \subseteq \Sigma^\omega$ to the greatest lower bound assigned by $\mathcal{A}$ to any word in L. For boolean value sets, QWAs define boolean properties of traces, and QLAs define boolean properties of sets of traces, i.e., hyperproperties. For more general value sets, QLAs serve as a specification language for a generalization of hyperproperties, called *quantitative hyperproperties*. A nonprobabilistic (resp. probabilistic) quantitative hyperproperty assigns a value to each set (resp. distribution) G of traces, e.g., the minimal (resp. expected) average response time exhibited by the traces in G. We give several examples of quantitative hyperproperties and investigate three paradigmatic problems for QLAs: evaluation, nonemptiness, and universality. In the *evaluation* problem, given a QLA $\mathbb{A}$ and an implementation G, we ask for the value that $\mathbb{A}$ assigns to G. In the *nonemptiness* (resp. *universality*) problem, given a QLA $\mathbb{A}$ and a value k, we ask whether $\mathbb{A}$ assigns at least k to some (resp. every) language. We provide a comprehensive picture of decidability for these problems for QLAs with common aggregators as well as their restrictions to ω-regular languages and trace distributions generated by finite-state Markov chains.

This presentation is based on a CONCUR 2025 paper [1].

Reference

1. Henzinger, T.A., Kebis, P., Mazzocchi, N., Saraç, N.E.: Quantitative language automata. In: CONCUR (2025). https://doi.org/10.4230/LIPICS.CONCUR.2025.21

Robust Identification of Hybrid Automata from Noisy Data

Niklas Kochdumper[1], Mohammed Aristide Foughali[2], Peter Habermehl[2], and Eugene Asarin[2(✉)]

[1] Technische Hochschule Ingolstadt, Ingolstadt, Germany
[2] Université Paris Cité, CNRS, IRIF, Paris, France
asarin@irif.fr

Abstract. In recent years, many different methods for identifying hybrid automata from data have been proposed. However, most of these methods consider clean simulator data, and consequently do not perform well for noisy data measured from real systems. We address this shortcoming with a new approach for the identification of hybrid automata that is specifically designed to be robust to noise. In particular, we propose a new high-level strategy consisting of the following three steps: clustering based on the dynamics identified from a local dataset, state space partitioning using decision trees, and conversion of the decision tree to a hybrid automaton. In addition, we introduce several new concepts for the realization of the single steps. For example, we propose an automated regularization of the dynamic models used for clustering via rank adaptation, as well as a new variant of the Gini impurity index for decision tree learning, tailored toward hybrid systems where different dynamics can be active within the same state space region. As our experiments on 19 challenging benchmarks with different characteristics demonstrate, in addition to being robust to both process and measurement noise, our approach avoids the need for extensive hyper-parameter tuning and also performs well for clean data without noise.

This presentation is based on a HSCC'25 paper [1].

Reference

1. Kochdumper, N., Foughali, M.A., Habermehl, P., Asarin, E.: Robust identification of hybrid automata from noisy data. In: HSCC (2025).https://doi.org/10.1145/3716863.3718030

Regular Model Checking for Systems with Effectively Regular Reachability Relation

Javier Esparza and Valentin Krasotin[✉]

Technical University of Munich, Munich, Germany
krasotin@in.tum.de

Abstract. Regular model checking is a well-established technique for the verification of *regular transition systems* (RTS): transition systems whose initial configurations and transition relation can be effectively encoded as regular languages. In 2008, To and Libkin studied RTSs in which the reachability relation (the reflexive and transitive closure of the transition relation) is also effectively regular, and showed that the recurrent reachability problem (whether a regular set L of configurations is reached infinitely often) is polynomial in the size of the RTS and the transducer for the reachability relation. We extend the work of To and Libkin by studying the decidability and complexity of verifying *almost-sure* reachability and recurrent reachability—that is, whether L is reachable or recurrently reachable w.p. 1. We then apply our results to the more common case in which only a regular overapproximation of the reachability relation is available. In particular, we extend recent complexity results on verifying safety using *regular abstraction frameworks*—a technique recently introduced by Czerner, the authors, and Welzel-Mohr—to liveness and almost-sure properties.

This presentation is based on a CONCUR 2025 paper [1].

Reference

1. Esparza, J., Krasotin, V.: Regular model checking for systems with effectively regular reachability relation. In: MFCS (2025). https://doi.org/10.4230/LIPICS.MFCS.2025.45

Galois Energy Games to Solve All Kinds of Quantitative Reachability Problems

Caroline Lemke[1](✉) and Benjamin Bisping[2]

[1] Department of Computing Science, Carl von Ossietzky Universität Oldenburg, Oldenburg, Germany
[2] Institute of Software Engineering and Theoretical Computer Science, Technische Universität Berlin, Berlin, Germany

Abstract. Most problems of how to reach a goal at minimal cost can be expressed as energy games. In energy games, one player fails if they run out of resources that they can gain or lose moving through the game. We provide a generic decision procedure for energy games with energy-bound reachability objective, moving beyond vector-valued energies and vector-addition updates. All we demand is that energies form well-founded bounded join-semilattices, and that energy updates have upward-closed domains and can be "undone" through Galois-connected functions.

Offering a simple framework to construct decidable games, we introduce the class of *Galois energy games*.

These can be instantiated to common energy games, declining energy games, multi-weighted reachability games, coverability on vector addition systems with states and shortest path problems. For such instantiations the running time of our simple algorithm is polynomial with respect to the game size and exponential in the dimension. We establish decidability of the (un)known initial credit problem for Galois energy games via energy-positional determinacy, an inductive characterization of winning budgets and properties of Galois connections. This result is supported by an Isabelle/HOL formalization [1].

This presentation is based on a CONCUR 2025 paper [2].

References

1. Lemke, C.: Galois energy games. Archive of Formal Proofs (2025). https://isa-afp.org/entries/Galois_Energy_Games.html
2. Lemke, C., Bisping, B.: Galois energy games: to solve all kinds of quantitative reachability problems. In: CONCUR (2025). https://doi.org/10.4230/LIPIcs.CONCUR.2025.29

Learning Deterministic One-Counter Automata

Prince Mathew[1(✉)], Vincent Penelle[2], and A. V. Sreejith[1]

[1] School of Mathematics and Computer Science, Indian Institute of Technology Goa, Goa, India
[2] UMR 5800, Univ. Bordeaux, CNRS, INP, LaBRI, 33400 Bordeaux, Bordeaux, France

Abstract. Active automata learning is a key technique in formal methods, particularly for model inference and verification. In the classical setting introduced by Angluins L* algorithm, a learner interacts with a teacher through *membership* and *equivalence* queries to infer a minimal deterministic finite automaton (DFA) for an unknown regular language $\mathcal{L}$. Membership queries check word inclusion in $\mathcal{L}$, while equivalence queries verify whether a proposed hypothesis DFA matches $\mathcal{L}$, returning a counterexample if not. This approach runs in time polynomial in the minimal DFA size and the longest counterexample. We extend this framework to deterministic one-counter automata (DOCA). DOCA are finite-state machines equipped with a non-negative integer counter that can be incremented, decremented, or reset. The presence of the counter allows DOCA to recognise certain non-regular, context-free languages such as $\{a^n b^n \mid n > 0\}$, making them a natural candidate for modelling some infinite state systems that arise in formal verification. We present OL*, a polynomial-time active learning algorithm for DOCA. Unlike previous exponential-time syntactic approaches for learning one-counter automata, OL* is semanticit learns the language of the target automaton without any knowledge of the DOCA structure. However, unlike L*, where the teacher can return any counter-example, we assume that the teacher always provides a *minimal counter-example* in response to an equivalence query. OL* can also be used to approximate DOCA minimisation, a problem whose exact version is NP-hard even for certain subclasses of DOCA. Future work includes improving the efficiency of equivalence checking and exploring richer query types to enhance practical applicability.

This presentation is based on a LICS 2025 paper. A preliminary version is available on arXiv [1].

Reference

1. Mathew, P., Penelle, V., Sreejith, A.V.: Learning deterministic one-counter automata in polynomial time. https://arxiv.org/abs/2503.04525

The authors would like to thank CEFIPRA for the project IFC/6602-1/2022/77. A.V. Sreejith would also like to thank the DST Matrics grant for the project MTR/2022/000788.

BT2Automata: Expressing Behavior Trees as Automata for Formal Control Synthesis

Ryan Matheu[✉], Aniruddh G. Puranic, John S. Baras, and Calin Belta

Institute for Systems Research, University of Maryland-College Park, College Park, USA

Abstract. The increasing complexity of robotic and cyber-physical systems (CPS) has made specifying task plans with explicit timing requirements a significant challenge. Temporal logics such as Linear Temporal Logic (LTL) and Metric Interval Temporal Logic (MITL) are widely used for expressing temporally evolving tasks, but encoding recovery behaviors and interdependent objectives often leads to intractable synthesis and verification problems. Behavior Trees (BTs), originally developed in the gaming industry and now widely adopted in robotics, offer a graphical, modular, and dynamic alternative. Their flexibility supports reconfiguration without complete redesign, making them suitable for dynamic environments. Temporal Behavior Trees (TBTs) extend BTs by embedding Signal Temporal Logic (STL) formulas at leaf nodes, enabling quantitative semantics; however, they remain limited to offline monitoring and synthesis capabilities needed for adaptive control.

To address this gap, we introduce **BT2Automata**, a framework that translates BTs into Timed Automata (TA), thereby bridging BTs' interpretability and modularity with the rigorous formal verification capabilities of temporal logic. While temporal logics excel at specifying temporally evolving behaviors but struggle with tractable synthesis, BTs provide dynamic adaptability but complicate safety analysis and control guarantees. By enabling falsification through UPPAAL, **BT2Automata** identifies inconsistencies, ensures language completeness under timing constraints, and supports monitoring of temporal properties. Furthermore, it enables both automaton-based and sampling-based control synthesis strategies that guarantee task satisfaction. This unified approach combines the intuitive specification advantages of BTs with the formal rigor of TA verification, advancing the development of adaptive, safe, and reliable control for CPS operating in safety-critical environments.

This presentation is based on a HSCC 2025 paper [1].

Reference

1. Matheu, R., Puranic, A.G., Baras, J.S., Belta, C.: Bt2automata: expressing behavior trees as automata for formal control synthesis. In: HSCC (2025). https://doi.org/10.1145/3716863.3718042

Model Checking as Program Verification by Abstract Interpretation

Paolo Baldan[1], Roberto Bruni[2], Francesco Ranzato[1], and Diletta Rigo[1(✉)]

[1] Department of Mathematics, University of Padua, Padua, Italy
diletta.rigo@phd.unipd.it
[2] Department of Computer Science, University of Pisa, Pisa, Italy

Abstract. Abstract interpretation offers a powerful toolset for static analysis, tackling precision, complexity and state-explosion issues. In the literature, state partitioning abstractions based on (bi)simulation and property-preserving state relations have been successfully applied to abstract model checking. Here, we pursue a different track in which model checking is seen as an instance of program verification. To this purpose, we introduce a suitable language—called MOKA (for MOdel checking as abstract interpretation of Kleene Algebras)—which is used to encode temporal formulae as programs. In particular, we show that (universal fragments of) temporal logics, such as ACTL or, more generally, universal μ-calculus can be transformed into MOKA programs. Such programs return all and only the initial states which violate the formula. By applying abstract interpretation to MOKA programs, we pave the way for reusing more general abstractions than partitions as well as for tuning the precision of the abstraction to remove or avoid false alarms. We show how to perform model checking via a program logic that combines under-approximation and abstract interpretation analysis to avoid false alarms. The notion of locally complete abstraction is used to dynamically improve the analysis precision via counterexample-guided domain refinement.

This presentation is based on a CONCUR 2025 paper [1].

Reference

1. Baldan, P., Bruni, R., Ranzato, F., Rigo, D.: Model Checking as Program Verification by Abstract Interpretation. CONCUR (2025). https://doi.org/10.4230/LIPIcs.CONCUR.2025.8

A Complexity Dichotomy for Semilinear Target Sets in Automata with One Counter

Yousef Shakiba[1](✉), Henry Sinclair-Banks[2], and Georg Zetzsche[1]

[1] Max Planck Institute for Software Systems, MPI-SWS), Kaiserslautern, Germany
yshakiba@mpi-sws.org
[2] University of Warsaw, Warsaw, Poland

Abstract. In many kinds of infinite-state systems, the coverability problem has significantly lower complexity than the reachability problem. In order to delineate the border of computational hardness between coverability and reachability, we propose to place these problems in a more general context, which makes it possible to prove complexity dichotomies. The more general setting arises as follows. We note that for coverability, we are given a vector t and are asked if there is a reachable vector x satisfying the relation $x \geq t$. For reachability, we want to satisfy the relation $x = t$. In the more general setting, there is a Presburger formula $\varphi(t, x)$, and we are given t and are asked if there is a reachable x with $\varphi(t, x)$. We study this setting for systems with one counter and binary updates: (i) integer VASS, (ii) Parikh automata, and (iii) standard (non-negative) VASS. In each of these cases, reachability is NP-complete, but coverability is known to be in polynomial time. Our main results are three dichotomy theorems, one for each of the cases (i)–(iii). In each case, we show that for every φ, the problem is either NP-complete or belongs to AC^1, a circuit complexity class within polynomial time. We also show that it is decidable on which side of the dichotomy a given formula falls. For (i) and (ii), we introduce novel density measures for sets of integer vectors, and show an AC^1 upper bound if the respective density of the set defined by φ is positive; and NP-completeness otherwise. For (iii), the complexity border is characterized by a new notion of *uniform quasi-upward closedness*. In particular, we improve the best known upper bound for coverability in (binary encoded) 1-VASS from NC^2 (as shown by Almagor, Cohen, Pérez, Shirmohammadi, and Worrell in 2020) to AC^1.

This presentation is based on a LICS 2025 paper. A preliminary version is available on arXiv [1].

Reference

1. Shakiba, Y., Sinclair-Banks, H., Zetzsche, G.: A Complexity Dichotomy for Semilinear Target Sets in Automata with One Counter. https://arxiv.org/abs/2505.13749

Quantifier Elimination for Regular Integer Linear-Exponential Programming

Mikhail R. Starchak[✉]

Max Planck Institute for Software Systems, Saarbrücken, Germany
mikhstark@gmail.com

Abstract. Regular integer linear-exponential programming (RegILEP) asks whether a system of inequalities $\sum_{i=1..n}(a_i \cdot x_i + b_i \cdot 2^{x_i}) \leq c$, where all coefficients are integers, has a solution in the integers whose binary representations belong to some regular set over the alphabet $\{0, 1\}$. RegILEP has recently been proved decidable in EXPSPACE using purely automata-theoretic techniques [1]. The main contribution of this work is a novel decision procedure for RegILEP, which works in a quantifier elimination fashion: after specifying a total order on the variables, the procedure gradually excludes the exponential occurrences of the leading variable (linearisation) and then eliminates the linear ones. Linearisation uses Chrobak normal form of unary sub-NFAs extracted from the NFAs associated with the systems produced during the quantifier elimination. These sub-NFAs describe the runs between the unique 1's in the binary expansions of consecutive exponentiated variables. This decision procedure meets the existing EXPSPACE upper bound for the problem. As a complementary result, it is shown that regular integer linear programming for the domain defined by the regular expression $(00 \cup 01)^*$ is PSPACE-complete.

This presentation is based on a LICS 2025 paper.

Reference

1. Draghici, A., Haase, C., Manea, F.: Semënov arithmetic, affine VASS, and string constraints. In: STACS (2024). https://doi.org/10.4230/LIPICS.STACS.2024.29

Flexible Catalysis

Mate Weisz[✉] and Sergii Strelchuk

Department of Computer Science, University of Oxford, Oxford, UK
mate.weisz@linacre.ox.ac.uk

Abstract. In quantum information and computation, a fundamental problem is determining which quantum states can be reached from a given initial state using a fixed set of free operations. Although some states are not directly reachable from the initial state, it may nevertheless be possible to devise a catalytic process that effectively results in the desired transformation. In chemistry, a catalyst is a substance that can be used to speed up a chemical reaction without being consumed in the process. In quantum information theory, catalysis refers to the phenomenon in which the presence of an auxiliary quantum state $|\eta\rangle$ enables an otherwise impossible state transformation $|\psi\rangle \to |\phi\rangle$. Here we introduce flexible catalysis, the concept of allowing the catalyst to change into a different catalyst while facilitating a transformation. The idea behind flexible catalysis is that in general the catalyst is not the main system of interest, and therefore it does not much matter whether it undergoes some change during the transformation. What does matter is that the modified auxiliary state that we get back at the end of the process instead of the original catalyst still has the ability to catalyse the transformation of interest. Alongside catalysis, there is another well-known way to simulate state transformations that cannot be realised directly: multiple copy transformations, that is, when n copies of a target state $|\phi\rangle$ are reached from n copies of an initial state $|\psi\rangle$. An interesting observation is that flexible catalysis can simulate multiple copy transformations. In fact, we find that flexible catalysis is at least as strong as the combination of catalytic and multicopy transformations, in the sense that if $|\psi\rangle^{\otimes n}|\eta\rangle \to |\phi\rangle^{\otimes n}|\eta\rangle$ can be implemented directly using our free operations, then there is a flexible catalytic process converting $|\psi\rangle$ into $|\phi\rangle$. Our main results demonstrate that we have obtained a useful generalisation of catalysis by allowing flexibility: we begin to explore what advantages flexibility can offer to catalytic transformations in quantum information and computation. Two of our main positive results are the following.

- When the set of free operations is LOCC (Local Operations and Classical Communication) and we have access to a fixed set of catalysts, flexible catalysis enables more transformations than traditional catalysis.
- When the set of free operations is LU (Local Unitaries) together with discard operations, there are transformations that cannot be implemented catalytically with any single catalyst, but can be realised by flexible catalysis.

Author Index

A
Ábrahám, Erika 202
Adamson, Duncan 68
Allamigeon, Xavier 233
Almagor, Shaull 126
Antal, László 202
Asarin, Eugene 112, 238

B
Baldan, Paolo 243
Baras, John S. 242
Barcelo, Pablo 17
Belta, Calin 242
Bisping, Benjamin 240
Bruni, Roberto 243
Brorholt, Asger Horn 97

C
Cadar, Cristian 3
Capetillo, Pascal 233
Carrasco, Manuel 3

D
Day, Joel D. 51
Degorre, Aldric 112
Demri, Stéphane 140
Di Cosmo, Francesco 156
Dima, Cătălin 112
Donaldson, Alastair F. 3
Doyen, Laurent 140
Dudey, Moritz 68

E
Esparza, Javier 239

F
Fervari, Raul 140
Fleischer, Lukas 234
Fleischmann, Pamela 68
Foughali, Mohammed Aristide 238

G
Gaubert, Stéphane 233

H
Habermehl, Peter 238
Hagihara, Fugen 236
Hartmanns, Arnd 186
Hasson, Itay 126
Hasuo, Ichiro 171
Henzinger, Thomas A. 237
Høeg-Petersen, Andreas Holck 97
Huch, Annika 68

I
Inclán, Bernardo Jacobo 112
Iorga, Dan 3

J
Jensen, Peter Gjøl 97
Junges, Sebastian 171

K
Kawamura, Akitoshi 236
Kebis, Pavol 237
Kochdumper, Niklas 238
Konefal, Matthew 51
Krasotin, Valentin 239

L
Larsen, Kim Guldstrand 97
Lemke, Caroline 240
Liew, Daniel 3
Lin, Anthony W. 17
Link, Franz 202

M
Mal, Soumodev 156
Matheu, Prince 242
Mathew, Prince 241
Mazzocchi, Nicolas 237

© The Editor(s) (if applicable) and The Author(s), under exclusive license to Springer Nature Switzerland AG 2026
P. Ganty and A. Mansutti (Eds.): RP 2025, LNCS 16230, pp. 247–248, 2026.
https://doi.org/10.1007/978-3-032-09524-4

Mikučionis, Marius 97
Modderman, Robert 186

P
Penelle, Vincent 241
Pilipczuk, Michał 126
Prince, Tephilla 156
Puranic, Aniruddh G. 242

R
Ranzato, Francesco 243
Randour, Mickael 31
Restanio, Jacob 217
Rigo, Diletta 243

S
Saraç, N. Ege 237
Schilling, Christian 97
Shakiba, Yousef 244
Sinclair-Banks, Henry 244
Sreejith, A. V. 241
Starchak, Mikhail R. 245

Strelchuk, Sergii 246
Stober, Florian 234
Subramani, A. 217
Subramani, K. 217

T
Thumm, Alexander 234
Tveretina, Olga 83

V
van der Vegt, Marck 171

W
Wasowski, Andrzej 97
Watanabe, Kazuki 171
Weisz, Mate 246
Weiß, Armin 234
Wickerson, John 3

Z
Zaslavski, Michael 126
Zetzsche, Georg 244

MIX
Papier aus verantwortungsvollen Quellen
Paper from responsible sources
FSC® C105338

If you have any concerns about our products,
you can contact us on
ProductSafety@springernature.com

In case Publisher is established outside the EU,
the EU authorized representative is:
**Springer Nature Customer Service Center GmbH
Europaplatz 3, 69115 Heidelberg, Germany**

Printed by Libri Plureos GmbH
in Hamburg, Germany